Science and Technology
in World History
*Volume 1*

# Science and Technology in World History

## *Volume 1: The Ancient World and Classical Civilization*

DAVID DEMING

McFarland & Company, Inc., Publishers
*Jefferson, North Carolina, and London*

The history of science is the history of mankind's unity,
of its sublime purpose, of its gradual redemption.
— George Sarton (1884–1956)

LIBRARY OF CONGRESS CATALOGUING-IN-PUBLICATION DATA

Deming, David, 1954–
Science and technology in world history / David Deming.
v. cm.
v. 1. The ancient world and classical civilization — v. 2. Early Christianity, the rise of Islam and the Middle Ages.
Includes bibliographical references and index.

**ISBN 978-0-7864-3932-4** (v. 1 : softcover : 50# alk. paper) —
**ISBN 978-0-7864-5839-4** (v. 2 : softcover : 50# alk. paper)

1. Science — History. 2. Technology — History. 3. World history. I. Title
Q125.D334 2010 509 — dc22 2010008935

British Library cataloguing data are available



On the cover: Friedrich Heinrich Füger, *Prometheus Bringing Fire to Mankind*, oil on canvas 16⅓" × 21½", 1817 (Museumslandschaft Hessen Kassel/Bridgeman Art Library); (background) Plate 1 of the *Harmonia Macrocosmica*, Andreas Cellarius, 1661

Manufactured in the United States of America

*McFarland & Company, Inc., Publishers*
*Box 611, Jefferson, North Carolina 28640*
*www.mcfarlandpub.com*

# Table of Contents

# Preface

The purpose of the present work is in part to define science by revealing through historical facts and analysis the origin of its methods. It was written primarily for those who want a better understanding of the advance of science and its relation to historical events, from the practicing scientist, to students, to the lay reader.

Most professional scientists today have been trained as narrowly focused technical specialists. They know many pertinent and specialized facts, but may lack an understanding of science as method. Consequently, their knowledge is not truly scientific, but as Aristotle put it, "merely accidental."[1]

Philosophy, history, and religion are foreign fields to many modern scientists, even with respect to something as crucial as recognizing common intellectual fallacies. The ability to practice science is necessarily hindered. And the ideal of a disinterested pursuit of truth has too often been lost as science has been professionalized and facts have been enrolled in the service of ideological causes. In the past, amateurs sometimes made important discoveries and contributions, being motivated by a sincere love of science.

Deficiencies originate in the way scientists are educated. In modern American universities the sciences are taught as bodies of accumulated fact. Method is not taught; history is mentioned only as a curiosity. Science has been surgically excised from its historical and philosophical underpinnings. As a result, students obtain the false idea that science is not only cumulative and progressive, but linear and inerrant. Blind to the mistakes of the past, modern scientists can only repeat them.

Understanding science must be accomplished through historical analysis because history deals with facts. Certainly, there are difficulties. Sources are biased. When Herodotus recorded that the Greeks at the Battle of Marathon (490 B.C.) had 192 casualties, while the Persians had 6,400, we understand that he wrote to glorify Greece, and that accordingly the body counts may have been exaggerated.[2] But there is no historian who doubts that the battle occurred, or that the Greeks won.

Artifacts are sparse and their interpretation can be contentious. Preservation can be selective and unrepresentative. Pottery survives, wood does not. Most history is unwritten, and our histories tend to be based largely upon written sources. The history of the practical arts and crafts is very important, but there is little written documentation of their development. Technological knowledge was mostly passed down by oral transmission, apprenticeship, and firsthand experience. The history of technology may be more important than the history of philosophy, but the history of philosophy is easier to reconstruct because philosophical knowledge was transmitted largely through books.

All sources are inadequate. Because only the exceptional tends to be recorded, it is easy to obtain the false idea that the extraordinary is the ordinary. All of these problems make it difficult to recover the past. But if we do not make the attempt, we will never learn from the immense amount of material that is available.

My approach has been chronological and conservative. I have diverged from a uniform chronological perspective when it made sense to do so, but the subject is best approached chronologically because science is a progressive and cumulative activity. Everyone read everyone else. The Greek physician Galen (c. A.D. 129–200) is an example. After Aristotle, Galen was the writer most commonly cited by Europeans during the High Middle Ages. It was from Galen, in part, that people such as Roger Bacon acquired an appreciation for the importance of empiricism. But Galen himself read voluminously and was apparently strongly influenced by writers such as Hippocrates (c. 460–370 B.C.), Aristotle (384–322 B.C.), Herophilus (4th century B.C.), and Erasistratus (3rd, 4th century B.C.). The methodology of modern science did not spring into being spontaneously, it developed painfully and gradually through an historical process.

Histories of science ought to be conservative. By "conservative," I mean not a political philosophy, but a treatment that shrinks from revisionism and novelty. By novelty, I mean a radical new interpretation that differs from those reached by all previous workers for hundreds of years. The inventor of the novelty believes that he alone has discovered a profound truth which heretofore was obscured by ignorance and prejudice, and that all previous writers have been wrong in their interpretations and conclusions. This is usually not the case, and novelties tend to pander to the human weakness to be attracted to the sensational. History is often rewritten and reinterpreted to meet current social or political fashions. Such treatments enjoy a short-lived popularity, but quickly become dated and have no lasting value. In their lifetime, the scalawag and fraud often receives more recognition than the honest and patient scholar.

Of course, if new data become available, that is another story. For example, our understanding of human evolution tends to be significantly revised every time an important fossil is found. This is because the complete database is very small. Interpretations must follow the data. This is quite different from constructing a radical new interpretation of a static body of accumulated evidence that others have systematically studied for more than a century.

The term "whig history" was introduced by Herbert Butterfield in his book, *The Whig Interpretation of History* (1931).[3] The Whigs and Tories were the two dominant political parties in England from about the late seventeenth through the nineteenth century. The Whigs were "regarded as the party of the great landowners, and of the merchants and tradesmen."[4] According to Butterfield, the whig historian did not try "to understand the past for the sake of the past," but studied "the past for the sake of the present."[5] The whig historian selectively filtered the chronicle of historical events, including those that demonstrate "throughout the ages the workings of an obvious principle of progress, of which the Protestants and whigs have been the perennial allies while Catholics and Tories have perpetually formed obstruction."[6]

The sin of a whig history is that the bias is implicit, not explicit. The picture of history that emerges has been predetermined before the chronicle emerges. History is prostituted to promote a political or cultural point of view. Whig histories have subsequently become anathema to professional historians.[7]

Butterfield asserted that it was a "fact that we can never assert that history has proved any man right in the long run."[8] As moral, social, political, and cultural values are relative, this statement is meaningful when dealing with most aspects of human history. The corollary is that all notions of progress must be chimerical.

But science is different. Should we accept Butterfield's absolute relativity in the field of science, we should have to accept that those men who thought the Earth was flat were just as correct as those who believed it to be spherical. The geocentric model of the solar system would have to be considered to be just as valid as the heliocentric. Newtonian physics would be no more "right" than Aristotelian. And we should have to accept that the germ theory of disease was no more valid than the Hippocratic theory that epidemics arose from tainted air. No one sincerely holds such views.

According to its own epistemological criteria of observation and reason, scientific knowledge is undeniably progressive.[9] It is "right" in modern times to believe that the Earth is spherical, "wrong" to think it flat. It is wrong to assert that the Sun moves around the Earth and not vice versa. Science constructs probable truths according to its own methods and criteria. Some scientific truths (e.g., the spherical nature of the Earth) have been strongly corroborated over time to the point of having universal acceptance. Scientific truths are not absolute like the deductive proofs of geometry, but they have become increasingly more reliable over history when judged by science's own methods of observation and reason.

It thus becomes valid, feasible, and interesting to trace the origin of scientific methods. Once it is accepted that it is possible to differentiate between true and false scientific knowledge, it also becomes possible to make judgments concerning effective and ineffective methods. For example, there are few who would now argue that divination and augury are as likely to lead to reliable knowledge as controlled and systematic experimentation. This justifies a selective search through the past for those developments and individuals that made the most significant contributions to modern science.

Ancient science should be described in terms of how it was conducted and understood by the people who did it. But if we want to understand the origins of modern science, we must necessarily focus on those activities and individuals that made the most significant contributions to it. There is only so much space that can be devoted to a discussion of blind alleys and errors. It is impossible to write a comprehensive history of human error, because "error is, of its very nature, infinite."[10] Anyone desiring a taste of average human behavior and aspirations is advised to close this book and open the pages of a current newspaper. All of history is a testament to the fact that "human wisdom is limited; human folly unlimited."[11] The failure to be selective is not a virtue, but an implicit confession of ignorance.

The assertion that scientific knowledge can be divided into "true" and "false" must be qualified. Science ultimately depends upon the methods of reason and experience. It is a well-known philosophic impossibility to establish absolute truth. Any statement with a pretension to truth must be tested against some criterion of truth. This criterion of truth must in turn be tested against some other criterion of truth, *ad infinitum*. Even geometry rests upon unprovable axioms. The only thing that may be demanded of any system of knowledge is internal consistency: it must satisfy its own criteria of truth.

Because observation is flawed and human reason unreliable, it must be conceded that scientific knowledge is always provisional and tentative. Some serious thinkers have entirely rejected the epistemological methods of science. The Islamic mystic and philosopher al-Ghazzali (A.D. 1058–1111) denied the existence of both natural law and cause and effect. Every event, he argued, was the immediate result of an omnipotent God. If the presence of fire ignited cotton, there was no guarantee that this would happen again. Every event proceeded from the immediate will of God. It was not possible to demonstrate the existence of natural law, only correlation.[12] This viewpoint is intellectually sound, but the person who applies it to human life will never learn from experience. They will never develop antibiotics, refrigeration, electronic digital computers, or any of the practical benefits to be derived from systematic experimentation combined with logical reasoning.

Is this a whig history? Acknowledging that the purpose of this history is to trace and understand the methodologies of modern science, the answer must be "yes." The book was written from the viewpoint of the present. But I did so with no predetermined outcome. The process of researching and writing this book and the volumes to follow changed most of my preexisting ideas regarding the development of scientific method.

Within my stated purpose of focusing on those individuals and events that have contributed most to the development of science, I have tried to be as objective as possible. I have made no intentional effort to distort the past by selecting only those facts that support some predetermined conclusion. To do so would have frustrated my own curiosity and defeated my purpose. But it is true that I have been primarily interested in those events, persons, and ideas that contributed to modern science. Most things did not. This history, like other histories, is therefore a history of the exceptional. Human error and folly are normal, unexceptional, and usually unrecorded.

No one questions that the work of Galileo was essential to the development of modern science. Galileo wrote that he considered Archimedes to possess "the greatest and most superhuman intellect."[13] By this statement, Galileo informs us that he read, and was heavily influenced by, Archimedes. Archimedes can hardly have been typical of Syracusans living in the third century B.C. But if we want to understand the development of science, we must focus on Archimedes and his work, not the average and unexceptional.

The *sine qua non* of historical research is original source material. Whenever possible, I have tried to cite and quote original sources, albeit in English translation. It must be understood however that this is a synthetic work of immense scope. It is not an analytical or focused work. Should anyone be interested in an in-depth knowledge of any of the individual topics discussed in this history, they should consult specialized works, many of which are listed in the bibliography.

There is a place for both synthetic and analytic works. Synthesis is necessary, because nothing takes place in a vacuum. Every event occurs in a broader context. In other words, focused analysis of an historical topic cannot be done without simultaneously understanding an array of related antecedents, as well as the historical development of a spectrum of cultural, social, and religious values. When Isaac Newton acknowledged that "if I have seen further it is by standing on the shoulders of giants,"[14] it becomes clear that Newton and his work cannot be fully understood without studying his predecessors. When Galileo praised Archimedes,[15] he revealed that his work was building on a long tradition dating back to the ancient Greeks. Thus understanding Galileo entails an acquaintance with Archimedes. Understanding the particulars involves comprehending the generals, and vice versa.

Given the synthetic nature and broad nature of this work, it was necessary for me to often rely upon secondary sources. Secondary sources in many cases were not only necessary, but advisable. The opinions and interpretations of a specialist who has spent his life studying a particular topic may be more reliable than those of a synthesist who can only afford two weeks of time.

My sources for this book and its companion volumes include many older works, including histories published during the nineteenth century. These include *Early Greek Philosophy* by Burnet, *Fifteen Decisive Battles* by Creasy, *History of Greece* by Grote, *Lectures on the History of Rome* by Niebuhr, *History of the Christian Church* by Schaff, *Life of Mahomet* by Muir, and *The Universities of Europe in the Middle Ages* by Rashdall. If I were writing a book on a technical field such as computer science, where knowledge was rapidly evolving, I would use recent sources almost exclusively. But in historiography our knowledge changes more slowly. Changes in interpretations and opinions over the last 150 years or so tend to be based more on political or cultural fashion than new discoveries. Do we really know significantly more about Aristotle, for example, than scholars did in the year 1850? Many nineteenth-century histories are superior to modern works in terms of their eloquence, lucidity, length, and elaboration.

My choice of sources has been determined by the usual factors. Did the author cite an extensive list of original sources? Was his prose intelligent and erudite? Did his interpretations smack of intelligence and insight? I am writing a history of science that extends at least over twenty-five hundred years. I am not going to ignore an impressive piece of scholarship simply because it might be a whig history or reflect cultural values that are no longer fashionable. Should I discard Herodotus because he wrote from a Hellenic perspective? If the study of history reveals nothing else, it is that people living at different times and places view the world differently. In fact, it is more dangerous to rely upon a recent piece of historical analysis because we are usually more oblivious to the biases of our own time. As Butterfield pointed out, the greatest danger is not bias, but the bias that "cannot be recognized."[16]

It has been cynically observed that the most creative author is the person who has been most successful at concealing his sources. In this book I have chosen to reveal my sources. I have tried to meticulously document every specific fact, giving full information including page numbers. There is nothing quite so infuriating as finding no page number given for a reference to a specific fact in an eight-hundred page work. I have found older works to be especially irritating in that the references are often cited in such an abbreviated form that only scholars in the field could possibly decipher the source. Scholars may disagree on both interpretations and facts. But every writer should document his sources. References, sources, and facts constitute the supporting structure of the work, interpretations are the surficial skin.

To write a proper history of science would require reading all original sources in the original languages. The knowledge acquired would have to be collated with historical reconstructions based on artifacts. To understand the human environment within which science developed would also require an understanding of all social, cultural, political, economic, and religious history. A deep appreciation for human nature and psychology would also be helpful. In other words, a comprehensive and accurate history of science is a history of all human knowledge. Such a project is an unachievable goal.

Thus this history, like others, is a work in progress. There simply is not enough time available for a truly comprehensive treatment. Like science itself, historiography is a patient and cumulative work that matures through the ages.

I would like to gratefully acknowledge Google Books. In the past few years, Google has digitized and placed online many millions of books, including one million published before 1923 that are now in the public domain. The entire database is searchable. This is a tremendous research tool, especially for a scholar attempting a broad synthesis. Hopefully, Google books is an embryo that will mature into a universal library of human knowledge.

# Introduction

## *Science, Religion and Epistemology*

From the time of Socrates, it has been understood that clear and unambiguous definitions are desirable and necessary antecedents to intelligible discussions. *Science* was defined by George Sarton as "the acquisition and systemization of positive knowledge."[1] By "positive," Sarton meant information derived empirically from the evidence of the senses. This definition has the virtue of being succinct, but does not explain the methodological procedures employed in science, how these methods differ from other systems of knowledge, the historical process by which these methods came to be adopted, or why other methods were ultimately rejected.

The *Oxford English Dictionary* defines science as "the state or fact of knowing,"[2] acknowledging a significant ambiguity inherent in the word and its history. In a historical treatment, it is impossible to pinpoint a definition of science that applies at more than one moment in time. Science is a living, organic activity. The definition and meaning of what is understood to be science has evolved incrementally and gradually over approximately the last twenty-five hundred years. Unless the process of history has stopped, it will continue to evolve in the future. At specific points in history, science has meant different things to different people. Perhaps the only definition of science that has universal application is that a *science* is a method designed to discover knowledge that is considered by some criteria to be reliable.

The traditional understanding is that modern science began with the natural philosophy of the ancient Greeks. However it is impossible to draw a line of temporal demarcation between natural philosophy and science. The methods employed in some sciences have remained the same for thousands of years. Astronomy has always been based on observation. And at least from the time of Eudoxus of Cnidus (c. 400–347 B.C.), astronomers have employed mathematical models to explain their observations. From the beginning, astronomy has essentially been science as it is understood today. A sharp line of differentiation between ancient and modern astronomy cannot be drawn, and it would be wrong to characterize ancient astronomy as anything other than science.

In contrast to astronomy, physics has evolved from a natural philosophy that employed reason and dialectic in the pursuit of final causes, to a science that collects empirical data, assigns efficient causes, and constructs mathematical models and explanations. A modern physicist would be baffled by many of the questions considered and discussed by Aristotle in *Physica* (*Physics*). At some point in history, physics divorced itself from metaphysics, ceased being natural philosophy, and became a science.

Science is not the only means of constructing knowledge, or even necessarily the most important. Science is method. It falls under the domain of *epistemology*, the study of how knowledge is acquired. There are four means of acquiring knowledge: observation, reason, revelation, and creative imagination.

Modern science is based on observation and reason, and incorporates a number of methodological assumptions and practices. The most important of these include naturalism, uniformity, induction, repeatability, and efficient causation.[3] The incorporation of these methodologies into what is recognized as modern science was a slow historical process.

Philosophy is based primarily on the sole exercise of reason, religion upon revelation. Technology, craft, and art rely upon observation, reason, and creative imagination. Of these different methods, creative imagination is the least understood but of undeniable importance.

The modern scientist embraces technology, grudgingly acknowledges philosophy, and tends to view religion as something that must be sharply demarcated from science. But as truth is a unity, it is difficult to disengage science from other epistemologies. This is especially true in a historical treatment. Attempts to construct barriers between science and other systems of knowledge must necessarily stunt and restrict our understanding of science and its historical development. In twelfth-century Europe, there were no "scientists," only theologians and philosophers. The first use of the word "scientist" in English appears to date from the year 1834.[4]

A historical appreciation of science necessarily entails a discussion of religion, technology, and philosophy. Of these, religion is perhaps the most important, technology the least appreciated, and philosophy the most acknowledged. Religion cannot be excluded. Christianity has been the single largest influence on Western Civilization, and Islam has been a major force in world history. In the introduction to his monumental work, *Introduction to the History of Science* (1931), George Sarton conceded that he was "obliged to devote much space to the history of religion."[5]

Religion is based upon revelation, an irrational and individual experience. Those who experience mystic communion claim that it transcends human reason. Saint Teresa of Avila (A.D. 1515–1582) described her ecstasy.

> I saw an angel close by me, on my left side, in bodily form.... He was not large, but small of stature, and most beautiful—his face burning, as if he were one of the highest angels, who seem to be all of fire.... I saw in his hand a long spear of gold, and at the iron's point there seemed to be a little fire. He appeared to me to be thrusting it at times into my heart, and to pierce my very entrails; when he drew it out, he seemed to draw them out also, and to leave me all on fire with a great love of God. The pain was so great that it made me moan; and yet so surpassing was the sweetness of this excessive pain that I could not wish to be rid of it. The soul is satisfied with nothing less than God.[6]

By the standards of modern science, the experience of St. Teresa was a psychological malfunction. But mystic communion is the root of religion, and religion establishes the moral codes and rules by which human civilizations are organized. Science does define morality or tell us how to order our civilizations; it is a human and social activity conducted within civilizations constructed by religion. If the development of science is to understood, the ecstasy of St. Teresa does not have to be accepted as endorsement of a supernatural reality, but it must be understood and appreciated as a historical force.

George Sarton characterized science as "a cumulative, a progressive activity."[7] That science is cumulative is self-evident, and does not require elaboration. The term "progressive" is unintelligible unless a specific goal is identified. If by "progressive," some moral quality is implied, then the alleged progressive nature of science becomes as arbitrary and relative as morality itself. But if science is judged by its own methods and goals, then it is as undeniably progressive as it is cumulative.

The goal of scientific activity is to construct knowledge considered to be reliable according to science's own criteria of reason and experience, as embodied in methodologies such as naturalism, uniformity, induction, repeatability, efficient causation, falsifiability, and multiple working hypotheses. The periodic table is a better model of matter than Empedocles' theory of the four elements because it has more predictive power and utility. The Hippocratic physicians c. 450 B.C. hypothesized that epidemics were caused by bad air. This theory relied upon observation and natural causation, and therefore represented a progression beyond the superstitious idea that diseases were punishments from the gods. But the modern germ theory of disease leads to cures in the form of antibiotics. Newtonian physics makes it possible to travel to the Moon in rocket ships. This cannot be done with Aristotelean physics.

## *Before the Greeks*

Genetic analysis suggests that human ancestors and chimpanzees diverged from a common ancestor between 4.6 and 6.2 million years ago.[8] The increasing complexity of stone tools is a proxy for the increasing brain size and development of the human species. The development of stone implements can be divided into three stages. The oldest stone tools (Lower Paleolithic) are about 2.5 to 2.6 million years old, and reveal "a sophisticated understanding of stone fracture mechanics and control."[9] About 250,000 years ago (Middle Paleolithic), tools became more complex. The most advanced tools of stone and bone date from about 50,000 years before present forward (Upper Paleolithic).[10]

In addition to constructing tools of stone, bone, and wood, early humans utilized fire to scare off predators, cook food, and provide warmth. The use of fire dates to at least one to two million years before present.[11] Fire occurs naturally when lightning starts a forest fire, or when sparks fly from a piece of flint. But the problem of how to create fire on demand required creative imagination, abstract reasoning, and experimentation. People may have inferred inductively that friction could be used to create fire. Friction creates heat, and heat is associated with fire. The technique that was developed involved rapidly twisting a stick in a piece of wood fast enough to set wood shavings on fire.

The most significant human invention was language, because it is language that allows the transmission of knowledge. It is language that makes us human, and only humans have language. "Monkeys and apes simply lack the ability to translate their thought into anything resembling human language."[12] Human being have an innate ability to acquire language. Charles Darwin (1809–1882) observed, "man has an instinctive tendency to speak, as we see in the babble of our young children."[13]

The beginning of language is difficult to date. It is possible that humans did not fully acquire the ability to utilize language until about 50,000 years ago, the time at which evidence suggests "behaviorally modern humans" emerged.[14]

Civilization, or sedentism, began about 10,000 years ago, but there is evidence that cereals were cultivated as early as 13,000 years ago. The onset of sedentism "happened independently and at different times in at least three parts of the world."[15] In a sedentary lifestyle, as opposed to foraging, food is largely acquired through agriculture and animal husbandry. A concomitant corollary of sedentism is a central organizing authority, or government. The current view is that the development of agriculture postdated, rather than preceded, the sedentary lifestyle. "Sedentary village life began several millenia before the end of the late glacial period, and the full-scale adoption of agriculture and stock rearing occurred much later.... It is now evident that agriculture was not a necessary prerequisite of sedentary life."[16]

The first large-scale civilizations were organized in river valleys in the Middle East around 8000 B.C. Egypt and the area between the Tigris and Euphrates Rivers (Mesopotamia or present-day Iraq) were the sites of the first sizable and complex human societies. In the arid climate of the Middle East, these locations were ideal for agriculture because the river valleys had rich soils and the rivers provided water for irrigation.

The primary advantage of sedentism over foraging was the ability to accumulate and store surplus food. An increased food supply led to an increase in population and made occupational specialization possible. This started a synergistic process that continues to the present day. The growth of knowledge led to more effective technologies that produced increases in wealth. More wealth allowed diversification and leisure time for systematic investigations that in turn led to the accumulation of yet greater knowledge. Occupational specialization was essential to the growth of knowledge. It is unlikely that the farmer would have ever figured out how to smelt copper from its ore. But the blacksmith would have had the time and inclination to solve the problem.

Egyptian and Mesopotamian civilizations developed complex societies whose needs spurred innovation.[17] Proclus (A.D. 411–485) argued that science began as a utilitarian search for practical ways of accomplishing goals and then naturally proceeded to abstraction. "Geometry was first invented by the Egyptians, deriving its origin from the mensuration of their fields ... nor ought it to seem wonderful, that the invention of this as well as of other sciences, should receive its commencement from convenience and opportunity ... whatever is carried in the circle of generations proceeds from the imperfect to the perfect. A transition, therefore, is not undeservedly made from sense to consideration, and from this to the nobler energies of intellect."[18]

The first science was astronomy; its study and growth were partly motivated by a practical application to agriculture: farmers needed to know when to plant and harvest crops.* In *Works and Days*, Hesiod (c. 700 B.C.) explained that the farmer should regulate his activities by the positions of the stars. "When Orion and Sirius are come into mid-heaven, and rosy-fingered Dawn sees Arcturus, then cut off all the grape-clusters, Perses, and bring them home ... but when the Pleiades and Hyades and strong Orion begin to set, then remember to plough."[19]

The movements of the sun, stars, and planets were studied and used to develop a calendar and predict the changing of the seasons. Astronomy was first developed as a science in Babylonia, where "the Babylonians acquired a remarkably accurate knowledge of the periods of the sun, moon, and planets."[20] Mesopotamian astronomers estimated the length of the year to be 360 days; this became the basis for our division of the compass into 360 degrees. In Egypt, a calendar of 365 days was used. In contrast, the Mesopotamians adopted a lunar calendar that had to be continually adjusted to the seasons. Marshall Clagett (1916–2005) suggested that the need for detailed calculations of the Moon's movements may have been the practical spur that provided the Mesopotamians with a superiority to the Egyptians in astronomy.[21]

*Arguably, medicine has at least an equal claim to "the first science."

Hunters and gatherers have no concept of individual land ownership. But to farmers, property rights are paramount. The need to irrigate crops demanded cooperation on a vast scale along with a command and control structure. Governments came into existence for the purpose of preserving and protecting property rights, regulating the distribution of common resources such as irrigation water, and fostering internal cooperation within a group. Tribal customs were replaced by written laws.

The earliest known set of written laws is the code of Hammurabi (2123–2081 B.C.). Hammurabi was a Mesopotamian king who governed the city of Babylon. His code was written on a stone tablet. In the prologue, Hammurabi explained his purposes. "[The gods] delighted the flesh of mankind by calling me, the renowned prince, the god-fearing Hammurabi, to establish justice in the earth, to destroy the base and the wicked, and to hold back the strong from oppressing the weak."[22]

Hammurabi's code contained 282 separate articles dealing with personal property, real estate, business, family, injuries, and labor. Laws dealing with irrigation illustrated the necessity of established law in complex societies where individual acts could affect others. "If a man has been too lazy to strengthen his dike, and has not strengthened the dike, and a breach has opened in the dike, and the ground has been flooded with water; the man in whose dike the breach has opened shall reimburse the corn [grain] he has destroyed."[23]

Although the code of Hammurabi is the oldest known example of written law, it is apparent from the context of the inscriptions that Hammurabi did not consider himself to be an innovator, but rather was promulgating long-standing legal traditions that existed among his people.[24]

Civilization both freed and bound the individual. Although tribes without written laws appear to allow more individual freedom, persons living in these societies were bound by innumerable customs and taboos. "No human being is so hidebound by custom and tradition as your democratic savage; in no state of society consequently is progress so slow and difficult. The old notion that the savage is the freest of mankind is the reverse of the truth. He is a slave ... [whose] lot is cast from the cradle to the grave in the iron mold of hereditary custom."[25] In a civilization, individuals tend to have a legal status recognized by the state, especially if they own land. Laws not only bind the individual, they also protect him.

The primary crops of the river civilizations were barley and wheat. These grains were ground into flour and baked into bread. The baking ovens were perhaps first nothing more than holes in the ground, but soon above-ground ovens were constructed of bricks. The hardening effect of fire on clay bricks was noted, and someone inductively inferred the possibility of fabricating clay vessels for the storage of both liquids and grains.[26] It is likely that this is the way in which pottery was invented.

Pottery facilitated the storage of grain surpluses, and the wealth of society increased. When potters began to impress their marks on wet clay, writing was invented. Writing was probably first used for creating commercial records. "The temple found it necessary to keep exhaustive records of its income and of its expenditures, done in cuneiform characters on clay tablets which could be fired to make them permanent. It can hardly be doubted that writing was invented for this purpose rather than for literary reasons."[27]

The widespread utilization of furnaces for firing pottery probably led to the discovery of how to manufacture glass. Glass is produced by melting high-silica sand in combination with a naturally alkaline mineral such as soda (sodium carbonate) or potash (potassium carbonate). Current evidence indicates that the first true glass was produced in Mesopotamia around 2500 B.C.[28] An illustration on the wall of an Egyptian tomb dating to 1900 B.C. shows glass-blowers.[29] Thus it appears that glass was being mass-produced in Egypt by this time. Different colors were obtained by mixing in trace amounts of certain metals or their oxides. Adding copper produced a light blue color, the addition of iron created a green tinge, and cobalt was responsible for the deep blue color that is still popular today.

The history of prehistoric cloth is difficult to reconstruct due to the fragile nature of artifactual evidence. But the technology to produce cloth by spinning and weaving appears to date from as early as 5000 B.C.[30] Silk cloth was manufactured in China in 3000 B.C.[31] The earliest samples of cotton cloth date from about 3000 B.C. in India.[32] Wool was common in Scandinavia c. 1000 B.C., but may have been spun earlier.[33] Linen, made from flax, was ubiquitous in the Mediterranean region, and fine linen cloth was manufactured in Egypt as early as 3000 B.C.[34] The robust nature of the prehistoric textile industry is shown by "eight thousand spindle-whorls" recovered from the ruins of Troy.[35]

The first metals to be utilized were those found in their native state that did not have to be refined from ore. These included copper, gold, silver, and iron in the form of metallic meteorites. By 3000 B.C., the Egyptians "were familiar with gold, copper, silver, lead, and iron."[36]

"The earliest metal worked by man was undoubtedly gold, followed closely by native copper and tin."[37] There is an apocryphal story in the *History* of Herodotus (c. 484–425 B.C.), that in Ethiopia gold was so common it was used for prisoners' chains, while copper was scarce and valuable. "The king led them to a prison, where the prisoners were all of them bound with fetters of gold. Among these Ethiopians copper is of all metals the most scarce and valuable."[38]

Diodorus Siculus (fl. 1st century B.C.) described the gold mining practices of the Egyptians. "In the confines of Egypt, and the neighboring countries of Arabia and Ethiopia, there is a place full of rich gold mines ... the earth which is hardest and full of gold they soften by putting fire under it, and then work it out with their hands: the rocks thus softened, and made more pliant and yielding, several thousands of profligate wretches break it into pieces with hammers and pickaxes."[39]

To be sentenced to work in the Egyptian gold mines was a fate worse than death. "The kings of Egypt condemn to these mines notorious criminals, captives taken in war, persons sometimes falsely accused, or such against whom the king is incensed ... [and] all are driven to their work with blows and cudgelling, till at length, overborn with the intolerable weight of their misery, they drop down dead in the midst of their insufferable labors; so that these miserable creatures ... long for death, as far more desirable than life."[40]

The Egyptians also had smelting techniques for refining gold ore, and separating it from other metals. "[They] put it into earthen urns, and, according to the quantity of the gold in every urn, they mix with it some lead, grains of salt, a little tin, and barley bran. Then covering every pot close, and carefully daubing them over with clay, they put them in a furnace, where they abide five days and nights together; then, after they have stood to cool a convenient time, nothing of the other matter is to be found in the pots, but only pure refined gold."[41]

The Greek geographer Strabo (c. 64 B.C.–A.D. 24) described the collection of placer gold with fleeces. "The winter torrents are said to bring down even gold, which the barbarians collect in troughs pierced with holes, and lined with fleeces; and hence the fable of the golden fleece."[42]

It is probable that silver was more common than gold.[43] In the *Iliad*, Homer described Achilles' armament as being ornamented with silver. "On his legs he set the fair greaves fitted with silver ankle-pieces, and next he donned the cuirass about his breast. Then round his shoulders he slung the bronze sword silver-studded."[44]

Strabo noted that Iberia (Spain) had rich mines. "Of metals, in fact, the whole country of the Iberians is full ... gold, silver, copper, and iron, equal in amount and of similar quality."[45] Strabo also described processes for refining ore. "When they have melted the gold, and purified it by means of a kind of aluminous earth, the residue left is electrum. This, which contains a mixture of silver and gold, being again subjected to the fire, the silver is separated and the gold left."[46]

Although gold was the most beautiful, the most common and useful of the metals was copper. It is plausible that the refining of copper from ore was discovered when a piece of malachite fell into a fire and generated a bead of copper. Malachite is a greenish-colored ore of copper that was also used as a cosmetic pigment; it is widely distributed throughout the Middle East.[47] Copper mines on Mount Sinai were worked as early as 5000 B.C.,[48] and the metal was in common use by 3000 B.C.[49]

Some copper ores naturally contain admixtures of tin. When tin is mixed with copper, the alloy of bronze is created. Bronze is twice as hard as copper, and therefore much more useful for many applications. Starting around 4000 B.C., copper and bronze tools started to replace wooden and stone implements. This was the beginning of a Bronze Age that lasted for nearly 3000 years.

The widespread use of copper and bronze fostered the creation of metallurgical technologies. Once the techniques for refining, casting, and shaping copper and bronze had been worked out, the processing of lead, tin, and silver followed in a straightforward way. But working with iron was more difficult, for a variety of reasons. Much higher temperatures were needed to smelt iron, so early experiments with ores would have failed. The physical properties of iron are also highly dependent on its carbon content, and this was difficult to control. Blacksmiths would have been initially misled by comparisons with familiar metals. Iron is hardened by quenching red-hot metal in water, but copper and bronze are softened by the same procedure. By about 1200 B.C., the techniques for mining, refining, and shaping iron had been developed, and the Iron Age began in the Middle East.

Engineers charged with building canals, irrigation systems, temples and defensive walls needed mathematics capable of calculating volumes, areas, and angles.[50] Herodotus described how the Egyptians developed geometry for the practical necessity of demarcating land boundaries.

> [The Pharaoh] ... made a division of the soil of Egypt among the inhabitants, assigning square plots of ground of equal size to all, and obtaining his chief revenue from the rent which the holders were required to pay him year by year. If the river carried away any portion of a man's lot, he appeared before the king, and related what had happened; upon which the king sent persons to examine, and determine by measurement the exact extent of the loss; and thenceforth only such a rent was demanded of him as was proportionate to the reduced size of his land. From this practice, I think, geometry first came to be known in Egypt, whence it passed into Greece.[51]

The pyramids and obelisks of Egypt are testimony to the intelligence and abilities of the ancient engineers. To this day, we do not know for sure how the Egyptians raised their monuments with the technologies that were available to them.

The most important early invention was writing. Potters had discovered they could make distinctive marks in wet clay, and the mark would be preserved by the firing process that turned the clay into pottery. The first writing consisted of pictures representing both nouns and verbs. The Mesopotamians kept records on clay tablets. Although the first use of writing was probably for the purpose of keeping commercial records, the ability to make permanent records "had the revolutionary effect of opening knowledge claims to the possibility of inspection, comparison, and criticism."[52] Thus writing was likely an indispensable adjunct to the development of philosophy, science, and serious inquiries into the nature of the world.

It became easier to write and distribute books after the Egyptians invented *papyrus*, a writing material made from the papyrus reed. Individual reeds were sectioned into long, narrow slices. A sheet was constructed by laying these slices side-by-side and gluing them together. Individual sheets of papyrus were then fastened end-to-end to make a long roll, and the roll was wound around a wooden spool.

Papyrus was a profitable export for the Egyptians, and its distribution facilitated the growth of knowledge. Egyptian mythology credits the invention of writing to the god Thoth, the god of wisdom and magic, patron of learning and the arts. Egyptian statues picture Thoth as a man with the head of an ibis, a large wading bird with a long, slender bill. The legend related by Plato (428–348 B.C.) in the dialogue *Phaedrus* is that Thoth presented the invention of writing to the Egyptian king, claiming that writing would "make the Egyptians wiser and give them better memories; it is a specific both for the memory and for the wit."[53] However, the king was unimpressed. He told Thoth that "this discovery of yours will create forgetfulness in the learner's souls, because they will not use their memories; they will trust to the external written characters and not remember of themselves."[54]

The most significant event in the history of writing occurred when pictograms were replaced by a phonetic system in which the written symbols represented not individual objects or actions, but sounds. The transition to a phonetic system probably occurred through the association of particular ideographic symbols with the sound of the object they represented. Presented for the first time with a new object for which no ideogram existed, a scribe naturally responded by writing it phonetically.[55] The origins of the first phonetic alphabet are obscure but appear to be Phoenician. However, the Phoenicians had only symbols for consonants, not vowels. The first fully functional phonetic writing was done by the Greeks around 900 B.C. (The word *alphabet* is derived by conjoining the first and second letters, *alpha* and *beta,* of the Greek alphabet). Hieroglyphic systems required the memorization of thousands of symbols, and literacy was largely confined to a scribal-priestly class. The adoption of a phonetic alphabet that consisted of fewer than thirty symbols made it possible for greater numbers of people to acquire literacy and thus fostered the transmission and preservation of knowledge.[56]

CHAPTER 1

# The Greeks (c. 600–300 B.C.)

The origin of science is traditionally ascribed to Greek natural philosophers who lived and wrote in the 6th century B.C. Greek philosophers borrowed from, and built upon, the work of Mesopotamians and Egyptians. But "the surviving literature [from Egypt and Mesopotamia] reveals a great emphasis on mythology and religion as the means of explaining the creation of the world and its operations ... what passed for natural philosophy among the ancient Egyptians was never distinct from religion and magic."[1]

John Burnet (1863–1928) explained, "It is an adequate description of science to say that it is thinking about the world in the Greek way. That is why science has never existed except among peoples who have come under the influence of Greece."[2] Burnet maintained that "rational science is the creation of the Greeks, and we know when it began. We do not count as philosophy anything anterior to that."[3]

The Mediterranean world was an ideal locus for the development of innovative ideas because it was both diverse and unified. The separation provided by the physical barrier of the Mediterranean Sea allowed individual civilizations to develop and flourish. The physical geography of the Mediterranean region made it more difficult to establish a single political and intellectual orthodoxy that would have strangled and suppressed innovation. But the partition was not so severe as to restrict the flow of information. There was an intellectual cross-fertilization, with different cultures exchanging technologies and ideas.[4]

Although there was a transmission of ideas between the Greeks and other cultures, Burnet argued that it was improbable that the Greeks imported science rather than invented it.

> If the Egyptians had possessed anything that could rightly be called mathematics, it is hard to understand how it was left for Pythagoras and his followers to establish the most elementary propositions in plane geometry; and, if the Babylonians had really any conception of the planetary system, it is not easy to see why the Greeks had to discover bit by bit the true shape of the Earth and the explanation of eclipses.... [I]f we mean by science what Copernicus and Galileo and Kepler, and Leibniz and Newton meant, there is not the slightest trace of that in Egypt or even in Babylon.... [M]odern Science begins just where Greek science left off, and its development is clearly to be traced from Thales to the present day.[5]

Before Greek civilization blossomed in the 6th century B.C., the Egyptians and other peoples had made important scientific contributions. But their cultures became stagnant. The zenith of Egyptian science was from the 20th through the 17th century B.C.[6] The Golenishchev papyrus documents that the Egyptians discovered the correct mathematical formula for the volume of a pyramid around 1900 B.C. George Sarton (1884–1956) described the solution as "breathtaking," but then went on to note that although the Egyptians worked for an additional three thousand years, they never found anything better.[7]

The point can be reinforced by comparing other aspects of Greek and Egyptian culture in the first millennium B.C. Greek culture was vital, inventive, and fertile, "it possessed no uniform and universally acknowledged dogmatic system."[8] But Egypt was stagnant and devoid of intellectual innovations. By the middle of the ninth century B.C., the Greeks were utilizing the world's first fully phonetic alphabet. The superiority of a phonetic system of writing is both obvious and overwhelming. The Egyptians had begun by incorporating into their writing a number of symbols that represented consonant sounds. However, while their neighbors, the Greeks, had made the transition to a fully phonetic system, the Egyptians insisted on retaining an antiquated and inferior system of hieroglyphics for thousands of years.[9] The Egyptians fared no better in mathematics. Their mathematical science was "grossly empirical" in nature, and their system of enumeration "awkward and cumbersome."[10]

The differences between the Greeks and Egyptians in this period of time can also be illustrated by comparing their architecture and sculpture. The Greeks surpassed the Egyptians in both areas. The Great Pyramid of Giza (c. 2650 B.C.) is monumental, but there is nothing in Egyptian architecture to compare with the beauty and symmetry of the Parthenon (completed in 432 B.C.).

Like their science and mathematics, Egyptian art became stagnant. "It was fettered by conventions and formulae."[11] Egyptian sculptures are impressive, but frozen into stereotypical forms. "On first entering an Egyptian museum, we are struck by the apparent resemblance between all the figures.... [T]hroughout its long career, Egyptian art never succeeded in casting off the trammels of certain conventions."[12]

The Greeks learned sculpture from the Egyptians, but went on to transcend their efforts. Greek sculpture is lifelike, it breathes and moves; it is undeniably beautiful, and invokes a range of emotions that Egyptian creations cannot. There is nothing in ancient Egyptian art to compare with the *Aphrodite of Melos* or the *Hermes of Praxiteles*. Greek sculpture from the Classical Period (c. 480–330 B.C.) was not equaled until Michelangelo finished *David* in A.D. 1504.

The Greeks were significantly influenced by surrounding cultures and civilizations. But Greece was undeniably the primogenitor of modern science.

## *Intellectual Freedom, Naturalism, and Demonstration*

There were three significant factors that contributed to the development of ancient Greek science and philosophy. First, Greece was an open society. "There was no dogmatic or systematic religious orthodoxy."[13] Philosophers enjoyed intellectual freedom, and there was a tradition of critical discussion.[14] Pupils were not only allowed to question their teachers, but were encouraged to do so.[15] Critical analysis was considered essential for the discovery of truth.

Personal feelings were secondary to the pursuit of truth. If someone's feelings were hurt by criticism, it was taken to be of no account. In the first book of Plato's *Republic*, the speakers attempted through critical dialogue to define the nature of justice. One of the participants became annoyed, at which point another explained to him, "if we were looking for gold, you can't suppose that we would willingly let mutual politeness hinder our search and prevent our finding it. Justice is much more valuable than gold, and we aren't likely to cramp our efforts to find it by any idiotic deference to each other."[16] If someone could offer a constructive criticism that would further the all-important search for truth, he was to be regarded as a friend, not an enemy. In Plato's *Timaeus*, it is noted that anyone who can present a better plan "shall carry off the palm, not as an enemy, but as a friend."[17]

Concomitant with intellectual freedom, the Greeks lived in relative political freedom. The Greek *polis* (city-state) "permitted freedom of thought and speech, especially in debate of public affairs, and it gave the individual ample freedom in his private life ... it [the *polis*] stimulated an excessive individualism."[18]

In the fifth century B.C., the Athenians "tolerated considerable variety of opinion and great license of speech."[19] Pericles (c. 495–429 B.C.) "governed by persuasion. Everything was decided by argument in open deliberation, and every influence bowed before the ascendancy of mind."[20]

The second factor contributing to the rise of science in Greece was naturalism. *Naturalism* is the explanation and interpretation of phenomena in terms of natural law rather than the interdiction of supernatural forces or beings. The Ionian Greeks were the first to propose natural explanations. "It is in the Greek writers that we must seek the inauguration of the scientific epoch. It is in them that, for the first time, appears the systematic effort to ascertain the relations of things objectively, to detect the causes of all changes as inherent in the things themselves, and to reject all supernatural or outlying agencies."[21]

Prior to the advent of naturalism, "the answers afforded by polytheism gave more satisfaction than could have been afforded by any other hypothesis.... The question asked was not, what are the antecedent conditions or causes of rain, thunder, or earthquakes, but, who rains and thunders? Who produces earthquakes? The Hesiodic Greek was satisfied when informed that it was Zeus or Poseidon. To be told of physical agencies would have appeared to him not merely unsatisfactory, but absurd, ridiculous, and impious."[22]

The origin of naturalism appears to be uniquely Greek. The Chinese did not develop the conception of a cosmos governed by impersonal natural law. "The Chinese world-view depended upon a totally different line of thought. The harmonious cooperation of all beings arose, not from the orders of a superior authority external to themselves, but from the fact that they were all parts in a hierarchy of wholes forming a cosmic pattern, and what they obeyed were the internal dictates of their own natures."[23]

The corollary to naturalism was uniformity. *Uniformity* is the supposition that the laws of nature are invariant throughout time and space. Uniformity is essential to naturalism. Without uniformity, nature is as arbitrary as the whims of the gods. With uniformity, the universe, formerly governed by the unpredictable and capricious notions of supernatural beings, became a *cosmos*, a place where cause and effect were regulated by inflexible natural law. The world became a place that could be understood.

The third factor in the development of Greek science was the idea of demonstration, or method. By developing mathematics from a set of empirical rules to an abstract science of demonstration, the Greeks found that it was possible to establish a method by which demonstrable truth could be established. *Opinion*, on which everyone disagreed, was replaced by *demonstration*, on which everyone agreed. Everyone who understood a geometrical proof necessarily accepted it. The science of geometry provided the idea that it was possible to establish a method by which certain truth could be found. The corollary to geometrical reasoning was verbal logic, the idea that there were correct ways of thinking and reasoning. Greek demonstration, and especially Aristotelean logic, evolved into the modern scientific method.

## *The Ionians (c. 600–400 B.C.)*

Around 1000 B.C., the Greeks started to spread eastward through the Aegean Sea to colonize the shores of Asia Minor and numerous islands off the coast. Strabo identified Athens as the source of the eastward migration. "The Ionians ... were descendants of the Athenians.... The Athenians sent out a colony of Ionians ... and the tract of country which they occupied was called Ionia after their own name ... and the inhabitants Ionians ... who were distributed among twelve cities."[24]

Herodotus claimed that Ionia had the best climate in the known world. "The Ionians of Asia ... have built their cities in a region where the air and climate are the most beautiful in the whole world."[25] In these lands, the summers were hot and dry, the winters mild and rainy. The main crops were wheat, barley, grapes, figs, and olives.

Some of the Greek colonists did not bother to bring women with them. According to Herodotus, they reproduced by the expedient of killing the male natives and marrying their daughters. "Even those who came from ... Athens, and reckon themselves the purest Ionians of all, brought no wives with them to the new country, but married Carian girls, whose fathers they had slain."[26] (Caria was a region of Anatolia located south of Ionia on the Aegean Coast and in Asia Minor.)

It was in Ionia that science was born in the 6th century B.C. In the introduction to *Early Greek Philosophy*, John Burnet testified to the intellectual ascendancy of the Ionians by noting, "it is a remarkable fact that every one of the men whose work we are about to study was an Ionian."[27] "These early Greek thinkers mark a dramatic break with all that went before in the Greek and non–Greek worlds."[28]

There were momentous intellectual developments throughout much of the world in the sixth century B.C. Monotheism was replacing polytheism. In India, Buddha (c. 563–483 B.C.) founded Buddhism, and in China, Confucius (c. 551–479 B.C.) founded the philosophy named after him. In Palestine, Hebrew prophets were writing many of the books of the Old Testament. Ionia was the center of communication between Western and Eastern civilizations.

The greatest city in Ionia was Miletus. Herodotus referred to Miletus as "the glory of Ionia."[29] It was there that the first school of Greek scientific thought arose. The Milesians were great traders and travelers, interacting with Persia, Egypt, and Phoenicia. At one time, Miletus had as many as eighty trading colonies and was one of the wealthiest cities in the world. According to Pliny the Elder (A.D. 23–79), "Miletus ... [was] the mother of more than ninety cities, founded upon all seas."[30] In addition to its advantageous position on the great east-west trading routes, Miletus was ideally suited for a seaport, possessing four separate harbors. The trading activities of the Milesians reached as far as India and the city was a fertile ground for the breeding of new ideas. Ionia was a "country without a past," owing no intellectual allegiance to older mythologies or Olympian polytheism.[31]

The emergence of Ionian natural philosophy in the 6th century B.C. appears to be without historical precedent. Earlier writings invoked supernatural explanations. Hesiod's (c. 700 B.C.) cosmogeny, *Theogony*,[32] is mythic, religious, and allegorical; "there is as yet no question of accounting for the origin of things by natural causes."[33] But "the spirit of the Ionians in Asia was ... thoroughly secular."[34] Ionian science was likely spawned in the interaction of cultures and ideas that takes place in a commercial trading center. The crossroads of trade are the breeding grounds for intellectual ferment and the exchange of ideas. It is "impossible to separate the interchange of wares from the interchange of knowledge and ideas."[35] When people of different cultures, religions, and traditions interact, the active intellect must conclude that there is "more than one way of looking at things and of solving problems."[36]

## *Thales (c. 624–547 B.C.)*

The Milesian Thales (c. 624–547 B.C.) is the Ionian generally credited with beginning Greek science. "He [Thales] attained note as a scientific thinker and was regarded as the founder of Greek philosophy because he discarded mythical explanations of things, and asserted that a physical element, water, was the first principle of all things."[37]

Thales is largely regarded as the first scientific thinker because he was the first to invoke naturalism. Of equal importance, Karl Popper (1902–1994) has argued that Thales is the founder of science because he not only tolerated criticism, but encouraged it. Rather than insist on the propagation of his ideas as dogma, Thales began the tradition of critical discussion and intellectual freedom. This "leads to the tradition of bold conjectures and of free criticism, the tradition which created the rational or scientific attitude, and with it our Western civilization."[38]

Thales left no surviving writings.[39] What we know of him and his ideas comes down to us from comments by other authors. In his youth, Thales visited Egypt and became acquainted with the mathematical discipline of geometry. Although the Egyptians invented geometry, their science was purely empirical. Egyptian geometry consisted of a series of rules of thumb. Thales generalized the special cases known to the Egyptians, and made original discoveries of his own.[40]

Proclus (A.D. 411–485) claimed that Thales first brought the empirical geometry of the Egyptians to Greece and made it a science. "Geometry was found out among the Egyptians from the distribution of land. When Thales, therefore, first went into Egypt, he transferred this knowledge from thence into Greece: and he invented many things himself, and communicated to his successors the principles of many. Some of which were, indeed, more universal, but others extended to sensibles."[41]

Thales demonstrated that the height of the Egyptian pyramids could be calculated geometrically by measuring the length of their shadows. All that was necessary was to realize that the shadow and height of a stick were in the same ratio as the shadow and height of a pyramid. "With little labor and no help of any mathematica, [or] instrument, you took so truly the height of one of the pyramids; for fixing your staff erect at the point of the shadow which the pyramid cast, and thereby making two triangles, you demonstrate, that what proportion one shadow had to the other, such the pyramid bore to the stick."[42]

In *Metaphysics*, Aristotle (384–322 B.C.) described Thales as the founder of natural philosophy, and attributed to him the belief that water was the elementary substance of the universe from which everything else was constituted. "Thales, the founder of this sort of philosophy, says that it is water ... [that] is the element and first principle of things."[43]

The nature of the fundamental substance was a primary concern of the Ionian natural philosophers. It was an interesting and fundamental question, and one not entirely removed from the realm of observation. The early philosophers must have observed, for example, that water evaporated, and thus apparently changed into air. Everything changed. Wood was consumed by fire, and solid substances such as salt apparently changed into water when they dissolved. The natural question that arose was if there was an irreducible and elementary substance whose permutations and changes gave rise to the observable phenomena.

The fundamental element of the universe is not water, but Thales and his Ionian successors had the correct approach. The entire point of science is to simplify. American geologist M. King Hubbert (1903–1989) maintained that the process of scientific discovery does not result in increased complexity, but in simplification. "The entire history of science has been the progressive reduction of one chaos of phenomena after another into a form that is within the powers of comprehension of an average human being."[44]

Although the idea of invoking water as a basic element may seem strange to the modern mind, it was a reasonable theory because it was consistent with the observations available to Thales. John Burnet explained:

> It is not hard to see how meteorological considerations may have led Thales to adopt the views he did. Of all the things we know, water seems to take the most various shapes. It is familiar to us in a solid, a liquid, and a vaporous form, and so Thales may well have thought he saw the world-process from water and back to water again going on before his eyes. The phenomenon of evaporation naturally suggests that the fire of the heavenly bodies is kept up by the moisture they draw from the sea.... Water comes down again in rain; and lastly, so the early cosmologists thought, it turns to earth.... Lastly, they thought, earth turns once more to water — an idea derived from the observation of dew, night-mists, and subterranean springs.[45]

According to Aristotle, Thales maintained that the Earth floated on water.[46] In the time of Thales, it was not yet recognized that the Earth was spherical. Aristotle referred to Thales successors, Anaximenes (c. 570–500 B.C.), Anaxagoras (c. 500–428), and Democritus (460–370 B.C.), as all believing in a flat Earth.[47] Accordingly, it seems that Thales envisaged the Earth as a round, flat disk floating on, and surrounded by, water.[48]

It has been claimed that Thales was the first Greek to study the science of astronomy. "Thales was the first of the Greeks to devote himself to the study and investigation of the stars, and was the originator of this branch of science."[49] Lucius Apuleius (c. A.D. 124–180) listed Thales' scientific discoveries, albeit with considerable ambiguity and lack of specificity.

> Thales of Miletus was easily the most remarkable of the famous seven sages. For he was the first of the Greeks to discover the science of geometry, was a most accurate investigator of the laws of nature, and a most skilful observer of the stars. With the help of a few small lines he discovered the most momentous facts: the revolution of the stars, the blasts of the winds, the wanderings of the stars, the echoing miracle of thunder, the slanting path of the zodiac, the annual turnings of the sun, the waxing of the moon when young, her waning when she has waxed old, and the shadow of her eclipse; of all these he discovered the laws. Even when he was far advanced into the vale of years, he evolved a divinely inspired theory concerning the period of the sun's revolution through the circle in which he moves in all his majesty.[50]

Plato related the story that Thales was so preoccupied with studying celestial phenomena that he did not watch where he was walking and fell into a well. "[There is a] jest which the clever witty Thracian handmaid is said to have made about Thales, when he fell into a well as he was looking up at the stars. She said, that he was so eager to know what was going on in heaven, that he could not see what was before his feet."[51]

According to Herodotus, Thales attained notoriety as a sage after he successfully predicted a solar eclipse in 585 B.C. "Just as the battle was growing warm, day was on a sudden changed into night. This event had been foretold by Thales, the Milesian, who forewarned the Ionians of it, fixing for it the very year in which it actually took place."[52] Modern astronomy fixes the exact date as May 28, 585 B.C. The eclipse interrupted a battle between the Medes and Lydians. The combatants interpreted the eclipse as divine disapproval of their warfare, and they reached a peace agreement.

There is evidence that Thales understood the cause of both solar and lunar eclipses.

"The eclipses of the sun take place when the moon passes across it in direct line.... Thales et al. agree with the mathematicians that the monthly phases of the moon show that it travels along with the sun and is lighted by it, and eclipses show that it comes into the shadow of the earth, the earth coming between the two heavenly bodies and blocking the light of the moon."[53]

But how Thales was able to successfully predict the exact date and location of a solar eclipse is unknown. The ability to make such predictions requires a knowledge of astronomy that is beyond what was known in his time. It is especially difficult to predict the relatively small area of the Earth's surface over which a total eclipse of the Sun will occur. It is possible that Thales only predicted the date of the eclipse.

Thales believed that all of nature was alive, and that not only men, but animals, plants, and stones possessed immortal souls. Everything contained a vital power that changed external forms, but itself remained unchanged. "Thales said that the mind in the universe is god, and the all is endowed with soul and is full of spirits; and its divine moving power pervades the elementary water."[54] According to Plutarch, "Thales said that the intelligence of the world was God."[55]

In *Politics*, Aristotle related that Thales wished to demonstrate that as a philosopher he was poor by choice, not by inability. Thales used his understanding of meteorology to predict a bountiful harvest of olives, and placed a deposit on all available olive presses, reserving their use for himself. By renting the presses out, he was thus assured of a significant profit.

> He [Thales] was reproached for his poverty, which was supposed to show that philosophy was of no use. According to the story, he knew by his skill in the stars while it was yet winter that there would be a great harvest of olives in the coming year; so, having a little money, he gave deposits for the use of all the olive-presses in Chios and Miletus, which he hired at a low price because no one bid against him. When the harvest-time came, and many were wanted all at once and of a sudden, he let them out at any rate which he pleased, and made a quantity of money. Thus he showed the world that philosophers can easily be rich if they like, but that their ambition is of another sort.[56]

But Aristotle insightfully pointed out that monopolizing the olive presses would have allowed Thales to set the price regardless of the size of the harvest. "He [Thales] is supposed to have given a striking proof of his wisdom, but, as I was saying, his device for getting wealth is of universal application, and is nothing but the creation of a monopoly."[57]

In *The Lives and Opinions of Eminent Philosophers**, Diogenes Laërtius (3rd century A.D.) attributed a number of apothegms to Thales. These include: "intellect is the swiftest of things, for it runs through everything: necessity is the strongest of things, for it rules everything: time is the wisest of things, for it finds out everything.... When he was asked what was very difficult, he [Thales] said 'To know one's self.' And what was easy, 'To advise another.'"[58]

According to Diogenes Laërtius, "Thales died while present as a spectator at a gymnastic contest, being worn out with heat and thirst and weakness, for he was very old."[59]

## *Anaximander (c. 610–546 B.C.)*

Anaximander (c. 610–546 B.C.) was a Milesian contemporary of Thales. Strabo also described Anaximander as a student of Thales. "Illustrious persons, natives of Miletus, were Thales ... [and] his disciple Anaximander."[60]

*Although Diogenes Laërtius' biographies are "devoid of all historical or philosophical discrimination ... [his writing] is yet distinguished by the circumstance that in his narrative the names of the earliest authorities still appear" (Blakesley, J. W., 1834, *A Life of Aristotle*: John W. Parker, London, p. 10).

Diogenes Laërtius stated that Anaximander was "the first discoverer of the gnomon,"[61] a vertical stick used as a sundial. But it is likely that Anaximander simply imported the invention from Babylonia. According to Herodotus, "the sun-dial ... and the gnomon ... were received by the Greeks from the Babylonians."[62]

Thales thought the Earth floated in water ("he declares that the Earth rests on water"[63]), but Anaximander believed the Earth was "freely suspended in space" without support.[64] The idea of an Earth that floats freely in space without support is also an indispensable corollary to heliocentrism. Karl Popper claimed, in support of his thesis that scientific theories can originate from logical or imaginative conjectures rather than systematic induction, that Anaximander's conception could not be based on experience or observation.[65] On the contrary, it can be argued that Anaximander's idea of an Earth floating freely in space was entirely inductive, based upon the observation that the Moon, Sun, planets, and stars appear to be suspended in free space without support.

Anaximander described the shape of the Earth as cylindrical, not spherical, "the form of it is curved, cylindrical like a stone column."[66] Anaximander was the "first to speculate on the relative distances of the heavenly bodies."[67] He made the conjecture that the Sun, stars, and Moon, were holes in rings filled with fire. "The heavenly bodies are a wheel of fire, separated off from the fire of the world, and surrounded by air. And there are breathing-holes, certain pipe-like passages, at which the heavenly bodies show themselves. That is why, when the breathing-holes are stopped, eclipses take place. And the moon appears now to wax and wane because of the stopping and opening of the passages. The wheel of the Sun is 27 times the size of (the Earth, while that of) the Moon is 18 times as large. The Sun is the highest of all, and lowest are the wheels of the stars."[68]

While Thales held that the primary substance was water, Anaximander instead proposed that matter had formed by differentiating itself from an ambiguous concept or principle that translates as the *infinite*.

Anaximander's system was described by Theophrastus (c. 371–287 B.C.), Aristotle's student and successor at the Lyceum. "Anaximander of Miletus, son of Praxiades, a fellow-citizen and associate of Thales, said that the material cause and first element of things was the *infinite*, he being the first to introduce this name of the material cause. He says it is neither water nor any other of the so-called elements, but a substance different from them which is infinite, from which arise all the heavens and the worlds within them."[69]

Anaximander recognized that water could not be the fundamental substance, because its properties were opposed to those of fire, another element. If any one element were primary, it would have destroyed the elements contrary to its nature by virtue of its primacy. In *Physics*, Aristotle explained, "they are in opposition one to another — air is cold, water moist, and fire hot — and therefore, if any one of them were infinite, the rest would have ceased to be by this time."[70]

As perhaps a natural corollary to his belief that everything originated in the *infinite*, both time and space were large in Anaximander's cosmology. He believed there were "innumerable worlds in the boundless [cosmos]."[71] In *De Natura Deorum* (*Of the Nature of the Gods*), Cicero (106–43 B.C.) wrote that Anaximander even conceived of the gods as being mortal. "It was Anaximander's opinion that the gods were born; that after a great length of time they died; and that they [there?] are innumerable worlds. But what conception can we have of a deity not eternal?"[72]

In ancient Greece, meteorology was not yet considered a separate science from astronomy.[73] Anaximander held the common view, that thunder and lightning "were all caused by the blast of the wind. When it is shut up in a thick cloud and bursts forth with violence, then the tearing of the cloud makes the noise, and the rift gives the appearance of a flash in contrast with the blackness of the cloud."[74] Pliny the Elder stated that Anaximander predicted an earthquake in Sparta. "Anaximander the Milesian ... warned the Lacedaemonians to beware of their city and their houses. For he predicted that an earthquake was at hand."[75]

Diogenes Laërtius credited Anaximander as "the first person ... who drew a map of the earth and sea."[76] Anaximander was also recognized by Strabo as being one of the first geographers. "If the scientific investigation of any subject be the proper avocation of the philosopher, geography ... is certainly entitled to a high place.... They who first ventured to handle the matter were distinguished men ... [and include] Anaximander the Milesian."[77]

Anaximander holds the distinction of being the person who first proposed a theory of biological evolution. He believed that human beings evolved from another form of life, and gave a reason for this belief. "Living creatures arose from the moist element as it was evaporated by the sun. Man was like another animal, namely, a fish, in the beginning.... [Anaximander] says that originally man was born from animals of another species. His reason is that while other animals quickly find food by themselves, man alone requires a lengthy period of suckling. Hence, had he been originally as he is now, he never would have survived."[78]

Diogenes Laërtius gave the time of Anaximander's death as the fifty-eighth Olympiad, at which time he was 64 years old.[79]

## *Heraclitus (c. 540–480 B.C.)*

The Ionian philosopher Heraclitus (c. 540–480 B.C.) was "renowned in antiquity for his obscurity."[80] Diogenes Laërtius further described him as "arrogant," and stated that toward the end of his life he became a "complete misanthrope."[81]

Heraclitus was the author of a book of which a number of fragments survive. He wrote "intentionally in an obscure style, in order that only those who were able men might comprehend it, and that it might not be exposed to ridicule at the hands of the common people."[82]

Heraclitus' philosophy was difficult, but apparently was based on the idea that there is an underlying unity in nature, a divine intelligence in nature termed *Logos*. "Not on my authority, but on that of truth, it is wise for you to accept the fact that all things are one."[83] "From all things arises the one, and from the one all things."[84] Heraclitus was the first to speak of a *cosmos*, an "ordered universe."[85]

But the unity of nature was hidden in a constant flux. "Nature loves to hide," and there is a "hidden harmony [that] is better than manifest."[86] Heraclitus' doctrine of continual and pervasive change in nature is exemplified in his statement that "you could not step twice in the same rivers."[87] In the dialogue *Kratylos*, Plato stated "all things are in motion, according to Heraclitus."[88] In Karl Popper's view, Heraclitus' philosophy of unceasing change meant "things are processes."[89]

Paradoxically, unity in nature was illustrated by pairs of opposites. "Cool things become warm, the warm grows cool; the wet dries, the parched becomes wet."[90] There was "harmony in contrariety, as in the case of the bow and the lyre."[91] "Opposition unites. From what draws apart results the most beautiful harmony. All things take place by strife."[92] "Good and bad are the same."[93] Aristotle described Heraclitus' philosophy in *Ethics*. "Heraclitus says that opposition unites, and that the most beautiful harmony results from opposites, and that all things come into being through strife."[94]

Heraclitus regarded fire as the fundamental element, because only fire is in a continual state of flux. "All things are exchanged for fire, and fire for all things."[95] All things began and ended in fire. "Fire is the source of all things, and ... all things are resolved into fire again."[96] The apparent conflict between Heraclitus considering both fire and *logos* to be the fundamental element of the cosmos can be reconciled by considering that "the *logos* on its material side is fire, and fire on its spiritual side is the *logos*."[97]

In *De Caelo*, Aristotle interpreted Heraclitus' cosmogeny as cyclic. "And others in their turn say that sometimes combination is taking place, and at other times destruction, and that this will always continue, as [says] ... Heraclitus."[98] Because Heraclitus regarded nature to be in a perpetual state of flux, it implied that philosophy or science based on observation was inadequate. In *Metaphysics*, Aristotle explained, "there can be no science of things in a state of flux."[99]

Diogenes Laërtius gave a brief summary of Heraclitus' views. "His doctrines are of this kind. That fire is an element, and that it is by the changes of fire that all things exist; being engendered sometimes by rarity, sometimes by density. But he explains nothing clearly. He also says, that everything is produced by contrariety, and that everything flows on like a river; that the universe is finite, and that there is one world, and that that is produced from fire, and that the whole world is in its turn again consumed by fire at certain periods, and that all this happens according to fate."[100]

Heraclitus explained the Sun, Moon, and stars as "solid bowls filled with fire."[101] The fire was fueled by moist exhalations from the sea that rose to the heavens. Eclipses occurred when the heavenly bowls somehow became inverted.[102]

In an apparent attempt to explain himself and his life, Heraclitus said "I inquired of myself,"[103] or "I have sought to understand myself."[104] In other apothegms, he said that "people ought to fight for their law as for a wall [of a city],"[105] "it is better to conceal ignorance than to put it forth into the midst,"[106] and "war is the father of all."[107]

A surviving fragment of Heraclitus' work records his criticism of other philosophers. He wrote "the learning of many things teacheth not understanding, else would it have taught Hesiod and Pythagoras, and again Xenophanes and Hekataios."[108] This is an indication that "the early Greek philosophers knew and criticized one another's ideas."[109] The emergence of Greek philosophy and science required intellectual freedom and an inquiring and critical attitude. G. E. R. Lloyd (b. 1933) theorized that this atmosphere of intellectual freedom likely was fostered in the political freedom and tradition of open debate in the Greek city-states.

Diogenes Laërtius related the story of Heraclitus' death at the age of seventy. "He was attacked by the dropsy, and so then he returned to the city, and asked the physicians, in a riddle, whether they were able to produce a drought after wet weather. And as they did not understand him ... he placed himself in the sun, and ordered his servants to plaster him over with cow-dung; and being stretched out in that way, on the second day he died."[110]

## *Leucippus (fl. 430 b.c.) and Democritus (460–370 b.c.)*

The Ionian Leucippus (fl. 430 B.C.) was "the inventor of the atomic system."[111] "We know next to nothing about his life,"[112] but it was Leucippus who answered the question first posed by Thales, as to the nature of the ultimate or fundamental substance.

Better known as the expositor of atomism was Leucippus' student, Democritus (460–370 B.C.). Democritus was born in Abdera in Thrace. According to Diogenes Laërtius, "he [Democritus] was a pupil of some of the Magi and Chaldeans whom Xerxes had left with his father as teachers.... From these men he, while still a boy, learned the principles of astronomy and theology. Afterwards, his father entrusted him to Leucippus."[113]

Democritus "traveled to Egypt to see the priests there, and to learn mathematics ... and he proceeded further to the Chaldeans, and penetrated into Persia."[114] He may have also traveled to India and Ethiopia. In reference to his own travels, Democritus said "of all my contemporaries I have covered the most ground in my travels, making the most exhaustive inquiries the while; I have seen the most climates and countries and listened to the greatest number of learned men."[115]

Diogenes Laërtius described Democritus as a polymath. "He was veritably a master of five branches of philosophy. For he was thoroughly acquainted with physics, and ethics, and mathematics, and the whole encyclic system, and indeed he was thoroughly experienced and skilful in every kind of art."[116] Democritus' contributions to mathematics were attested to by Archimedes (287–212 B.C.), who credited him with being the first to "to state the important propositions that the volume of a cone is one third of that of a cylinder having the same base and equal height, and that the volume of a pyramid is one third of that of a prism having the same base and equal height."[117]

Democritus was known as the "laughing philosopher."[118] The Latin poet Horace (65–8 B.C.) wrote, "Democritus, if now on Earth, had laughed to see the gaping vulgar stare."[119] In his *Epistles*, the Roman Emperor Julian (A.D. 332–363), related the story of how the Persian king, Darius, had hired Democritus to resurrect his wife from the dead. Democritus informed the king that he was ready to proceed. Democritus only required that Darius "inscribe on the tomb of your wife the names of three who have never known affliction.... Darius hesitating, and not being able to recollect any one who had not experienced some sorrow, Democritus laughed, as usual, and said to him, and are not you the absurdest of men, ashamed still to lament, as if you alone were involved in such distress?"[120]

In the atomic theory originated by Leucippus and expounded by Democritus, material reality is composed of atoms and the *void*. The *void* is a complete vacuum. Space is infinite, as are the number of atoms.[121] An atom is an indivisible, and fundamental particle that is too small to be seen. Atoms were described by Aristotle in *On Generation and Corruption*. "The primary bodies of the Atomists— the primary constituents of which bodies are composed, and the ultimate elements in which they are dissolved — are indivisible."[122]

Although fundamental and indivisible, the atoms of the ancient Greeks were not uniform. Differences in their attributes and the way they combined led to diverse elements such as the familiar earth, air, water, and fire. In *Physics*, Aristotle said "Democritus ... speaks of differences in position, shape, and order."[123] Aristotle also discussed atomism in *Metaphysics*. "Leucippus and his associate Democritus ... say the differences in the elements are the causes of all other qualities. These differences they say, are three — shape and order and position."[124] Aristotle criticized the atomists for not being more specific in explaining how differences in atomic attributes could result in diverse substances. "They have never explained in detail the shapes of the various elements, except so far as to allot the sphere to fire."[125]

Diogenes Laërtius summarized Democritus' philosophy. "His principal doctrines were these. That atoms and the vacuum were the beginning of the universe; and that everything else existed only in opinion. That the worlds were infinite, created, and perishable. But that nothing was created out of nothing, and that nothing was destroyed so as to become nothing. That the atoms were infinite both in magnitude and number, and were borne about through the universe in endless revolutions. And that thus they produced all the combinations that exist; fire, water, air, and earth; for that all these things are only combinations of certain atoms."[126]

Because the only things that truly existed were atoms and the void, it followed as a necessary consequence that all human sensory perceptions in reality were a form of touch or contact, and that nothing could be perceived directly. Democritus wrote, "In reality there are atoms and the void. That is, the objects of sense are supposed to be real and it is customary to regard them as such, but in truth they are not. Only the atoms and void are real."[127]

Democritus believed in the existence of the human soul, but regarded the soul as material, "composed of atoms like everything else."[128] The atomistic philosophy of Leucippus and Democritus was necessarily materialistic. "Creation is the undesigned result of inevitable natural processes."[129] The single surviving fragment of Leucippus' writing reads "nothing comes into being without a reason, but everything arises from a specific ground and driven by necessity."[130]

The atomic theory of the ancient Greeks differed in several respects from modern atomic theory but it was, nevertheless, a forerunner of modern chemistry. The theory was reductionist, mechanical, and deterministic. The eternal nature of the constituent atoms foreshadowed the modern concepts of the conservation of matter and energy.

The atomists did not know the Earth was spherical. They thought "the Earth was shaped like a tambourine, and floated on air. It was inclined towards the south because the heat of that region made the air thinner, while the ice and cold of the north made it denser and more able to support the Earth."[131]

However Democritus had an intuitive conception of the universe as a place that was large and evolving. His view was closer to the modern one than the small and unchanging cosmos Europeans believed in nearly two millennia after his death. According to St. Hippolytus (c. A.D. 165–235),

> [Democritus] said that the ordered worlds are boundless and differ in size, and that in some there is neither sun nor moon, but that in others both are greater than with us, and in yet others more in number. And that the intervals between the ordered worlds are unequal, here more and there less, and that some increase, others flourish and others decay, and here they come into being and there they are eclipsed. But that they destroyed by colliding with one another. And that some ordered worlds are bare of animals and plants and of all water. And that in our cosmos the earth came into being first of the stars and that the moon is the lowest of the stars, and then comes the sun and then the fixed stars: but that the planets are not all at the same height. And he laughed at everything, as if all things among men deserved laughter.[132]

Democritus was evidently interested in ethics and had a number of "golden sayings" which offered pithy advice to the receptive. The apothegms included "not from fear but from a sense of duty refrain from your sins," "fools learn wisdom through misfortune," "it is better to correct your own faults than those of another," and "a sensible man takes pleasure in what he has instead of pining for what he has not."[133]

Democritus evidently lived to an old age. Diogenes Laërtius stated "he died ... after having attained the age of more than a hundred years,"[134] but later in the same brief biography claimed "[Democritus] expired, without any pain ... having lived a hundred and nine years."[135]

## THE TUNNEL OF EUPALINOS (C. 530 B.C.)

In addition to natural philosophers, there were also engineers among the Ionians. On the Ionian island and city of Samos there is indisputable evidence of the Ionians' engineering prowess in the form of a thousand-meter long tunnel cut through solid rock. The tunnel was commissioned by the ruler of Samos, Polycrates (reigned c. 535–522 B.C.), and executed by the engineer, Eupalinos.

Polycrates had risen to power through a coup d'etat at an opportune moment. During the festival of Hera, most of the population of the town was outside the city walls and Polycrates was able to make himself and his two brothers masters of the city with the aid of only a handful of men. Once in control, Polycrates rapidly consolidated his power. According to Herodotus, he did not care to share power with his brothers, so he killed one and exiled the other. "At the outset he [Polycrates] divided the state into three parts, and shared the kingdom with his brothers, Pantagnôtus and Syloson; but later, having killed the former and banished the latter, who was the younger of the two, he held the whole island."[136]

Nearly all of the important coastal trade passed through a narrow strait between Samos and the Asian mainland. Polycrates built a fleet of a hundred ships and seized control of the trade route, enriching himself through extortion and piracy. "He conquered several of the neighboring islands, and even some towns on the mainland.... Alike terrible to friend and foe by his indiscriminate spirit of aggression, he acquired a naval power which seems at that time to have been the greatest in the Grecian world."[137]

The chief town on the island of Samos was also named Samos, and occupied the slopes of Mount Castro. Polycrates fortified the city of Samos through a series of public works. In *Politics*, Aristotle cynically noted that the construction of public works was a device employed by tyrants to maintain their power, and specifically mentioned the works of Polycrates. "[The tyrant] should impoverish his subjects; he thus provides against the maintenance of a guard by the citizens, and the people having to keep hard at work, are prevented from conspiring ... the great Polycratean monuments at Samos ... were intended to occupy the people and keep them poor."[138]

One of Polycrates' projects was the tunnel of Eupalinos. Polycrates wanted a secure supply of water for Samos so the city could survive a protracted siege. There was a suitable spring on the other side of Mount Castro, but the problem was how to bring the water around the side of the mountain without building an aqueduct that would be vulnerable to attack. Eupalinos' solution was to cut a tunnel, 1,036 meters in length, through a hill composed of solid limestone.

The tunnel of Eupalinos was originally described by Herodotus.

> I have dwelt the longer on the affairs of the Samians, because three of the greatest works in all Greece were made by them. One is a tunnel, under a hill one hundred and fifty fathoms [274 meters] high, carried entirely through the base of the hill, with a mouth at either end. The length of the cutting is seven furlongs [1408 meters]—the height and width are each eight feet [2.4 meters]. Along the whole course there is a second cutting, twenty cubits [10 meters] deep and three feet [0.9 meters] broad, whereby water is brought, through pipes, from an abundant source into the city. The architect of this tunnel was Eupalinos.[139]

The tunnel of Eupalinos was rediscovered during an excavation in 1882. It is still largely intact and open to inspection today.[140] The tunnel was "dug through solid limestone by two separate teams advancing in a straight line from both ends, using only picks, hammers, and chisels."[141] Eupalinos evidently designed the tunnel to be cut through the hill at a level elevation. This is shown by the fact that the six-foot (1.8 meter) offset in elevation between the two branches occurs at the junction where they meet under the hill.[142]

At the site of the spring, an underground reservoir was constructed and hidden from view. Water from the reservoir traveled through pipes to the mouth of the tunnel and emerged on the other side inside the city walls. Water was carried through the tunnel in a separate channel, excavated on one side. The water-bearing trench is about 4 feet (1.2 meters) deep at the entrance of the tunnel, but the depth at the exit is about 25 feet (7.6 meters). "Similar underground conduit systems in Syracuse, Acragas, and Athens show the same double-tunnel construction ... [thus the construction method] was intentional and customary."[143]

Polycrates' wisdom in fortifying Samos was confirmed in 525 B.C. Angry at repeated losses to Polycrates' piracy, Sparta sent a fleet of ships to attack Samos. The city was under siege for forty days, but the Spartans were unable to break in. Herodotus said that eventually they had to simply turn around and go home. "The Lacedaemonians arrived before Samos with a mighty armament, and forthwith laid siege to the place ... [they] besieged Samos during forty days, but not making any progress before the place, they raised the siege at the end of that time, and returned home to the Peloponnese."[144]

Polycrates was said to be a man of unusual luck. In Herodotus' words, he was "fortunate in every undertaking."[145] When his ally, the Egyptian Pharaoh Amasis, heard of Polycrates' phenomenal fortune he was concerned. Amasis wrote a letter to Polycrates stating that if his unusual luck continued he would evoke the jealousy and wrath of the gods. Amasis counseled Polycrates to rid himself of his most valued possession so that his luck would not be entirely good. "Bethink thee which of all thy treasures thou valuest most and canst least bear to part with; take it, whatsoever it be, and throw it away, so that it may be sure never to come any more into the sight of man."[146]

Polycrates thought about Amasis' counsel for a while and finally decided the Pharaoh was correct. Accordingly, he boarded a ship and sailed out to deep water. He took his signet ring, a gold ring with an emerald stone that was his most precious possession, and "flung it into the deep."[147]

A few days later a fisherman "caught a fish so large and beautiful that he thought it well deserved to be made a present of to the king. So he took it with him to the gate of the palace, and said that he wanted to see Polycrates."[148] Polycrates was pleased with the fisherman's present, and had the fish taken to his kitchens to be prepared for dinner. When the cooks cut the fish open, they found Polycrates' ring inside. In great excitement, they brought it to the tyrant of Samos.

Polycrates wrote to Amasis describing what had happened. When the Egyptian Pharaoh read the letter, he concluded that it was wrong for one man to try and change the fate of another. "He felt certain that Polycrates would end ill, as he prospered in everything, even finding what he had thrown away."[149] Accordingly, Amasis "sent a herald to Samos, and dissolved the contract of friendship."[150]

Amasis' insight was correct, Polycrates did incur the wrath of the gods. The Persian governor of the Ionian island of Sardis, Oroetes, became resentful of Polycrates' good fortune and independence. He resolved to seize control of Samos and kill Polycrates. To accomplish this goal he resorted to subterfuge. Knowing Polycrates' greed, Oroetes sent him the following message. "I hear thou raisest thy thoughts high, but thy means are not equal to thy ambition. Listen then to my words, and learn how thou mayest at once serve thyself and preserve me. King Cambyses is bent on my destruction — of this I have warning from a sure hand. Come thou, therefore, and fetch me away, me and all my wealth—share my wealth with me, and then, so far as money can aid, thou mayest make thyself master of the whole of Greece. But if thou doubtest of my wealth, send the trustiest of thy followers, and I will show my treasures to him."[151]

Polycrates sent a trusted advisor to Oroetes to ascertain the size of Oroetes' fortune. Oroetes again resorted to trickery. "He filled eight great chests almost brimful of stones, and then covering over the stones with gold ... held them in readiness."[152] Polycrates traveled to Sardis, eager to claim the fortune. Upon his arrival, Oroetes killed him and hung his body on a cross. Herodotus said Oroetes "slew him [Polycrates] in a mode which is not fit to be described."[153] Thus the tyrant of Samos perished. This apocryphal story of unbalanced fortunes is typical of Grecian morals, for above all else the Greeks favored moderation and balance.

## *Pythagoras (c. 569–475 B.C.)*

### *The First Philosopher*

One of the most enigmatic of the ancient Greek philosophers was Pythagoras. Pythagoras was a mathematician, natural philosopher, and the founder of a religious cult. The Pythagorean philosophy was an integrated view of human existence that unified science and religion. "Religion and science were ... to Pythagoras ... two inseparable factors in a single way of life."[154]

Pythagoras was born around 582 B.C. on the Ionian island of Samos. He "wrote nothing,"[155] and there is little information on his life that can be considered to be authentic or certain. Legendary stories surround his life, most derived from unreliable and worshipful biographies written by faithful followers hundreds of years after Pythagoras died.

According to Iamblichus (c. A.D. 250–330), Pythagoras' father, Mnesarchus, visited the oracle at Delphi. The oracle foretold that his wife would bear "a son surpassing in beauty and wisdom all that ever lived, and who would be of the greatest advantage to the human race in every thing pertaining to the life of man."[156] The youth Pythagoras "was most venerable, and his habits most temperate, so that he was even reverenced and honored by elderly men; and converted the attention of all who saw and heard him speak, on himself, and appeared to be an admirable person to every one who beheld him. Hence it was reasonably asserted by many, that he was the son of a god."[157]

Iamblichus linked Pythagoras with both Thales and Anaximander in Miletus,[158] and John Burnet concluded, "it is stated that he [Pythagoras] was a disciple of Anaximander, which is no doubt a guess, but probably right."[159] As a young man, Pythagoras reportedly educated himself by traveling. Diogenes Laërtius stated "he went to Egypt ... and he associated with the Chaldeans and with the Magi."[160] Iamblichus assures us that Pythagoras "frequented all the Egyptian temples with the greatest diligence."[161]

In 525 B.C., the king of Persia, Cambyses II, invaded Egypt[162] and Pythagoras was taken prisoner. Iamblichus described Pythagoras' internment at Babylon as an intellectual sabbatical. "He [Pythagoras] gladly associated with the Magi, [and] was instructed by them in their venerable knowledge.... After associating with them twelve years, he returned to Samos about the fifty-sixth year of his age."[163]

Upon his return to Samos, Pythagoras "established a school ... called the semicircle."[164]

Iamblichus attests that Pythagoras lived in a cave outside of the city where "he spent the greatest part both of the day and night; employing himself in the investigations of things useful in disciplines, [and] framing intellectual conceptions."[165]

Because Pythagoras did not write anything himself, it is questionable how many of his original conceptions can be authentically attributed to him, and how many to his disciples. Aristotle was evidently aware of the difficulty, because he always referred to "the Pythagoreans," rather than Pythagoras himself.[166] John Burnet argued "all great advances in human knowledge have been due to individuals rather than to the collective work of a school, and so it is better to take the risk of ascribing a little too much to the founder than to lose sight of him among a crowd of disciples."[167] But Kirk and Raven (1957) took the opposite point of view, maintaining "such was the respect paid to its founder that later discoveries made by members of the fraternity seem not to have been claimed as individual achievement but rather attributed indiscriminately to Pythagoras himself."[168]

Although Pythagoras was respected in his native city of Samos, he evidently felt he could fulfill his philosophical mission more completely in another land. Around 518 B.C. (or earlier), he embarked for southern Italy, settling in the Greek city of Croton. Strabo related that Pythagoras' travels were motivated in part by his distaste for Polycrates' tyrannical government of Samos. "Pythagoras, observing the growing tyranny, left the city [Samos], and traveled to Egypt and Babylon, with a view to acquire knowledge. On his return from his travels, perceiving that the tyranny still prevailed, he set sail for Italy, and there passed the remainder of his life."[169]

If Thales can be called the first scientist, "Pythagoras was the first who called himself a philosopher."[170] Indeed, the very word *philosophy* was an invention of Pythagoras. He rejected the Greek word *sophia* (wisdom) as pretentious, and introduced the term *philosophia* (love of wisdom) and referred to himself as a *philosopher*, or lover of wisdom.[171]

Pythagoras compared human life to a gathering. In this gathering, there were three classes of people, just as there were three classes of people who attended the Olympics. Some came to the Olympic games to buy and sell, seeking riches. Others come to compete, seeking honor. But the best came merely to observe and learn. Accordingly, in Pythagorean philosophy men were ranked from base to noble. The lowest were those who loved wealth and material possessions. Intermediate were the seekers of status and honor among men. But the noblest type of man was the philosopher who sought wisdom.[172] In describing the Pythagorean tradition, Iamblichus observed, "the most pure and unadulterated character is that of the man who gives himself to the contemplation of the most beautiful things, and whom it is proper to call a philosopher."[173]

The Pythagorean brotherhood practiced communal living. "His disciples used to put all their possessions together into one store, and use them in common."[174] The Pythagoreans also held their doctrines and teachings secret. "The strictness of their secrecy is astonishing."[175] The need for secrecy arose from the Pythagorean recognition that knowledge could be used both for man's liberation and for his destruction. Therefore the degree of knowledge entrusted to an acolyte was proportional to the degree to which he had purified his body and mind. It was forbidden to transmit any teaching outside of the Pythagorean community.

## *All Things Are Numbers*

The central idea of the Pythagorean philosophy was that the ultimate basis of reality and the universe is numbers. "The doctrine of Pythagoras was that all things are numbers."[176] Aristotle noted, "[the Pythagoreans] supposed the elements of numbers to be the elements of all things."[177] Pure numbers were considered to be the final, irreducible basis of existence because only numbers are eternal, everything else changes. Therefore, it must be the goal of science to interpret and understand nature in terms of mathematics.

An important discovery of Pythagoras was that the pitch of a musical note depends on the length of the string that produces it; thus music is fundamentally mathematical in nature. The apocryphal story[178] of Pythagoras' discovery is that one day he was walking by a blacksmith shop and noticed that hammers of different weights produced different tones when they struck the anvil. "[Pythagoras] found by various experiments, that the difference of sound arose from the magnitude of the hammers."[179] From this observation, Pythagoras was led to discover the mathematical basis of music. "It was ... the first step towards the mathematization of human experience—and therefore the beginning of science."[180]

In *Metaphysics*, Aristotle said that the Pythagoreans even sought to describe abstract psychological principles such as justice and reason in terms of numbers.

> Contemporaneously with these philosophers and before them, the so-called Pythagoreans, who were the first to take up mathematics, not only advanced this study, but also having been brought up in it they thought its principles were the principles of all things. Since of these principles numbers are by nature the first, and in numbers they seemed to see many resemblances to the things that exist and come into being—more than in fire and earth and water (such and such a modification of numbers being justice, another being soul and reason, another being opportunity—and similarly almost all other things being numerically expressible).[181]

Prior to Pythagoras and Thales, mathematics had been largely an empirical science. Specific rules and relations were found and recorded for their practical significance. The Greeks made mathematics a theoretical and abstract field of study by generalizing rules-of-thumb into precise theorems that could be proved. They invented *demonstration*, the method of logical reasoning that produced proofs on which all men were bound to agree. Proclus (A.D. 411–485) claimed that it was Pythagoras who transformed geometry into a mathematical science. "Pythagoras changed that philosophy, which is conversant about geometry itself, into the form of a liberal doctrine, considering its principles in a more exalted manner; and investigating its theorems immaterially and intellectually."[182]

The chief mathematical contribution for which Pythagoras is remembered today is the *Pythagorean Theorem*, the geometrical theorem which states that in a right triangle the square of the length of the hypotenuse ($c$) is equal to the sums of the squares of the lengths of the two shorter sides ($a$, $b$), or $a^2 + b^2 = c^2$. Although the Egyptians and others had previously known that this was true for the special case of a right triangle with sides of length 3, 4, and 5, Pythagoras was the first to mathematically prove that this was the case for all right triangles. However it must be conceded that "no really trustworthy evidence exists that it [the theorem] was actually discovered by him [Pythagoras]."[183]

Pythagoras also discovered the five regular or perfect solids.[184] A regular solid is a three-dimensional geometrical figure in which all faces are identical. The simplest is the tetrahedron or pyramid, whose four faces are composed of identical equilateral triangles. The next level of complexity is occupied by the cube, whose six faces are each occupied by a square. The remaining four solids are the octahedron (composed of eight equilateral triangles), the dodecahedron (composed of twelve pentagons), and the icosahedron (composed of twenty equilateral triangles).

Although Pythagoras is given credit for discovering the regular solids, it was the geometer Euclid (c. 325–265 B.C.) who later proved that these five regular solids are the only ones that can exist in three dimensional space. It is mathematically impossible to construct any other symmetrical bodies. Nearly two thousand years after the life of Pythagoras, contemplation of the five regular solids led Johannes Kepler (A.D. 1571–1630) to discover the laws of planetary motion.

Assessing the mathematical contributions of Pythagoras, G. J. Allman (1824–1904) concluded that Pythagoras formed a secret society for the purpose of transmitting his knowledge and philosophy.

> In establishing the existence of the regular solids he showed his deductive power; in investigating the elementary laws of sound he proved his capacity for induction; and in combining arithmetic with geometry ... he gave an instance of his philosophic power. These services, though great, do not form, however, the chief title of this sage to the gratitude of mankind. He resolved that the knowledge which he had acquired with so great labor, and the doctrine which he had taken such pains to elaborate, should not be lost ... so Pythagoras devoted himself to the formation of a society *d'elite*, which would be fit for the reception and transmission of his science and philosophy; and thus became one of the chief benefactors of humanity, and earned the gratitude of countless generations.[185]

## *Cosmology and the Music of the Spheres*

The Pythagoreans are generally credited as the first to propose that the Earth and heavenly bodies are spheres. Pythagoras was described by Diogenes Laërtius as "the first person ... who called the Earth round,"[186] but there are other attributions as well.[187] The Pythagorean recognition of a spherical Earth may not have been based on empirical reasoning, but on their conception that the sphere was a perfect geometrical form.[188] In Pythagorean cosmology, heavenly bodies and their motions were eternal, divine, and incorruptible. Therefore, their motion must be both constant and circular, because only circular motion is perfect.

Although the Pythagoreans postulated a spherical Earth, their cosmology was not geocentric. In a critical passage in *De Caelo* (*On the Heavens*), Aristotle noted that the Pythagoreans cosmos had "fire" at its center, that they postulated the existence of a "another earth," and that their ideas were not in accordance with observation or "appearances."

> Most of those who hold that the whole universe is finite say that it [the Earth] lies at the center, but this is contradicted by the Italian school called the Pythagoreans. These affirm that the center is occupied by fire, and that the earth is one of the stars, and creates night and day as it travels in a circle around the center. In addition they invent another earth, lying opposite our own, which they call by the name of "counter-earth," not seeking accounts and explanations in conformity with the appearances, but trying by violence to bring the appearances into line with accounts and opinions of their own.[189]

The Pythagorean concept of perfect circular motion in the heavens became an axiom of astronomy that was not abandoned until Kepler recognized the elliptical form of planetary orbits in the seventeenth century.

In his text *Elements of Astronomy*, Geminus (c. 10–60 B.C.) explained that circular motion was axiomatic in astronomy, and that the concept had originated with the Pythagoreans. "It is a fundamental assumption in all astronomy that the sun, the moon, and the five planets move in circular orbits at uniform speed in a sense contrary to that of the universe. For the Pythagoreans, who were the first to apply themselves to investigations of this kind, assumed the movements of the sun, the moon, and the five planets to be circular and uniform."[190]

It is Pythagoras who originated the concept of the "music of the spheres." According to Pythagoras, as it is natural for a moving body to create sound, the Sun, Moon, and planets generated music as they moved through the heavens. It was inconceivable that this music could be disharmonious, therefore the distances between the planets must be governed by the same mathematical rules that established the lengths necessary for the concordant tones of musical instruments.

In his *Commentary on Aristotle's Metaphysics*, Alexander of Aphrodisias (c. 200 B.C.) described the harmony of the spheres. "They [the Pythagoreans] said that the bodies which revolve round the center have their distances in proportion, and some revolve more quickly, others more slowly, the sound which they make during this motion being deep in the case of the slower, and high in the case of the quicker; these sounds then, depending on the ratio of the distances, are such that their combined effect is harmonious."[191]

Pythagoreans believed that Pythagoras alone was able to hear the music of the spheres. Iamblichus claimed:

> Employing a certain ineffable divinity, and which it is difficult to apprehend, he [Pythagoras] extended his ears, and fixed his intellect in the sublime symphonies of the world, he alone hearing and understanding, as it appears, the universal harmony and consonance of the spheres, and the stars that are moved through them, and which produce a fuller and more intense melody than any thing effected by mortal sounds.... For he conceived that by him alone, of all the inhabitants of the Earth, the mundane sounds were understood and heard, and this from a natural fountain itself and root.[192]

Of course, Iamblichus also attributed a number of other miraculous powers to Pythagoras. These included the ability to be present simultaneously in two places, the talent of making "infallible predictions of earthquakes," and the power to make hail stop falling and to calm waves on the sea and rivers.[193]

Apparently unimpressed by Pythagoras' attainment of semi-divine status, in *De Caelo* (*On the Heavens*) Aristotle dismissed the idea of occult heavenly sounds. "When it is asserted that the movement of the stars produces harmony, the sounds which they make being in accord, the statement, although it is a brilliant and remarkable suggestion on the part of its authors, does not represent the truth."[194] Aristotle went on to explain that the reason the planets do not produce sounds is that they move simultaneously with the medium in which they are embedded. "What makes a noise is that which is moving in a stationary medium."[195] "Things which are themselves in motion create noise and impact, but whatever is fixed or otherwise contained in something moving, as the different parts are in a ship, cannot create noise; nor can the ship itself, if it is moving down a river."[196]

One day, the Pythagoreans made a discovery that must have shocked them. They discovered the existence of irrational numbers, numbers that could not be expressed as either whole numbers or the ratio of whole numbers. An example of an irrational number is the square root of two.[197] Irrational numbers can only be expressed as indeterminate decimals of infinite length. Because the Pythagoreans believed that "all things are numbers," the existence of numbers which could never be represented in a comprehensible way must have implied to them that parts of the cosmos were forever beyond the capacity of the human mind to understand. According to one legend, a member of the sect who revealed the existence of irrational numbers to the outside world was punished by being taken out to sea and drowned. "Hippasus ... was drowned at sea ... for making known the discovery of the irrational."[198]

## *Transmigration, Music, and Taboos*

One of Pythagoras' doctrines was transmigration of the soul, the belief that human beings have an immortal soul that can be reincarnated in either human or animal form. Pythagoras personally claimed to be aware of "the former lives he had lived."[199] According to Diogenes Laërtius, "[Pythagoras stated] that he had received as a gift from Mercury the perpetual transmigration of his soul, so that it was constantly transmigrating and passing into whatever plants or animals it pleased."[200] Xenophanes of Colophon (c. 570–475 B.C.) related that Pythagoras once asked a man to stop beating a puppy because he recognized the soul of a friend in the puppy's cries.[201]

The doctrine of transmigration did not originate with the Greeks. Herodotus stated that the idea originated in Egypt. "They [the Egyptians] were also the first to broach the opinion, that the soul of man is immortal, and that, when the body dies, it enters into the form of an animal which is born at the moment, thence passing on from one animal to another, until it has circled through the forms of all the creatures which tenant the earth, the water, and the air, after which it enters again into a human frame, and is born anew. The whole period of the transmigration is three thousand years."[202]

Pythagoreans used music to heal the body. The body was the instrument of the soul; like a musical instrument, it was strung to a certain pitch. If a body was ill, it was out of tune. The body could be purified and healed through music. "Through certain melodies and rhythms ... the remedies of human manners and passions are obtained."[203] According to Porphyry (c. A.D. 233–304), "by singing ... he [Pythagoras] cured the sick."[204] The soul was best purified through the study of philosophy, especially mathematics. Mathematics was the highest music.

The Pythagorean brotherhood observed a number of taboos. The path to purification started with the observance of the taboos and ended with the mathematical contemplation and interpretation of nature. The taboos included, "do not stir the fire with a sword," "do not devour your heart," and "when you are traveling abroad, look not back at your own borders."[205] Diogenes Laërtius explained that the sayings were symbolical. "The precept not to stir fire with a sword meant, not to provoke the anger or swelling pride of powerful men.... By not devouring one's heart, he intended to show that we ought not to waste away our souls with grief and sorrow. In the precept that a man when traveling abroad should not turn his eyes back, he recommended those who were departing from life not to be desirous to live, and not to be too much attracted by the pleasures here on earth."[206]

Other rules promulgated by Pythagoras included vegetarianism.

> He [Pythagoras] ordered his familiars to abstain from [eating] all animals, and farther still from certain foods, which are hostile to the reasoning power, and impede its genuine energies. He likewise enjoined them continence of speech, and perfect silence, exercising them for many years in the subjugation of the tongue, and in a strenuous and assiduous investigation and resumption of the most difficult theorems. Hence also, he ordered them to abstain from wine, to be sparing in their food, to sleep little, and to have an unstudied contempt of, and hostility to glory, wealth, and the like.[207]

## *Demise*

The details of Pythagoras' death, like his life, are obscure. Diogenes Laërtius said "Pythagoras ... died when he was eighty years of age ... but according to the common account, he was more than ninety."[208] Iamblichus stated that Pythagoras incurred the wrath of a powerful noble in Croton named Cylon, fled to the city of Metapontium, and died there.

Cylon the Crotonian held the first place among the citizens for birth, renown, and wealth; but otherwise, he was a severe, violent, and turbulent man, and of tyrannical manners. He had, however, the greatest desire of being made a partaker of the Pythagoric life, and having applied himself to Pythagoras, who was now an elderly man, for this purpose, was rejected by him on account of the above-mentioned causes. In consequence of this, therefore, he and his friends exercised violent hostilities against Pythagoras and his disciples. So vehement likewise and immoderate was the ambition of Cylon, and of those who arranged themselves on his side, that it extended itself to the very last of the Pythagoreans. Pythagoras, therefore, for this cause went to Metapontum, and there is said to have terminated his life.[209]

By the fourth century B.C., "Pythagorean clubs [in Italy] ... had grown to be political,"[210] and there was a violent uprising against them. Polybius (c. 203–120 B.C.) said "the burning of the Pythagorean clubs in Magna Grecia was followed by great constitutional disturbances ... and the Greek cities in that part of Italy became the scene of murder, revolutionary warfare, and every kind of confusion."[211]

## *Parmenides (c. 520–450 B.C.) and the Eleatics*

### *Eleatic Metaphysics*

Parmenides was a citizen of the town of Elea (present day Velia) on the west coast of Italy. Elea was founded c. 546 B.C. by refugees from the Ionian city of Phocaea.[212] Rather than submit to the Persians under the rule of Cyrus the Great, the Phocaeans decided to flee.

In Elea, Xenophanes of Colophon (c. 570–475 B.C.) founded the Eleatic school of philosophy. Xenophanes is primarily remembered for enunciating philosophical monotheism. A fragment of his surviving writings refers to "one god, the greatest god among gods and men, neither in form like unto mortals nor in thought."[213]

According to one tradition, Parmenides was a student of Xenophanes, but is more accurately characterized as a Pythagorean. Diogenes Laërtius stated, "though he [Parmenides] was a pupil of Xenophanes, he was not afterwards a follower of his; but he attached himself to Aminias, and Diochartes the Pythagorean."[214] In the dialogue titled *Parmenides*, Plato said that Parmenides "came to Athens ... [with Zeno ... and] at the time of his visit ... [was about] 65 years old, very white with age.... Socrates, then a very young man, came to see them."[215]

Diogenes Laërtius credited Parmenides with being "the first person who asserted that the earth was of a spherical form; and that it was situated in the center of the universe."[216] However other traditions give credit to Pythagoras. Thomas Heath (1861–1940) noted, "certain astronomical innovations are alternatively attributed to Parmenides and Pythagoras."[217] For example, "Parmenides is thought to have been the first to recognize that the Evening and Morning Stars [Venus] are one and the same, while others say that it was Pythagoras."[218] Parmenides may also have been the first, or one of the first, Greek astronomers to recognize that the planets followed different paths than the stars, and therefore belonged to a different category of celestial objects.[219]

Parmenides astronomical theories are difficult to understand, and appear to be derived from Anaximander's concept of hoops filled with fire. "These hoops were rings of compressed air filled with fire which burst out in flame at the outlets, thereby producing what we see as the sun, moon, and stars."[220] This is difficult to reconcile with the claim that Parmenides recognized that the Moon is illuminated by the Sun.[221] An existing fragment of his writing appears to refer to the Moon, when it states "shining by night with borrowed light, wandering round the earth."[222] It is clear that in the fifth century B.C., the Greeks had not yet fully developed the idea of a Earth-centered universe surrounded by concentric, rotating spheres.

Parmenides is primarily noted for his metaphysical conceptions, which have been characterized as "exceedingly obscure,"[223] but nevertheless influential. Parmenides' metaphysics was centered or fixated on his claim that "what does not exist" does not exist. "It is not possible for what is nothing to be."[224] Although this statement superficially appears to be a trivial tautology, its meaning can be illustrated by example. We are used to thinking of contrasts, for example, between light and dark. But Parmenides would maintain that this is a mistake. Dark, being the absence of light, is "that which is not," and therefore does not exist.

> It follows that that which exists "cannot have arisen out of nothing; for there is no such thing as nothing. Nor can it have arisen from something; for there is no room for anything but itself. What *is* cannot have beside it any empty space in which something else might arise; for empty space is nothing, nothing cannot be thought, and therefore cannot exist. What *is* never came into being, nor is anything going to come into being in the future.... That which is, *is*; and it cannot be more or less. There is, therefore, as much of it in one place as in another, and the world is a continuous, indivisible *plenum*. From this it follows at once that it must be immovable. If it moved, it must move into an empty space, and there is no empty space.... It must be finite, and can have nothing beyond it.... Hence, too, it is spherical."[225]

John Burnet summarized Parmenides' metaphysics. "What *is*, is a finite, spherical motionless corporeal *plenum*, and there is nothing beyond it. The appearances of multiplicity and motion, empty space and time, are illusions."[226] It follows irresistibly that no reliable or true knowledge can be constructed by observation, or by relying upon the senses. The only way to find truth is by pure cognition. According to Diogenes Laërtius, Parmenides "used to say that argument was the test of truth; and that the sensations were not trustworthy witnesses."[227]

## *Zeno's Paradoxes*

Parmenides' student,[228] Zeno of Elea (c. 490–425 B.C.), constructed a number of logical paradoxes to show that motion was indeed impossible. As an Eleatic, Zeno believed that reality was an indivisible unity. If the contrary was true, if reality was divided into separate entities, then these things must be infinite in number. "If things are a many, they must be just as many as they are, and neither more nor less. Now, if they are as many as they are, they will be infinite in number."[229]

Having established that multiplicity required space to be infinitely divisible, Zeno then attempted to show that this led to a contradiction, and thus reality was indivisible and motion was impossible. He argued, "You cannot cross a race-course. You cannot traverse an infinite number of points in a finite time. You must traverse the half of any given distance before you traverse the whole, and the half of that again before you can traverse it. This goes on *ad infinitum*, so that there are an infinite number of points in any given space, and you cannot touch an infinite number one by one in a finite time."[230]

Thus a paradox exists. It is mathematically impossible for a runner to move any given distance, yet this is routinely observed to happen. The resolution of the paradox is apparent when we realize that Zeno has failed to realize that the sum of an infinite series may be finite. Points have no dimension, and thus an infinite number of points *can be* traversed in a finite time.

A related paradox offered by Zeno was the argument that it is impossible for a fast runner to overtake a slow runner. "Achilles will never overtake the tortoise. He must first reach the place from which the tortoise started. By that time the tortoise will have got some way ahead. Achilles must then make up that, and again the tortoise will be ahead. He is always coming nearer, but he never makes up to it."[231] But of course a fast runner can overtake a slow one, so a paradox apparently exists. The paradox is again resolved by the recognition that the faster runner can in fact traverse an infinite number of points in a finite time.

The Eleatics offered the paradoxes to support their philosophy. But Zeno's paradoxes accomplish the opposite. The fact that what is argued to be logically impossible does in fact occur simply illustrates that Eleatic logic was fundamentally flawed.

Eleatic metaphysics illustrated a fundamental division in ancient natural philosophy. Parmenides (and later Plato) believed that the senses were unreliable, and that true knowledge could only come from logical cogitation. Other philosophers (e.g., Epicurus) gave primacy to observation and sensory information. Aristotle took an intermediate position. It would take two thousand years of trial-and-error to ultimately demonstrate the futility of attempting to construct reliable knowledge from pure reason unconstrained by experience.

Parmenides' contention that "all being is one and self-contained, and has no place in which to move"[232] provoked criticisms. Aristotle said that the Eleatic philosophy spurred Leucippus to develop the atomic theory in which both the void and motion exist. "Some of the older philosophers [e.g., Parmenides] thought that 'what is' must of necessity be 'one' and immovable. The void, they argue, 'is not'.... Leucippus, however, thought he had a theory which harmonized with sense-perception and would not abolish either coming-to-be and passing-away or motion and the multiplicity of things."[233]

It is likely that the Greeks themselves took Eleatic metaphysics with a grain of proverbial salt. In Plato's *Sophist*, one of the speakers observed "I think that Parmenides, and all who ever undertook to determine the number and nature of existence, talked to us in rather a light and easy strain."[234]

In *Greek Science*, Benjamin Farrington noted "there is nothing wrong with this [Parmenides'] argument except that it flouts all experience."[235] One must wonder if Parmenides would have objected to being hit in the head with a hammer, since he believed that motion was an illusion. The fact that intelligent men could have made serious arguments so completely at odds with common sense is testimony to the Greeks' recognition of the value of logical demonstration.

## *Empedocles (c. 492–432 B.C.)*

### Theory of the Four Elements

In an attempt to reduce the complexity of the natural world to a primal substance, the Ionians had speculated as to the ultimate nature of matter. Thales put forth a conjecture that the primal substance was water, and Heraclitus claimed that it was fire. The chemical system that was finally adopted was Empedocles' system of the four elements: earth, air, fire, and water.

The four elements were hypothetical abstractions, with physical substances representing heterogeneous amalgamations of the four primaries. For example, common dirt or stone was not identical with the element "earth," although this element was probably regarded as predominant in the composition. The pure element "earth" was "an extremely cold and dry substance."[236] Empedocles' theory of four elements was adopted by Aristotle, and remained the dominant chemical model in Western thought for eighteen-hundred years.

Empedocles was not an Ionian, but was born in the Greek colony of Agrigentum in Sicily. In 406 B.C., Agrigentum was "in the highest state of prosperity and magnificence," with an aggregate male population between 200,000 and 800,000.[237] Agrigentum's "temples and porticos, especially the spacious temple of Zeus Olympius, its statues and pictures, its abundance of chariots and horses, its fortifications, its sewers, its artificial lake of nearly a mile in circumference, abundantly stocked with fish — all these placed it on a par with the most splendid cities of the Hellenic world."[238]

As is the case for other presocratic philosophers, what little we know of Empedocles life is mixed with legend and fantastic anecdotes of questionable authenticity.[239] Diogenes Laërtius related that "he was a pupil of Pythagoras ... [but] was afterwards convicted of having divulged his doctrines."[240] Listing the successors of Pythagoras, Iamblichus included "of the Agrigentines, Empedocles."[241]

Empedocles was "not only a philosopher but a poet, a seer, a physicist, a physician, [and] a social reformer."[242] "He [Empedocles] claimed to be a god."[243] In his lost work, *Sophist*, Aristotle said that Empedocles "first discovered rhetoric."[244] Empedocles composed two poems; only fragments of these survive in the form of quotations by later authors. The first poem was titled *On Nature*, and was about two to three thousand lines long. The second poem had the appellation *Purifications*, and was typically Pythagorean in its theme of spiritual cleansing through philosophy.

Empedocles adopted the Pythagorean doctrine of transmigration.[245] In *Purifications*, he wrote, "I have been born as boy, girl, plant, bird, and dumb sea-fish."[246] In an apparent indication that he embraced monotheism, and the recognition of a single God as a higher spiritual principle, Empedocles described God as "only mind, sacred and ineffable mind, flashing through the whole universe with swift thoughts."[247]

A political reformer, Empedocles reportedly instituted democratic reform in Agrigentum. He "abolished the assembly of a thousand, and established a council in which the magistrates were to hold office for three years, on such a footing that it should consist not only of rich men, but of those who were favorers of the interests of the people."[248]

Empedocles is generally recognized as the originator of the theory of the four elements: earth, air, fire, and water. In *De Generatione et Corruptione* (*On Generation and Corruption*), Aristotle attributed the hypothesis to Empedocles, noting "and some say at once that there are four elements, as Empedocles."[249] In Empedocles' theory, all things are composed of a combination of four primary elements. Ideally, these elements were eternal and unchanging, and thus truly "elementary." But other philosophers proposed variations on Empedocles' model, in which the elements could transform into each other, or only one element was truly elemental. Ancient philosophic thought was not monolithic. Empedocles' model of the four elements was accepted as the best extant until chemists in the eighteenth century began to discover the true nature of the elements and how they combine to form compounds.

In Empedocles chemistry, material substances composed of the four elements are acted upon by two basic forces that exist in the universe: *love* and *strife* (or hate). Love tends to make matter aggregate, while strife leads to its dispersion. Change in the natural world results from the forces of love and strife acting upon matter in varying intensities.

Empedocles' cosmology was cyclic. He stated that in the beginning there was a unity, followed by a divergence and then again by a unity. "At one time things grew to be one alone out of many; and then again fell asunder so that there were many from the one,— fire and water and earth and the endless height of the air; and, apart from these, baneful strife, with equal weight throughout, in their midst love, equally distributed in length and breadth."[250] Aristotle referred to the cyclicity of Empedocles' worldview in *De Caelo* (*On the Heavens*). "Some say that alternately at one time there is coming into being, at another time there is perishing, and that this always continues to be the case; so says Empedocles of Agrigentum."[251]

Adumbrating the modern concepts of conservation of mass and energy, Empedocles described the four elements and the forces of love and strife as eternal and indestructible. "Nothing comes into being besides these [the four elements], nor do they pass away; for, if they had been passing away continually, they would not be now."[252]

## *Experiment Demonstrating the Materiality of Air*

Empedocles demonstrated that air was a material substance by conducting an experiment. He showed that water will not flow into an empty *clepsydra** unless the air inside is allowed to escape. "When a girl, playing with a water-clock of shining brass, puts the orifice of the pipe upon her comely hand, and dips the water-clock into the yielding mass of silvery water — the stream does not then flow into the vessel, but the bulk of the air inside ... keeps it out."[253]

The significance of Empedocles' experiment is controversial. John Burnet concluded "the rise of the experimental method dates from the time when the medical schools began to influence the development of philosophy, and accordingly we find that the first recorded experiment of a modern type is that of Empedocles with the clepsydra."[254] Kirk and Raven (1957) did not agree, and described Burnet's conclusion as "an exaggerated view."[255] But Benjamin Farrington (1891–1974) concluded that Empedocles' experiments were an important step toward the eventual development of an experimental method. Empedocles "demonstrated the fact that the invisible air was something that could occupy space and exert power."[256] Thus, Empedocles showed that the senses could provide information on the hidden parts of nature, and therefore were an important tool in the acquisition of knowledge.

## *Astronomy, Geology, and Medicine*

Anticipating the first measurement of the speed of light (in A.D. 1676) by two thousand years, Empedocles maintained that the speed of light must be finite, not infinite. In *De Sensu* (*On Sense*), Aristotle said "Empedocles says that the light from the sun first enters the intermediate space before it comes to vision or to the earth."[257]

The Greek physician Galen (c. A.D. 129–200) who practiced in Rome, described Empedocles as "the founder of the Italian school of medicine."[258] Empedocles' "fundamental

*A *clepsydra* is a water clock, a graduated vessel with a hole in the bottom. The passage of time is marked by the rate at which the volume of water inside the clepsydra diminishes.

[medical] doctrine was the identification of the four elements with the hot and the cold, the moist and the dry.... The heart, not the brain, was regarded as the organ of consciousness."[259] He "is said to have written on medicine not less than six thousand verses, and he was nearly contemporary with Hippocrates [c. 460–370 B.C.]."[260]

The existence of hot springs evidently led Empedocles to the conclusion that the interior of the Earth was hot. In *Natural Questions*, the Roman author Seneca (c. 4 B.C.–A.D.65), wrote "Empedocles is of [the] opinion that as there are fires concealed in many places beneath the earth, water is heated when they happen to lie beneath the ground through which it has to flow."[261]

There are parts of Empedocles' astronomy that are "scarcely intelligible,"[262] at least to a modern mind confronted with the fragments of original material available to us. "Empedocles considered the finite universe to be spherical, solid, [and] made of condensed air after the manner of crystal. To this sphere the fixed stars, which were formed of fiery matter pressed upward by the air, are attached, while the planets wander freely in space. The moon is air rolled together and mixed with fire, it is flat like a disc and illuminated by the sun.... The earth was held in its place by the rapid spinning round of the rotating heavens."[263] In *De Caelo* (*On the Heavens*), Aristotle referred to this idea when he explained "others, with Empedocles, [say] that the motion of the heavens, moving about it [the Earth] at a higher speed, prevents movement of the earth, as the water in a cup, when the cup is given a circular motion."[264]

Empedocles said "the sun is, in its nature, not fire, but a reflection of fire similar to that which takes place from the surface of water."[265] Thomas Heath explained this theory as "the sun which we see is a concentration of rays reflected from the earth upon the crystal sphere [of the heavens]."[266] But the theory becomes obscure and difficult at the point where Empedocles refers to a second sun. "There are two suns; one is the original sun which is the fire in one hemisphere of the world ... the other is the apparent sun which is a reflection in the other hemisphere."[267] Heath explained, one problem is "[the sun] can hardly be in the other hemisphere because that hemisphere is night."[268]

Diogenes Laërtius related several apocryphal versions of Empedocles demise. According to one story, "he rose up and went away as if he were going to mount Etna; and when he arrived at the crater of fire he leaped in, and disappeared, wishing to establish a belief that he had become a god. But afterwards the truth was detected by one of his slippers having been dropped. For he used to wear slippers with brazen soles.... Timaeus contradicts all these stories; saying expressly, that he [Empedocles] departed into Peloponnesus, and never returned at all, on which account the manner of his death is uncertain."[269]

## *Hippocrates of Cos (c. 460–370 B.C.)*

### *The Asclepiadae*

While the natural philosophers relied primarily upon reason, the practice and science of medicine from the earliest times was based largely upon observation and experience. Medicine was not a speculative or idle occupation, it was a practical art with observable results. Physicians were craftsmen, not philosophers.[270]

The best known of the Greek physicians was Hippocrates of Cos. Little is known of the details of Hippocrates' life, but it is reasonably certain that he was a real person, famous in his time.[271] References to Hippocrates in the writings of Plato and Aristotle imply that Hippocrates was recognized as the leading physician of his time. In Plato's *Protagoras*, Hippocrates of Cos was referred to as a "physician."[272] In *Phaedrus*, Hippocrates was quoted as holding the view that "then nature even of the body can only be understood as a whole."[273] In *Politics*, Aristotle also mentioned Hippocrates, the "physician."[274]

Strabo described Hippocrates native town, Cos, as a fertile island that produced "excellent wine." According to the geographer, Hippocrates was "one of the illustrious natives of Cos," and "learned and practiced the dietetic part of medicine from the narrative of cures suspended in the temple [of Asclepius]."[275]

In *Natural History*, Pliny the Elder stated that Hippocrates had considerable knowledge of herbal remedies. "Hippocrates, it is well known, was the first to compile a code of medical precepts, a thing which he did with the greatest perspicuity, as his treatises, we find, are replete with information upon the various plants."[276] Pliny also related the apocryphal story that Hippocrates burned the temple of Asclepius in Cos so that he might profit from being in sole possession of the medical records and knowledge otherwise destroyed. It was "the practice for persons who had recovered from a disease to describe in the temple of that god [Asclepius] the remedies to which they owed their restoration to health, that others might derive benefit therefrom in a similar emergency; Hippocrates, it is said, copied out these prescriptions, and, as our fellow-countryman, Varro will have it, after burning the temple to the ground, instituted that branch of medical practice which is known as 'clinics.' There was no limit after this to the profits derived from the practice of medicine."[277] In another version of this tale, Hippocrates burned "the library of Cnidus."[278] Both versions are doubtful.

"From his own account, Hippocrates was the eighteenth in descent from Asclepius."[279] Asclepius was the Greek god of healing, "son of Apollo and healer of sickness."[280] According to Greek mythology, Asclepius' mother, Koronis, was unfaithful to Apollo. The news of her infidelity was brought to the god by a crow. Upon hearing the bad tidings, Apollo was so angry that he punished the messenger by changing the color of the crow from white to black.[281]

Apollo turned his son, Asclepius, over "to the centaur Cheiron to be brought up. The child was named Asclepius, and acquired, partly from the teaching of the beneficent leech [physician] Cheiron, partly from inborn and superhuman aptitude, a knowledge of the virtues of herbs and a mastery of medicine and surgery, such as had never before been witnessed. He not only cured the sick, the wounded, and the dying, but even restored the dead to life."[282]

In the *Iliad*, Homer stated that Asclepius' two sons, Podaleirios and Machaon, were physicians for the Greeks during the Trojan War. "Asclepius' two sons were leaders, the cunning leeches Podaleirios and Machaon."[283] The Trojan War is believed to have occurred c. 1250 B.C.[284]

It is not clear "whether Asclepius was originally a god, or whether he was first a man and then became afterwards a god."[285] The fact that Hippocrates boasted of his lineal descent from Asclepius is an indication that "the legendary element pervade[d] even the most philosophical and positive minds of historical Greece."[286]

Whatever the case may have been concerning the divinity and origin of Asclepius, ancient Greece contained an "order of priest-physicians"[287] known as the Asclepiadae who derived their authority as physicians in part from alleged genealogical connections with Asclepius. The Asclepiadae presided over "temples of health, called Asclepia ... erected in various parts of Greece, as receptacles for the sick, to which invalids resorted in those days for the cure of diseases, under the same circumstances as they go to hospitals and spas at the present time.... A large proportion of these temples were built in the vicinity of thermae, or medicinal springs."[288] Healing shrines were ubiquitous in ancient Greece, and the Asclepia were especially prominent in the fifth century B.C.[289]

It is not clear to what extent the practice of the Asclepiadae was scientific, and what part relied upon the invocation of superstitious rituals.[290] The Asclepiadae "endeavored to cure the sick partly by superstitious modes of working upon the imagination, and partly by more rational means, suggested by observation and a patient study of the phenomena of disease."[291]

The Asclepiadae were not the only healers in ancient Greece. In *Theaetetus,* Plato referred to midwives as "respectable women."[292] "By the use of potions and incantations they are able to arouse the pangs and to soothe them at will; they can make those bear who have a difficulty in bearing, and if they think fit they can smother the embryo in the womb."[293]

Superstition and religion in early medicine were inevitable, because many diseases had no apparent cause. Absent any observable natural cause, people were compelled to infer that a supernatural influence was at work. "Within a primitive or ancient culture internal diseases are usually believed to be strange events which cannot be understood as merely natural ones."[294] "Gods as disease-causers are to be found in every passage of the *Iliad* and *Odyssey* ... for Homer as well as the culture represented by him, internal diseases are usually caused by supernatural beings."[295]

Diseases were sent by the gods as punishment for sins.[296] In Book 1 of the *Iliad,* Apollo dispersed disease by means of arrows. "And the arrows clanged upon his shoulders in his wrath, as the god moved; and he descended like to night. Then he sate him aloof from the ships, and let an arrow fly; and there was heard a dread clanging of the silver bow ... and the pyres of the dead burnt continually in multitude."[297]

There is significant evidence that Hippocrates was not literally "the first to cast superstition aside, and to base the practice of medicine on the principles of inductive philosophy."[298] To the extent that Hippocrates deserves the title of "Father of Medicine,"[299] this recognition derives from his eminence in his own time, the body of writings he left, and the school that he founded. However in the sixth century B.C., magic and medicine were intertwined. Healers routinely invoked charms, chants, incantations, prayers, and other superstitious rituals. But by the fourth century B.C., after the influence of Hippocrates and his colleagues defined a standard of scientific medicine, magic was no longer considered orthodox.[300]

## *The Plague of Athens, 430 B.C.*

Hippocrates' reputation among his contemporaries in part derived from his extensive travels. He "visited Thrace, Delos, Thessaly, Athens, and many other regions.... He practiced, and probably taught, his profession in all these places."[301] It is likely that Hippocrates was present at the great plague that struck Athens in 430 B.C. He reportedly attempted to extinguish the plague "by kindling fires."[302]

The epidemic at Athens was described with merciless objectivity by Thucydides. "The plague broke out at Athens for the first time ... there is no record of such a pestilence occurring elsewhere, or of so great a destruction of human life. For a while physicians, in ignorance of the nature of the disease, sought to apply remedies; but it was in vain, and they themselves were among the first victims, because they oftenest came into contact with it. No human art was of any avail."[303]

"The malady took a form not to be described, and the fury with which it fastened upon each sufferer was too much for human nature to endure. There was one circumstance in particular which distinguished it from ordinary diseases. The birds and animals which feed on human flesh, although so many bodies were lying unburied, either never came near them, or died if they touched them."[304]

"Appalling too was the rapidity with which men caught the infection; dying like sheep if they attended on one another; and this was the principal cause of mortality.... The dead lay as they had died, one upon another, while others hardly alive wallowed in the streets and crawled about every fountain craving for water. The temples in which they lodged were full of the corpses of those who died in them; for the violence of the calamity was such that men, not knowing where to turn, grew reckless of all law, human and divine."[305]

Thucydides' description of the symptoms of the disease was so meticulous, factual, and free of hypothesizing, that it enabled the plague to be identified as typhoid fever.[306] "His simple and precise summary of observed facts carries with it an imperishable value, and even affords grounds for imagining that he was no stranger to the habits and training of his contemporary, Hippocrates."[307]

## *The Hippocratic Corpus*

The Hippocratic Corpus consists of about "sixty medical works," dating from the last part of the fifth century B.C. and "the first half of the fourth."[308] Attribution is uncertain, but it seems very probable that many of the most important works were authored by Hippocrates himself.[309] Titles of some of the better-known works include the Hippocratic *Oath*, *On Ancient Medicine*, the *Canon*, *Tradition in Medicine*, *Epidemics*, *Prognosis*, *Aphorisms*, *The Sacred Disease*, *Regimen*, and *Airs, Waters, and Places*.[310]

Nothing comparable to the Hippocratic Corpus has yet been found in Babylonian or Egyptian writings. There was certainly a long history of medical practice in Egypt, but the Greek innovations of naturalism and the rejection of supernaturalism appear to have been unique, not imported. In Greece, medical ideas and theories were openly discussed and debated. Greek medicine was "radically different from that known from elsewhere."[311] Although Hippocrates may not have been the genuine author of most of the works in the Corpus, this body of works defined medicine and its practice in Western Civilization.[312]

The treatise *Epidemics* contains the origin of the famous medical aphorism, "do no harm." Hippocrates explained, "the physician must ... have two special objects in view with regard to disease, namely, to do good or to do no harm."[313] Nearly of equal fame is the aphorism, "life is short, science is long."[314] In this pithy statement, Hippocrates affirmed the cumulative and progressive nature of medical science, recognizing that it was an activity that depended on the patient accumulation of generations of experience.

## *Empiricism and Naturalism*

That Hippocrates was not a revolutionary innovator, but building on a scientific tradition, is evidenced by his written remarks. He recognized the existence of scientific method by noting that scientific knowledge differed from mere opinion. "Science and opinion govern the world: the one points out our knowledge—the latter our deficiency."[315] And he acknowledged that the science of medicine had been established through a long period of practice and experience. "[Medicine] is an art of long existence, of sure principles, and certain regulations, through which, for a long period, numerous discoveries have been made, and which are confirmed by experience, unmixed with hypothesis."[316]

In such a crucial art or science as medicine, the important of experience had been recognized for a long time. Herodotus related that the Babylonians had "no physicians, but when a man is ill, they lay him in the public square, and the passers-by come up to him, and if they have ever had his disease themselves or have known anyone who has suffered from it, they give him advice."[317]

Written records of disease symptoms and treatments were kept at the temples of the Asclepiadae.[318] "Sick visitors ... were numerous and constant, and the tablets [were] hung up to record the particulars of their maladies, the remedies resorted to, and the cures operated by the god."[319] "It is very certain that many of the remedies employed by Hippocrates had been in common use long before him."[320] Hippocrates' *Epidemics* contains numerous case histories of diseases, each detailed and specific.[321] It is evident that Greek physicians had a tradition of making careful observations and keeping meticulous records.

Although there was a long empirical tradition in ancient Greek medicine, Hippocrates also employed theory (e.g., the four humors) and recognized that raw empiricism must be tempered with reason. In an aphorism, he noted that "the occasion [is] fleeting, experience fallacious, and judgment difficult."[322] Hippocrates "never exempts the apparent results of experience from the strict scrutiny of reason. Above all others, Hippocrates was strictly the physician of experience and common sense. In short, the basis of his system was a rational experience, and not a blind empiricism."[323]

Hippocrates was a contemporary of Democritus (460–370 B.C.), and evidently respected his work.[324] Celsus (fl. c. A.D. 25) described Hippocrates as Democritus' student, but also noted that Hippocrates was "the first to separate this study [medicine] from that of philosophy."[325] Hippocrates had a disdain for applying speculative philosophy to a practical art like medicine. Theory had to be tempered and subordinated to empiricism. In *The Art [or Science] of Medicine*, Hippocrates acclaimed, "facts are far superior to reasoning."[326]

In *On Ancient Medicine*, Hippocrates appeared to explicitly criticize his contemporary, Empedocles.[327] "Those who have undertaken to treat of medicine, have manifestly been deceived in most particulars, by attempting to found this doctrine on the hypothetical notions of cold and hot, of dry and moist, thus reducing to one or two principles the causes of death and of disease."[328]

Hippocrates explained that the natural philosophers had no established criterion of truth. "Recourse to hypothesis should therefore be avoided in medicine ... [because] there is no established rule of truth."[329] In contrast to the philosophers, physicians had long ago established experience as their "rule of truth." "Such a rule [of truth], however, exists in medicine; it is an art of long existence, of sure principles, and certain regulations, through which, for a long period, numerous discoveries have been made, and which are confirmed by experience, unmixed with hypothesis."[330]

Hippocrates noted that the natural philosophers could also not be believed, because in their primary task of reducing the materials of the universe to one primary substance, they had all reached conclusions that contradicted one another. "They [the natural philosophers] first advance the assertion that every thing existing is a unit, and that this unity is the universal whole; but then they disagree as to what this universal unit is. One affirms it to be air, another that it is fire, a third that it is water, and a fourth that it is earth; and each one grounds his assertion on reasoning and testimony of no value. Now, that they should agree at setting off, in one opinion, and then differ in what they say, is an evidence of their ignorance of the whole subject."[331]

The Hippocratic school rejected supernaturalism, embraced naturalism, and believed in cause and effect. In *The Art* [or *Science*] of *Medicine*, Hippocrates maintained that there was no such thing as chance: every effect had a cause. "Chance, when we come to examine the phrase, means absolutely nothing. Every event has a certain cause, which is, itself, the effect of some preceding one. Chance, therefore, cannot be said to have existence. It is a term employed by ignorance for what it does not comprehend."[332]

In a treatise on epilepsy, *On the Sacred Disease*, the writer maintained that epilepsy, like other diseases, had a natural cause. "With regard to the disease called sacred: it appears to me to be nowise more divine nor more sacred than other diseases, but has a natural cause from which it originates like other affections. Men regard its nature and cause as divine from ignorance and wonder."[333]

To the extent that the origin of epilepsy was divine, all disease processes were divine. "This disease seems to me to be nowise more divine than others; but it has its nature such as other diseases have, and a cause whence it originates, and its nature and cause are divine only just as much as all others are, and it is curable no less than the others."[334]

Impostors and charlatans who invoked supernaturalism were rebuked. "They who first referred this disease to the gods, appear to me to have been just such persons as the conjurors, purificators, mountebanks, and charlatans now are, who give themselves out for being excessively religious, and as knowing more than other people."[335]

Naturalism was also emphasized in the Hippocratic treatise, *Airs, Waters, and Places*. "To me it appears that such affections are just as much divine as all others are, and that no one disease is either more divine or more human than another, but that all are alike divine, for that each has its own nature, and that no one arises without a natural cause."[336]

By maintaining that "all [diseases] are alike divine,"[337] Hippocrates was able to differentiate explicitly between superstition and theism. It was not impious to rebuke superstition, nor did naturalism necessarily imply atheism.

## *Epidemiology*

The primary health concerns in modern, industrialized civilizations are degenerative diseases, such as cancer and atherosclerosis. But the world in which the Greek and Roman physicians operated was different. Degenerative diseases were uncommon, because most people did not live long enough for them to develop. Life expectancy at birth was between twenty and thirty years. Food supplies varied, and could be affected drastically by inclement weather and other variables. Infant mortality was high. Only about five percent of the population reached the age of sixty.[338]

The Hippocratic tradition emphasized observation, but the ancient Greek physicians had no microscopes and could not possibly have known that infectious diseases were caused by microorganisms. The germ theory of disease, championed by Louis Pasteur (1822–1895) and Robert Koch (1843–1910), was not confirmed and accepted until 1879.[339]

Unaware of the existence of microorganisms, Hippocratic physicians concentrated on natural environmental factors they could observe. Primary among these was diet. Making adjustments to a patient's diet was the central element in Hippocratic medicine.[340] After diet, the most important factors were considered to be common environmental factors, such as air, water, climate, locality, and seasons.

The Hippocratic physicians recognized that obesity and inactivity were unhealthy. In *Airs, Waters, and Places*, the physician was advised to investigate "the mode in which the inhabitants live, and what are their pursuits, whether they are fond of drinking and eating to excess, and given to indolence, or are fond of exercise and labor, and not given to excess in eating and drinking."[341]

The nature of the winds in specific localities were considered to be a factor contributing to health or disease.[342] Water was also important, because it "contributes much towards health."[343] "The best [waters] are those which flow from elevated grounds, and hills of earth; these are sweet, clear, and can bear a little wine; they are hot in summer and cold in winter, for such necessarily must be the waters from deep wells."[344]

The seasons were regarded as obvious environmental factors that affected human health, "for with the seasons the digestive organs of men undergo a change."[345] By carefully observing the weather, including patterns of wind, temperature, and precipitation, the skilful physician should be able to "judge whether the year will prove sickly or healthy."[346] The perspicuous doctor was advised to be vigilant and "particularly guarded during the greatest changes of the seasons."[347]

The Hippocratic physicians believed that climate determined racial temperaments. "The principal reason the Asiatics are more unwarlike and of gentler disposition than the Europeans is, the nature of the seasons, which do not undergo any great changes either to heat or cold, or the like."[348] This was a common view in classical Mediterranean civilization. Vitruvius (c. 90–20 B.C.), for example, admitted that the peoples of the "southern nations have the keenest wits, and are infinitely clever in forming schemes," but "the moment it comes to displaying valor, they succumb because all manliness of spirit is sucked out of them by the sun."[349]

## *The Four Humors*

Although the mode in which Hippocrates practiced, and the tradition from which he descended, emphasized empiricism, it was necessary to have a theoretical structure to explain how changes in environmental conditions caused disease. Hippocrates utilized the theory of the four humors. The four humors were four bodily fluids: blood, phlegm, and yellow and black bile. To divide the essential fluids of the human body into a quartet was naturally analogous to other divisions in Greek natural philosophy. The Greeks recognized four elements (earth, air, fire, water), four primary qualities (hot, cold, moist, dry), and four winds (east, west, north, south).

Disease resulted when some environmental factor caused the four humors to be imbalanced. "We continue in a state of health so long as they [the four humors] continue in a natural state, and in due proportion as to quantity, quality, and mixture."[350] Thus bloodletting was an attempt to restore balance by removing an excess of this fluid. The existence of the menstrual cycle showed that nature itself regarded the process as beneficial.[351]

The theory was explained in the treatise *On the Nature of Man*. "The body of man contains blood, pituita [phlegm], and two kinds of bile — yellow and black; and his nature is such that it is through them that he enjoys health, or suffers from disease. He enjoys the former when each is in due proportion of quantity and force, but especially when properly commingled. Disease takes place if either is in excess or deficient, or if not duly united."[352]

The theory of the four humors was derived from the tradition of careful observation. The humors were both "common and perceptible."[353] The Greek physicians noted, for example, that phlegm "abounds in man more largely in the winter."[354] "Bile ... appears in vomit and diarrhea."[355] The doctrine of the four humors remained the ruling medical theory for more than two thousand years.

## *Regimen and Remedies*

The theory of the four humors could not explain all diseases. An epidemic could only be explained by a common factor among all victims: bad air.[356] "Whenever, in the same place, many persons are attacked with the same disease, at the same time, we must attribute this to some common cause. Now this is the air. It is evident it cannot be the diet, because the disease attacks all."[357]

Treatment began with a proper regimen of diet designed to maintain health and prevent disease. By adjusting and changing diet to match the requirements of the changing seasons, the four humors could be kept in balance. The humors "increase or diminish, each according to the season."[358] Diets had to be "conditioned by age, the time of year, habit country and constitution."[359]

Recommended physical exercises included walking, running, and wrestling.[360] The physicians learned from the athletes. In *On Ancient Medicine*, the author noted "those who devote themselves to gymnastics and training, are always making some new discovery, by pursuing the same line of inquiry, where, by eating and drinking certain things, they are improved and grow stronger than they were."[361]

In *The Republic*, Plato noted that the practice of dietary regime had been introduced by an athletic trainer, Herodicus. "Before the time of Herodicus, the guild of Asclepius did not practice our present system of medicine, which may be said to educate diseases. But Herodicus, being a trainer, and himself of a sickly constitution, by a combination of training and doctoring found out a way of torturing first and chiefly himself, and secondly the rest of the world."[362] It is possible that Hippocrates was a student of Herodicus.[363]

The preoccupation of the modern physician is not so much to maintain health, as it is to prescribe powerful remedies after a disease process has developed. In contrast, Hippocrates was primarily concerned with maintaining health through a proper regimen of diet and exercise. In *A Regimen for Health*, Hippocrates explained "a wise man ought to realize that health is his most valuable possession."[364]

But if the regimen failed, the Hippocratic physicians had an array of medicines and procedures that could be called upon. Even in Hippocrates' time, herbal medicine was hundreds or thousands of years old. Writing c. A.D. 25, Celsus noted "even the most uncivilized tribes have some knowledge of herbs."[365] The *Materia Medica* of Pedanius Dioscorides, written c. A.D. 50–70, contained descriptions of more than a thousand remedies, of which seven hundred are herbs or plants.[366]

Hippocrates prescribed laxatives, narcotics, diuretics, and sudorifics.[367] Medicinal herbs utilized by the Hippocratic physicians included elaterium, colocynth, hellebore, mandragora (mandrake), and henbane (*hyoscyamus niger*).[368] It is possible that Hippocrates also derived substances with narcotic effects from poppies.[369] "Blood-letting was known, but not greatly practiced."[370] Hippocrates and his colleagues even resorted to surgery. They opened veins, "perforated the skull," and "opened the chest."[371]

Hippocrates and his colleagues developed the craft of medicine into a science, based on observation and reason. Theory was employed, but theoretical speculation was subordinated and constrained by stubborn facts. But it is doubtful if the methods and experiences of the physicians had much effect upon the natural philosophers. The philosophers tended to have a disdain for the practical arts and applied sciences, even a craft as indispensable as medicine.

## *The Greco–Persian Wars (c. 499–448 B.C.)*

### *Battle of Marathon (490 B.C.)*

In the sixth century B.C., the greatest power in the eastern Mediterranean region was the Persian Empire. From 559 through 500 B.C., the Persians extended and consolidated their conquests. By 490 B.C., ancient Persia controlled an area that stretched east to India, north to Russia, and west to Egypt. Greek Ionia was under Persian rule. The Persian King, Darius, controlled virtually the entire world known to the ancient Greeks. Only Europe remained free from Persian domination.

In his classic book, *The Fifteen Decisive Battles of the World from Marathon to Waterloo* (1851), British history professor Sir Edward Shepherd Creasy (1812–1878), described the extent of the Persian Empire and their aurora of invincibility.

> With the exception of the Chinese Empire ... all the great kingdoms which we know to have existed in Ancient Asia, were, in Darius' time, blended with the Persian. The northern Indians, the Assyrians, the Syrians, the Babylonians, the Chaldees, the Phoenicians, the nations of Palestine, the Armenians, the Bactrians, the Lydians, the Phrygians, the Parthians, and the Medes,— all obeyed the scepter of the Great King: the Medes standing next to the native Persians in honor, and the empire being frequently spoken of as that of the Medes, or as that of the Medes and Persians. Egypt and Cyrene were Persian provinces; the Greek colonists in Asia Minor and the islands of the Aegean were Darius' subjects; and their gallant but unsuccessful attempts to throw off the Persian yoke had only served to rivet it more strongly, and to increase the general belief: that the Greeks could not stand before the Persians in a field of battle. Darius' Scythian war, though unsuccessful in its immediate object, had brought about the subjugation of Thrace and the submission of Macedonia. From the Indus to the Peneus, all was his.[372]

The Persians maintained their rule through the satrapy system. A *satrap* was essentially a provincial governor who was loyal to the Persian king; the area he controlled was a *satrapy*. "The satrap was the head of the administration of his province; he collected the taxes, controlled the local officials and the subject tribes and cities, and was the supreme judge of the province."[373] Conquered lands and peoples were enrolled in the Persian Empire through the satrapy system. They were free to retain their own religion and manage internal affairs. However, they had to pay tribute to Persia and make their men available for military service. The Persian monarch had absolute authority in war; disobedience was harshly punished.

Around 500 B.C., the Persian King, Darius, decided to conquer Europe. If he had succeeded, the nascent Greek culture that formed the basis of Western Civilization might have been crushed out of existence. Creasy described the nature and culture of Persian governance as inimical to intellectual activity. Philosophy and science were restricted by authoritarianism and the establishment of state religions.

> The governments of all the great Asiatic empires have in all ages been absolute despotisms.... We should bear in mind, also, the inseparable connection between the state religion and all legislation which has always prevailed in the East, and the constant existence of a powerful sacerdotal

> body, exercising some check, though precarious and irregular, over the throne itself, grasping at all civil administration, claiming the supreme control of education, stereotyping the lines in which literature and science must move, and limiting the extent to which it shall be lawful for the human mind to prosecute its inquiries.[374]

In Greece, the priesthood was not able to rise to power because it was preempted by "a military aristocracy."[375] Burnet argued it was the rise of science itself that checked the development of a religious stranglehold over the intellectual world, explaining "it was not so much the absence of a priesthood as the existence of the scientific schools that saved Greece."[376]

Creasy described Greek culture as innovative and original; Persian civilization as authoritarian and stagnant. "In literature and science the Greek intellect followed no beaten track, and acknowledged no limitary rules.... Versatile, restless, enterprising, and self-confident, the Greeks presented the most striking contrast to the habitual quietude and submissiveness of the Orientals."[377]

At the beginning of the fifth century B.C., Darius' plans of conquest were delayed by a Greek revolt in Ionia. Herodotus recorded that "all of Ionia has revolted from the king."[378] In 499 B.C., a constitutional government was declared in the Ionian city of Miletus and tyrants were expelled from the other Ionian states. The Ionians sought aid from Greece, and "the Athenians now arrived with a fleet of twenty sail, and brought also in company five triremes* of the Eretrians."[379]

Upon the arrival of the triremes in 498 B.C., the Ionians "came down upon Sardis and took it ... the whole city fell into their hands.... Sardis was burnt, and, among other buildings, a temple of the native goddess Cybelé was destroyed; which was the reason afterwards alleged by the Persians for setting on fire the temples of the Greeks."[380] The Athenians and Eretrians withdrew their ships, but the success sparked revolts in other occupied Greek cities and colonies.

Although it took a few years for the Persians to respond effectively to the revolt, in 495 B.C. they won a decisive battle at sea. The Ionian "cities fell one after another."[381] "The Persians, when they had vanquished the Ionians in the sea-fight, besieged Miletus both by land and sea ... until at length they took both the citadel and the town, six years from the time when the revolt first broke out.... After killing most of the men, [the Persians] made the women and children slaves."[382] In other Ionian cities, "the most beautiful Greek youths and virgins were picked out, to be distributed among the Persian grandees as eunuchs or inmates of the harems."[383] By the succeeding year, the Ionian revolt was completely crushed.

Darius made a liberal settlement with the subdued Ionian states. Although the leaders of the rebellion were either executed or deported, Ionian cities were allowed to institute democratic governments and the tribute rate that was imposed was only moderate. Darius did not want to unnecessarily alienate the Ionians, because he was already planning on using their navy in an attack on Greece.

Darius reserved and focused his desire for revenge on the Athenians who had assisted the Ionians.

> King Darius received tidings of the taking and burning of Sardis by the Athenians and Ionians; and at the same time he learnt that the author of the league, the man by whom the whole matter had been planned and contrived was Aristagoras the Milesian. It is said that he no sooner understood what had happened, than, lying aside all thought concerning the Ionians, who would, he was sure, pay dear for their rebellion, he asked, "Who the Athenians were?" and, being informed,

*A *trireme* was a warship with three tiers of oars on each side.

> called for his bow, and placing an arrow on the string, shot upward into the sky, saying, as he let fly the shaft—"Grant me, Jupiter, to revenge myself on the Athenians!" After this speech, he bade one of his servants every day, when his dinner was spread, three times repeat these words to him—"Master, remember the Athenians."[384]

Darius prepared a large army for an invasion of the Grecian mainland. While war preparations were ongoing, he also sent envoys to all the city-states in Greece and nearby islands. Darius' emissaries demanded submission, which was to be signified by the token delivery of "earth and water" in acceptance of vassalage. "Darius resolved to prove [test] the Greeks, and try the bent of their minds, whether they were inclined to resist him in arms or prepared to make their submission. He therefore sent out heralds in divers directions round about Greece, with orders to demand everywhere earth and water for the king."[385]

Many of the mainland Greek cities submitted, as did nearly all of the island colonies. They had witnessed how the Persians had handily put down the Ionian revolt, and were terrified of the consequences of resisting Darius. However, the two largest and strongest Greek cities, Athens and Sparta, refused to submit. "[Darius'] messengers ... were thrown, at Athens, into the pit of punishment, at Sparta into a well, and bidden to take there from earth and water for themselves, and carry it to their king."[386]

Darius placed the Median general, Datis, in command of the Grecian invasion forces. Datis was instructed by Darius to completely conquer and subjugate Greece. Special orders were issued regarding the inhabitants of the Greek cities of Athens and Eretria. These were the cities that had aided the Ionian revolt. The Athenians and Eretrians were to be taken captive and brought back as slaves before Darius.

In the late summer or early fall of 490 B.C., the Persians set sail with six hundred ships. They first landed at Euboea, a large island immediately off the Greek coast, and attacked the city of Eretria. The Eretrians successfully defended their city for six days. On the seventh day, they were betrayed by two of their own chiefs. The Persians "no sooner entered within the walls than they plundered and burnt all the temples that there were in the town, in revenge for the burning of their own temples at Sardis; moreover, they did [so] according to the orders of Darius, and carried away captive all the inhabitants."[387] The victorious Persians anticipated that the defeated Eretrians would soon be joined by the Athenians.

In September of 490 B.C., the Persians landed unopposed on the plain at Marathon, northeast of Athens. Europe was on the precipice. If Athens fell, the only other Greek city that could fight Persia was Sparta. Although the Spartans would have fought to the last man, by themselves they would have been unable to prevail against the Persian horde. If the Persians were able to defeat Athens, other Greek cities might have enlisted in their cause rather than face disaster. The momentum of the Persian attack could have been irresistible. Western civilization might have been strangled in the cradle.

> Nor was there any power to the westward of Greece that could have offered an effectual opposition to Persia, had she once conquered Greece, and made that country a basis for future military operations. Rome was at this time in her season of utmost weakness.... Had Persia beaten Athens at Marathon, she could have found no obstacle to prevent Darius, the chosen servant of Ormuzd*, from advancing his sway over all the known Western races of mankind. The infant energies of Europe would have been trodden out beneath universal conquest; and the history of the world, like the history of Asia, would have become a mere record of the rise and fall of despotic dynasties, of the incursions of barbarous hordes, and of the mental and political prostration of millions beneath the diadem, the tiara, and the sword.[388]

**Ormuzd* was the chief deity of Zoroastrianism, a religious system founded in Persia by Zoroaster.

The Athenians sent a runner to Sparta requesting aid. But the Spartans refused to come. Because of a religious scruple, they would not be able to set march until the moon was full. "The Spartans wished to help the Athenians, but were unable to give them any present succor ... so they waited for the full of the moon."[389]

Athens hurriedly rushed an army to the high ground overlooking the plains where the Persians were encamped at Marathon. "The Athenians were 10,000 hoplites either including, or besides, the 1,000 who came from Plataea ... the numbers of the Persians we cannot be said to know at all, nor is there anything certain except that they were greatly superior to the Greeks."[390] The size of the Persian army was probably in the neighborhood of 50,000.[391]

The Greek army was commanded by ten generals and one war-ruler, Callimachus. Opinion as to how to proceed was divided. Five of the generals voted to attack immediately, the other five argued that it was better to wait on reinforcements from Sparta. One of the generals was Miltiades; he had formerly fought the Persians in the east, and was knowledgeable of their tactics. "[Miltiades] was a man, not only of the highest military genius, but also of that energetic character which impresses its own type and ideas upon spirits feebler in conception. Miltiades was the head of one of the noblest houses at Athens: he ranked the AEacidae among his ancestry, and the blood of Achilles flowed in the veins of the hero of Marathon."[392]

Miltiades argued for an immediate attack. He and the other Athenians knew that the Eretrians had lost because of treachery. The longer they waited, the greater the chance that certain traitorous factions would view the Persian conquest as inevitable and seek to advance themselves by facilitating the defeat of Athens. The Persians, in turn, were also aware of this possibility, and were content to wait in the hope that Athenian resolve would crumble.

Among the Persians was a man named Hippias. Hippias was a native Athenian and former tyrant of Athens. In 506 B.C., he had been deposed at Athens and fled to Persia. There he had entreated Darius to help him regain his power at Athens. In return, Hippias offered to govern Athens as a satrapy in the Persian Empire. Now that the Persian army had landed on the shores of Greece, the traitor Hippias was eagerly anticipating a return to power.

In council, Miltiades argued for an immediate attack.

> It now rests with you, Callimachus, either to enslave Athens, or, by assuring her freedom, to win yourself an immortality of fame, such as not even Harmodius and Aristogeiton have acquired. For never, since the Athenians were a people, were they in such danger as they are in at this moment. If they bow the knee to these Medes, they are to be given up to Hippias, and you know what they then will have to suffer. But if Athens comes victorious out of this contest, she has it in her to become the first city of Greece. Your vote is to decide whether we are to join battle or not. If we do not bring on a battle presently, some factious intrigue will disunite the Athenians, and the city will be betrayed to the Medes. But if we fight, before there is anything rotten in the state of Athens, I believe that, provided the Gods will give fair play and no favor, we are able to get the best of it in the engagement.[393]

Miltiades convinced Callimachus, and the vote was narrowly decided in favor of attack. Miltiades assumed tactical command, and had to confront the question of how to attack an superior force. The Greek hoplites were covered with heavy bronze armor and carried a formidable eight-foot spear as their primary weapon. The normal tactic was to construct a phalanx eight men deep and advance in an irresistible forward march. However, due to the inequality in numbers, Miltiades had to alter his tactics. In order to lengthen his line of attack and avoid being flanked, he deliberately weakened his center while maintaining normal strength on the left and right flanks.

Instead of the usual slow march, Miltiades ordered his men to run approximately 1.6 kilometers to close with the Persian army. The Athenians apparently accomplished this handily, although each carried a weight in the neighborhood of fifty to seventy pounds (23–32 kilograms) in the form of spear, short sword, shield, and armor.

The Greeks closed with the Persians. "The Athenians, so soon as they were let go, charged the barbarians at a run. Now the distance between the two armies was little short of eight furlongs [1600 meters]. The Persians, therefore, when they saw the Greeks coming on at speed, made ready to receive them, although it seemed to them that the Athenians were bereft of their senses, and bent upon their own destruction; for they saw a mere handful of mean coming on at a run without either horsemen or archers."[394]

At first, the Persians drove the weak center of the Greek line back. However the Greeks were able to turn and reform their line. The stronger Athenian flanks meanwhile began to decisively beat back the Persian troops. As the flanks started to wheel and close upon the Persian center, the Persians panicked and retreated in a rout. The Athenians chased them back to the shoreline and sought to burn the Persian ships, but most of the vessels escaped.

Instead of resting after the victory, Miltiades immediately ordered the army to march all night in order to reach Athens by daybreak. He had anticipated that the Persians would seek to sail to Athens and destroy the city behind their backs. When the Persian fleet arrived the next morning, it saw the same army that had defeated it the previous day. Frustrated, the Persians returned home.

According to Herodotus, the total body count at Marathon consisted of 6400 Persian and 192 Greek casualties.[395] Two factors were decisive in the Greek victory at Marathon. First, the Greek hoplite possessed superior armament and weaponry. The Persian infantry carried wicker shields, lacked body armor, and had short swords. The Greek hoplite was encased in bronze armor, carried a heavy wooden shield reinforced with iron, and employed a formidable eight-foot spear as his primary weapon.

The second reason for the Greek rout of the Persian troops was psychological. The Persians were mostly conscript troops from various nations, divided by nationality, language, and culture. The Persian army was "composed of subject soldiers of very uneven worth."[396] In contrast, the Greek habit was to fight shoulder-to-shoulder with men with whom they had trained since childhood. The Athenians had recently expelled their dictator, and were flush with the light of democratic freedom. They knew that defeat meant slavery. It may seem paradoxical that the Greeks valued freedom so highly since they themselves practiced slavery. However the endemic presence of slavery in the ancient world only made freedom that much more valued.

In assessing the importance of the battle at Marathon, Creasy concluded that it was critical for the future of Western Civilization. "The day of Marathon is the critical epoch in the history of two nations. It broke for ever the spell of Persian invincibility, which had paralyzed men's minds.... It secured for mankind the intellectual treasures of Athens, the growth of free institutions, the liberal enlightenment of the Western world, and the gradual ascendancy for many ages of the great principles of European civilization."[397]

## *Sparta*

Ten years would pass before the Persians sought to avenge their defeat at Marathon. This time, hoplites from the Greek city of Sparta (Lacedaemon) would face the initial brunt of the attack.

The Spartans were descended from Dorians who had invaded Greece around 1000 B.C., conquering and enslaving the native peoples. There were three classes of people in Sparta. The heirs of the Dorian invaders became Spartan citizens and lived as free men. The descendants of the native inhabitants were known as *helots*, and were held as slaves by the Spartans. A third group were called *perioeci* ("dwellers around the city"[398]), and lived in Sparta as free men, but did not hold citizenship. The perioeci played a critical role in supporting Sparta as they undertook the manual and commercial occupations the Spartans themselves disdained.

Unlike Athens, the glory of Sparta did not lie in art, literature, philosophy, or architecture. In *History of the Peloponnesian War*, the Greek historian Thucydides (c. 460 B.C.–c. 395 B.C.) noted the lack of architectural edifices in Sparta. "Their city is not regularly built, and has no splendid temples or other edifices; it rather resembles a straggling village like the ancient towns of Hellas, and would therefore make a poor show."[399] Sparta was famed for the stability of its government, its military ascendancy, and the loyalty and dedication of the citizens to the Spartan system.

The Spartan system was originated by Lycurgus (c. 800–730? B.C.). Prior to the Lycurgean reforms, the Spartans were "the very worst governed people in Greece."[400] Writing circa fourth century B.C., Xenophon (c. 431–355 B.C.) attributed Sparta's power to its political and cultural institutions. "Sparta, one of the least populous of states, had proved the most powerful and celebrated city in Greece, I wondered by what means this result had been produced. When I proceeded, however, to contemplate the institutions of the Spartans, I wondered no longer."[401]

According to Plutarch (c. A.D. 46–120), some of Lycurgus' political ideas were derivative, others original. "He first arrived at Crete, where, having considered their several forms of government, and got an acquaintance with the principal men among them, some of their laws he very much approved of, and resolved to make use of them in his own country; a good part he rejected as useless."[402]

The Lycurgean system was not democratic, but a unique combination of different political systems. When questioned on the advisability of democracy, Lycurgus replied "begin, friend, and set it up in your family."[403]

The Spartans had two hereditary kings of equal power, and a senate of twenty-eight composed of men over sixty years of age. At times, the entire population of male citizens over thirty years of age were assembled to vote on certain questions. Above all of these was an executive council of five *ephors* elected by popular vote of the citizens. Most power resided in the council of *ephors*; given a sufficient pretext, they could even depose a king.

It is difficult to classify the Spartan political system. In the view of one historian, "it was in substance a close, unscrupulous, and well-obeyed oligarchy."[404] In *Politics*, Aristotle described the Spartan system as a hybrid. "It is made up of oligarchy, monarchy, and democracy, the king forming the monarchy, and the council of elders the oligarchy, while the democratic element is represented by the ephors; for the ephors are selected from the people."[405]

The most powerful element in the government was the council of ephors. Originally constituted as "nothing more than subordinates and deputies of the kings,"[406] over time the ephors became so powerful that they "reduced the kings to a state of intolerable humiliation and impotence."[407] "Both the internal police and the foreign affairs of the state are in the hands of the ephors, who exercise an authority approaching to despotism, and altogether without accountability."[408]

Instituted in the 8th or 9th century B.C., the Lycurgean system in Sparta generated four or five centuries of ascendancy. Thucydides noted, "[Sparta] has preserved the same form of government for rather more than four hundred years, reckoning to the end of the Peloponnesian War [404 B.C.]. It was the excellence of her constitution which gave her power."[409]

But by the third century B.C., the Spartan system had deteriorated. "Among all the Grecian states, Sparta had declined the most; her ascendancy was totally gone, and her peculiar training and discipline had degenerated in every way."[410]

It was forbidden to write down any Spartan law. Plutarch related, "Lycurgus would never reduce his laws into writing; nay there is a *rhetra* [law] expressly to forbid it."[411] Rather, the young were required to memorize the entire body of laws by rote. Lycurgus believed that for the law to be effective it had to live in the mind and hearts of the people. The highest duty of every Spartan was to obey the law. "At Sparta the citizens pay the strictest obedience to the magistrates and laws."[412] Xenophon (c. 430–355 B.C.) concluded, "Sparta is naturally superior in virtue to all other states, as it is the only one that engages in a public cultivation of honor and virtue."[413]

The Lycurgean system was designed to eliminate the moral degradations brought by civilization and the pursuit of wealth. The virtues of self-discipline, austerity, and service to the state were emphasized. Lycurgus "realized his project of creating in the 8,000 or 9,000 Spartan citizens unrivalled habits of obedience, hardihood, self-denial, and military aptitude."[414] All gold and silver money was eliminated from Spartan society. "He [Lycurgus] commanded that all gold and silver coin should be called in ... in a country which had no money ... at once a number of vices were banished from Lacedaemon; for who would rob another of such a coin? Who would unjustly detain or take by force, or accept as a bribe, a thing which it was not easy to hide, nor a credit to have, nor indeed of any use to cut in pieces? ... luxury, deprived little by little of that which fed and fomented it, wasted to nothing and died away of itself."[415]

Plutarch stated that at the institution of the Lycurgian reforms, the state seized land that was held primarily by a few wealthy individuals and divided it equally among citizens. "[Lycurgus] obtained of them to renounce their properties, and to consent to a new division of the land, and that they should live all together on an equal footing; merit to be their only road to eminence, and the disgrace of evil, and credit of worthy acts, their one measure of difference between man and man."[416] But George Grote (1794–1871) questioned the authenticity of Plutarch's account, noting "all the historical evidences exhibit decided inequalities of property among the Spartans."[417]

The law was made that all meals had to be eaten in common messes.

> The third and most masterly stroke of this great lawgiver [Lycurgus], by which he struck a yet more effectual blow against luxury and the desire of riches, was the ordinance he made, that they should all eat in common, of the same bread and same meat, and of kinds that were specified, and should not spend their lives at home, laid on costly couches at splendid tables, delivering themselves up into the hands of their tradesmen and cooks, to fatten them in corners, like greedy brutes, and to ruin not their minds only but their very bodies which, enfeebled by indulgence and excess, would stand in need of long sleep, warm bathing, freedom from work, and, in a word, of as much care and attendance as if they were continually sick.... The rich, being obliged to go to the same table with the poor, could not make use of or enjoy their abundance, nor so much as please their vanity by looking at or displaying it.[418]

Xenophon explained, "When they take their meals together in this manner, how can any one ruin either himself or his family by gluttony or drunkenness?"[419]

Foremost in the Lycurgian reform was the total dedication of the state to the education of children, including girls. Girls as well as boys had to perform sports publicly in the nude. These public exhibitions were not in any sense lascivious. The purpose was to inculcate a physical culture of the body. Knowing that their nude body was to be put on public display, Spartans were motivated to exercise it, and avoid obesity and physical degeneration.

In *The Republic*, Plato said that "the Cretans and then the Lacedaemonians introduced the custom" of exercising in the nude.[420] When the practice began, it was regarded as "ridiculous and improper," but "experience showed that to let all things be uncovered was far better than to cover them up, and the ludicrous effect to the outward eye vanished before the better principle which reason asserted."[421]

Children of both sexes had to learn to run, wrestle, and throw the discus and javelin. According to Plutarch, "he [Lycurgus] ordered the maidens to exercise themselves with wrestling, running, throwing the quoit*, and casting the dart ... and he [Lycurgus] ordered that the young women should go naked in the processions, as well as the young men, and dance, too, in that condition."[422] The Spartan youths may not have been completely naked. George Grote concluded, "they seem to have worn a light tunic, cut open at the skirts, so as to leave the limbs both free and exposed to view — hence Plutarch speaks of them as completely uncovered."[423]

Women in Sparta had more freedom and greater status than any other state in ancient Greece. Unlike other Greek states, girls were taught to read and write. Female literacy meant that Sparta had the highest overall literacy rate of any Greek community. Women could also own property and accumulate wealth. In *Politics*, Aristotle noted "nearly two-fifths of the whole country are held by women."[424]

Because most men were preoccupied with their military duties, the management of the family estate and household was left to the wife. Aristotle complained of the pernicious influence of women on the government of Sparta. "What difference does it make whether women rule, or the rulers are ruled by women? The result is the same. Even in regard to courage, which is of no use in daily life, and is needed only in war, the influence of the Lacedaemonian women has been most mischievous."[425] Gorgo, wife of King Leonidas, was asked why Spartan women were the only ones who could rule men. She replied, because "we are the only women who bring forth men."[426]

Spartan mothers told their sons to "return either with your shield or upon it."[427] In 371 B.C., Thebes broke Sparta's power by badly defeating it at the battle of Leuctra. According to Plutarch, the families of those who had been slain in battle publicly celebrated, while the relatives of those who had survived remained hidden in their homes, laden with shame.

> The fathers, relatives, and friends of the slain came out rejoicing in the market-place, saluting each other with a kind of exultation; on the contrary, the fathers of the survivors hid themselves at home among the women. If necessity drove any of them abroad they went very dejectedly, with downcast looks and sorrowful countenances. The women outdid the men in it; those whose sons were slain openly rejoicing, cheerfully making visits to one another, and meeting triumphantly in the temples; they who expected their children home being very silent and much troubled.[428]

The Spartans practiced an unabashed eugenics. Newborn children were brought before a council of elders and examined. If the child was deemed to be weak or defective, it was thrown off a cliff.[429] Husbands were encouraged to let their wives be inseminated by men deemed to be superior; this was not considered to be adultery. "He [Lycurgus] made it ...

*A ring thrown at a peg in the ground.

honorable for men to give the use of their wives to those whom they should think fit, that so they might have children by them."[430]

Music flourished among the warlike Spartans. Their favorite instrument was the flute; they favored it for marching rhythms, and were said to stride into battle singing.

> When their army was drawn up in battle array, and the enemy near, the king sacrificed a goat, commanded the soldiers to set their garlands upon their heads, and the pipers to play the tune of the hymn to Castor, and himself began the paean of advance. It was at once a magnificent and a terrible sight to see them march on to the tune of their flutes, without any disorder in their ranks, any discomposure in their minds, or change in their countenances, calmly and cheerfully moving with the music to the deadly fight. Men, in this temper, were not likely to be possessed with fear or any transport of fury, but with the deliberate valor of hope and assurance, as if some divinity were attending and conducting them.[431]

The key to Spartan society was the upbringing of male children. At the age of seven, boys were taken from their mothers and subjected to continuous and rigorous military training. "As soon as they were seven years old they were to be enrolled in certain companies and classes, where they all lived under the same order and discipline."[432] Male youth learned to read and write, but the physical arts were emphasized. "Their chief care was to make them good subjects, and to teach them to endure pain and conquer in battle."[433] Boys were required to undergo pain, hardship, and misfortune without complaint. They bathed infrequently, and slept in the open, both summer and winter. The Spartans believed that water and oils made the body soft, but cold air and soil made it hard and resistant.[434]

Teachers had to demonstrate to elders and magistrates that their punishments were neither too harsh nor too lenient. The entire education of male youth was designed to inculcate the virtues and strengths that would bring victory in battle. Military supremacy was necessary for Spartan survival. Not only were they incessantly at war with other sovereignties, but the citizens of Sparta were greatly outnumbered by the helots, the slaves who tilled their soil. Thucydides explained, "remember that in the cities from which you come, not the many govern the few, but the few govern the many, and have acquired their supremacy simply by successful fighting."[435]

Boys were maintained on a frugal diet, but encouraged to supplement their diet by stealing food. It was thought that the skills needed for successful thievery would be useful in a military occupation. "Lycurgus ... gave them [boys in military training] liberty to steal certain things to relieve the cravings of nature [hunger]; and he made it honorable to steal as many cheeses as possible ... it is evident that he who designs to steal must be wakeful during the night, and use deceit, and lay plots; and, if he would gain anything of consequence, must employ spies. All these things, therefore, it is plain that he taught the children from a desire to render them more dexterous in securing provisions, and better qualified for warfare."[436]

Anyone caught in the act of stealing was severely punished by being whipped. The punishment was not for the act of stealing, but for failure to carry out the theft successfully. Plutarch wrote that he had personally witnessed Spartan youths being whipped to death. "What is practiced to this very day [c. A.D. 100] in Lacedaemon is enough to gain credit to this story, for I myself have seen several of the youths endure whipping to death."[437]

Although Sparta was renowned for physical training, there may have been a significant intellectual accompaniment. In the dialogue *Protagoras*, Plato described Spartans as "perfectly educated" men.[438]

> There is a very ancient philosophy which is more cultivated in Crete and Lacedaemon than in any other part of Hellas, and there are more philosophers in those countries than anywhere else in the world. This, however, is a secret which the Lacedaemonians deny; and they pretend to be

ignorant, just because they do not wish to have it thought that they rule the world by wisdom ... [because] ... if ... their superiority were disclosed, all men would be practicing their wisdom. And [this is how] you may know that I am right in attributing to the Lacedaemonians ... excellence in philosophy and speculation: If a man converses with the most ordinary Lacedaemonian, he will find him seldom good for much in general conversation, but at any point in the discourse he [the Spartan] will be darting out some notable saying, terse and full of meaning, with unerring aim; and the person with whom he is talking seems to be like a child in his hands. And many of our own age and of former ages have noted that the true Lacedaemonian type of character has the love of philosophy even stronger than the love of gymnastics; they are conscious that only a perfectly educated man is capable of uttering such expressions.[439]

Although it is widely believed that homosexuality and the exploitation of male children was rampant in Sparta, the evidence is ambiguous. In his book *Greek Homosexuality* (1989), Kenneth James Dover concluded that there was a variety of evidence indicating that "overt and unrepressed homosexuality ... [was] a conspicuous feature of Greek life."[440] But Xenophon, whose description of Sparta can be characterized as a panegyric, said that pederasty was expressly forbidden and considered to be as immoral as incest.

If any man showed that his affections were fixed only on the bodily attractions of a youth, Lycurgus, considering this as most unbecoming, appointed that at Lacedaemon suitors for the favors of boys should abstain from intimate connection with them, not less strictly than parents abstain from such intercourse with their children, or children of the same family from that with one another. That such a state of things is disbelieved by some, I am not surprised; for in most states the laws are not at all adverse to the love of youths; but Lycurgus, for his part, took such precautions with reference to it.[441]

In Plato's *Laws*, the charge was made that the institutions of Sparta "have had a tendency to degrade the ancient and natural custom of love below the level, not only of man, but of the beasts. The charge may be fairly brought against your city [Sparta] above all others ... the intercourse of men with men, or of women with women, is contrary to nature."[442] Plato, of course, wrote from the perspective of an Athenian after Athens had lost the long and bitter Peloponnesian War to Sparta.

Spartan training was not mindlessly dedicated to abuse; its severity was imposed gradually upon youths as their capacities grew. Humfrey Michell (1883–1970) noted that Spartan boys did not engage in competitive sports until they were ten years old and observed "we must credit the Spartans with some wisdom in their treatment of children."[443]

In this age, there were no machine guns, bombs, or poison gas. The greatest force was the discipline of cold iron. Conflict was a personal affair, managed by mental fortitude and muscle, accompanied by blood and sweat. The greatest qualities that were necessary for victory were nerve and steadfastness. The psychological factor was more important than the physical. If the phalanx broke, defeat was inevitable.[444] Strength and skill were desirable, but worthless without resolve. The key to victory was maintaining an ordered formation; the man who broke and ran would spread fear and destroy hope.

In battle, the Spartans were highly disciplined. They fought alongside men with whom they had lived since the age of seven. Their entire life was devoted to duty; to falter or exhibit cowardice was unthinkable. When they had defeated an enemy, it was their practice to never pursue and slaughter foes who fled from them. "Knowing that they [the Spartans] killed only those who made resistance, and gave quarter to the rest, men generally thought it their best way to consult their safety by flight."[445] Conversely, they were known to be unrelenting in those who stood against them. As these practices became known, it encouraged their enemies to flee from them.

From the time of Xenophon (c. 400 B.C.), writers have expressed admiration for the

Spartan system. However it would be circumspect to realize that "we can afford to praise Sparta, for we do not have to live in her."[446]

## *Battle of Thermopylae (480 b.c.)*

After his defeat at Marathon, the Persian king, Darius, was eager for revenge. However his plans were delayed by a revolt in Egypt and he died in 494 B.C. Darius' son, Xerxes, inherited his father's empire and ambitions for an invasion of Greece. According to Herodotus, Xerxes "called together an assembly of the noblest Persians ... to lay before them his own designs."[447] Xerxes spoke: "My intent is to throw a bridge over the Hellespont and march an army through Europe against Greece, that thereby I may obtain vengeance from the Athenians for the wrongs committed by them against the Persians and against my father ... I ... pledge myself not to rest till I have taken and burnt Athens."[448]

To minimize the possibility of defeat, Xerxes assembled a huge army and made diligent and careful preparations for the invasion of Greece. "Xerxes spent four full years in collecting his host, and making ready all things that were needful for his soldiers. It was not till the close of the fifth year that he set forth on his march, accompanied by a mighty multitude. For of all the armaments whereof any mention has reached us, this was by far the greatest.... All [other] expeditions ... are as nothing compared to this. For was there a nation in Asia which Xerxes did not bring with him against Greece? Or was there a river, except those of unusual size, which sufficed for his troops to drink?"[449]

Herodotus estimated the size of Xerxes' army as exceeding five million men. "This will give 5,283,220 as the whole number of men brought by Xerxes."[450] Historians generally consider Herodotus' estimate to be an exaggeration. But there is no good way to know what the size was with any accuracy. Modern estimates of the size of Xerxes' army, based on various methods, vary between 50,000 and 1,000,000.[451]

When Greek spies were discovered among the Persian Troops, Xerxes declined to have them executed. Instead, he led them about, exhibited the huge size and resources of his army, and had the Greeks released. "'Had the spies been put to death,' he [Xerxes] said, 'the Greeks would have continued ignorant of the vastness of his army, which surpassed the common report of it; while he would have done them a very small injury by killing three of their men. On the other hand, by the return of the spies to Greece, his power would become known; and the Greeks,' he expected, 'would make surrender of their freedom before he began his march, by which means his troops would be saved all the trouble of an expedition.'"[452]

Accompanying Xerxes was Demaratus, a Spartan king in exile. When asked by Xerxes for an assessment of the Greek resistance, he warned that the Spartans would fight even if greatly outnumbered.

> Brave are all the Greeks who dwell in any Dorian land; but what I am about to say does not concern all, but only the Lacedaemonians. First then, come what may, they will never accept thy terms, which would reduce Greece to slavery; and further, they are sure to join battle with thee, though all the rest of the Greeks should submit to thy will. As for their numbers, do not ask how many they are, that their resistance should be a possible thing; for if a thousand of them should take the field, they will meet thee in battle, and so will any number, be it less than this, or be it more.[453]

In 480 B.C., Xerxes' huge army was on the move. It had been ten years since the Athenians defeated the Persians at Marathon. Knowing that invasion was imminent, the Greek cities that had not allied themselves with the Persians met in council to plan a defensive strategy. It was decided to hold the mountain pass at Thermopylae. The pass was narrow, and it offered the best chance for a smaller army to defend itself against a larger one. The Spartan king, Leonidas, was placed in command of the forces that were to assemble there. "The opinion which prevailed was, that they should guard the pass of Thermopylae.... Considering that in this region the barbarians could make no use of their vast numbers, nor of their cavalry, they resolved to await here the invader of Greece."[454]

Upon his return to Sparta, Leonidas was refused permission by the ruling council of ephors to advance the army during an approaching religious holiday. Furthermore, the consensus in Sparta favored a defensive position closer to Sparta, not the more distant Thermopylae. Leonidas, however, was morally committed — he had promised the other Greek cities in council that the Spartans would lead the united forces at Thermopylae. Although he could not take the entire Spartan army, he was allowed to advance with 300 Spartans chosen as his personal bodyguard.

In choosing men for the march to Thermopylae, Leonidas picked only men with living male heirs. He knew from the beginning that it was a suicidal mission, but he also knew that the other Greek cities would not fight if the Spartans did not lead them. Someone questioned the wisdom of fighting the Persians with only 300 men. Leonidas replied, "if you esteem number, all Greece is not able to match a small part of that army; if courage, this number is sufficient."[455]

As Leonidas marched with his 300 Spartans to Thermopylae, troops from other Greek cities swelled the numbers to 7000. When they arrived at the mountain pass, the Greeks busied themselves rebuilding a crumbled stone wall that had served as a defensive bastion in a forgotten age.

The Persian army arrived, but Xerxes delayed his attack. Xerxes thought the Greeks would surely flee in fear when they saw the hopelessness of trying to fight his enormous force. "Four whole days he [Xerxes] suffered to go by, expecting that the Greeks would run away."[456]

Xerxes sent an envoy to the Spartans to request their surrender. The note carried by the emissary told Leonidas that if he surrendered, Greece would be assimilated into the Persian Empire and Leonidas would be its governor. Leonidas replied, "if you understood wherein consisted the happiness of life, you would not covet other men's: but know that I would rather die for the liberty of Greece than be a monarch over my countrymen."[457] The Persian messenger warned Leonidas that the sky would be darkened when the arrows of the innumerable Persian archers blotted out the sun. But Leonidas merely observed, "it will be pleasant for us to fight in the shade."[458]

Exasperated, Xerxes finally launched an attack. The Persians made no headway against the resolute Greeks. "The Medes rushed forward and charged the Greeks, but fell in vast numbers ... it became clear to all, and especially the king, that though he [Xerxes] had plenty of combatants, he had but very few warriors."[459] "The position was one in which bows and arrows were of little avail: a close combat hand to hand was indispensable, and in this the Greeks had every advantage of organization as well as armor. Short spears, light wicker shields, and tunics, in the assailants, were an imperfect match for the long spears, heavy and spreading shields, steady ranks, and practiced fighting of the defenders."[460]

Herodotus claimed that when the Spartans decided they were not killing Persians fast enough, they would fall back, tricking the enemy into moving forward. After luring the Persians into their clutches, they would turn and slaughter them. "The Lacedaemonians fought in a way worthy of note, and showed themselves far more skillful in fight than their adversaries, often turning their backs, and making as though they were all flying away, on which the barbarians would rush after them with much noise and shouting, when the Spartans at their approach would wheel round and face their pursuers, in this way destroying vast numbers of the enemy. Some Spartans likewise fell in these encounters, but only a very few."[461]

On the dawn of the second day, Xerxes sent another envoy to Leonidas. He praised the Spartans for their courage, and said that he had never seen such skilled and resolute warriors. He promised the Spartans that if they would lay down their arms, they could join the vanguard of his army and share in the spoils of victory. Leonidas' reply was exceedingly terse and to the point. In answer to the demand to lay down his weapons, he said "come and take them."[462]

Xerxes responded by attacking with 10,000 of his elite troops known as "immortals." "They were called 'the Immortals,' for the following reason. If one of their body failed either by the stroke of death or of disease, forthwith his place was filled up by another man, so that their number was at no time either greater or less than 10,000."[463]

The "immortals" were Xerxes' personal bodyguard, and commanded by his brother. At the end of the day, the result was the same. The pass was still held by the Greeks and Xerxes' brother was dead. The "immortals" had been decimated, and the Persian army was demoralized to see their best troops so easily defeated.

Seemingly, the small Greek force could hold the world's greatest army at bay indefinitely, but on the evening of the second day they were betrayed. A traitorous Greek goat herder named Ephialtes showed Xerxes an obscure path through the mountains that the Persians could use to circle around and attack the Greeks from behind.

Leonidas had been aware of the existence of the path from the beginning, and had it watched and guarded. As the Persians started to infiltrate, he was informed of their action. The tactical situation had changed. The pass could not be held from a simultaneous attack on two sides. Knowing that defeat was now certain, Leonidas ordered the non–Spartans to retreat so that they could live to fight another day. The Thebans and Thespians refused to withdraw and remained with the Spartans. About 1,000 Greeks were left.

"To Leonidas the idea of retreat was intolerable. His own personal honor, together with that of his Spartan companions and of Sparta herself, forbade him to think of yielding to the enemy the pass which he had been sent to defend. The laws of his country required him to conquer or die in the post assigned to him, whatever might be the superiority of number on the part of the enemy."[464]

On the morning of the third day, Leonidas told his men to eat breakfast with the expectation of having dinner in Hades. There were three youths that Leonidas wanted to save from destruction, so he sent them back to Sparta on the pretext of needing someone to carry an urgent message. He tried to do the same with three older men, but they saw through the ruse and refused to leave. One said, "I came, sir, to be a soldier, and not a courier."[465] A man who had gone blind from infection demanded to be led to the coming battle, sword in hand.

Only one of the Spartans, a man named Aristodêmus, chose to return home after being grievously wounded. "He returned only to scorn and infamy among this fellow-citizens. He was denounced as 'the coward Aristodêmus;' no one would speak or communicate with him ... after a year of such bitter disgrace, he was at length enabled to retrieve his honor at the battle of Plataea, where he was slain, after surpassing all his comrades in heroic and even reckless valor."[466]

As the Persians started to close the pass behind them, Leonidas ordered his troops to rush forward and attack. "The Greeks fought with reckless bravery and desperation against this superior host, until at length their spears were broken, and they had no weapon left except their swords. It was at this juncture that Leonidas himself was slain."[467]

The Spartans had to drive the enemy off four times before they could retrieve Leonidas' body. The few Greeks that were left retreated to a small hillock and formed a circle. Rather than see more of his infantry killed, Xerxes had the remaining Greeks killed by his archers. By the time the battle was finally over, the Greeks had lost about 4,000 men, including helots, but the Persians had 20,000 dead.[468] According to Herodotus, a monument was erected at Thermopylae with the inscription, "Go, stranger, and to Lacedaemon tell, that here, obeying her behests, we fell."[469]

Greece lost the battle at Thermopylae, but the circumstances turned the defeat into a moral victory. In the next year, Greece won decisive sea and land battles against Persia. In the Straits of Salamis, 300 Greek warships defeated 800 Persian ships. Herodotus attributed the Greek victory to better tactics and seamanship. "As the Greeks fought in order and kept their line, while the barbarians were in confusion and had no plan in anything that they did, the issue of the battle could scarce be other than it was."[470]

The Persian defeat was likely due to a combination of factors, both physical and psychological. "Their signal defeat was not owing to any want of courage, but, first, to the narrow space which rendered their superior number a hindrance rather than a benefit: next, to their want of orderly line and discipline as compared to the Greeks: thirdly, to the fact that when once fortune seemed to turn against them, they had no fidelity or reciprocal attachment, and each ally was willing to sacrifice or even to run down others, in order to effect his own escape."[471]

After his defeat at Salamis, Xerxes returned to Persia for the winter but left a large force of men in Greece under the command of his general, Mardonius. In the Spring of 479 B.C., an allied force of about 40,000 Greeks faced 100,000 Persians at Plataea. The Greeks were victorious and the Persian general Mardonius was killed.

Herodotus described the battle, attributing victory to the superior armament of the Greek hoplites. "A fierce contest took place by the side of the temple of Ceres, which lasted long, and ended in a hand-to-hand struggle. The barbarians many times seized hold of the Greek spears and brake them; for in boldness and warlike spirit the Persians were not a whit inferior to the Greeks; but they were without bucklers, untrained, and far below the enemy in respect of skill of arms.... Their light clothing, and want of bucklers, were of the greatest hurt to them: for they had to contend against men heavily armed, while they themselves were without any such defense."[472]

The Persian army withdrew and never again invaded Greece. Sporadic fighting continued for several years until the Greco-Persian wars ended with a peace treaty in 448 B.C. The treaty forced the Persians to recognize the liberty of the Greek states in Europe and Asia and kept the Persian fleet out of the Aegean Sea.

Following the defeat of the Persians at Salamis and Plataea, Sparta lapsed into agriculture, seclusion, and stagnation. However Athens turned her navy into a fleet of merchant ships and became the hub of a Greek world that stretched from Italy in the west to the shores of Asia Minor in the east. There was peace for fifty years, and Athens became the intellectual, commercial, and cultural capital of Greece.

Athens had been burned by the Persians, but the Athenians had faith in the future. They rebuilt the city, and constructed the Parthenon, a building of legendary grace and beauty. The arts flourished during a period that "called forth creative genius in oratory, in dramatic poetry, and in philosophical speculation."[473] It was the golden age of Athens, one in which the contacts fostered by trade led to comparisons of cultures, ideas, and theories.

## *Socrates (470–399 B.C.)*

### *Personal Characteristics*

The most famous philosophers of ancient Greece were Socrates, Plato, and Aristotle. These three are related by direct academic descent. Socrates was the teacher of Plato, and Plato in turn was the teacher of Aristotle.

Socrates began an intellectual revolution by establishing that the process of inquiry is more important than specific answers. Science is a method of generating reliable knowledge. Although the practice of science leads to the accumulation of a body of knowledge, this knowledge is not science itself, it is the product of the practice of science. Karl Popper explained, "[Some people] are still possessed by the pre–Socratic magical attitude towards science, and towards the scientist, whom they consider as a glorified shaman, as wise, as learned, initiated. They judge him by the amount of knowledge in his possession, instead of taking, with Socrates, his awareness of what he does not know as a measure of his scientific level as well of his intellectual honesty."[474]

Socrates left no writings behind, so we cannot be absolutely sure that he even existed. However the question of existence does not diminish Socrates' importance. John Burnet argued, "if he is a fictitious character, he is nevertheless more important than most men of flesh and blood."[475]

One of the few primary sources that describes the life and philosophy of Socrates are the writings of Socrates' friend, the historian Xenophon (c. 431–355 B.C.). But the most important primary sources are the *Dialogues* of Plato. In these books, Plato himself does not appear, but the chief character is always Socrates, who explores philosophical questions by carrying on dialogues on various topics.

The eternal question concerning the *Dialogues* is whether Plato was a mere raconteur who recorded the true words of Socrates in his works, or a philosopher who used Socrates as a mouthpiece for his own ideas. The truth may lie somewhere in between. In George Grote's opinion, this "is a point not to be decided with certainty or rigor."[476]

It is likely that the early dialogues were written by Plato as a historical narrative of his beloved teacher. As Plato's literary career progresses, it becomes apparent that Socrates has become a literary tool for Plato to present his own views.[477] It is unlikely, for example, that the historical Socrates would have endorsed the totalitarian views put in his mouth in *The Republic*. This raises the question as to how much of the historical Socrates we can infer from Plato's dialogues. It is probable that the dialogues provide an accurate picture of Socrates' methods and character, if not his views on specific issues. John Burnet concluded, "if Plato's Socrates is not meant for the real Socrates, I find it very hard to imagine what he can be meant for."[478]

"Of the circumstances of his [Socrates'] life we are almost wholly ignorant."[479] Socrates was born in 470 B.C., ten years after the battle at Thermopylae. He was a citizen of Athens, and during his lifetime he fought as a hoplite in the Peloponnesian War, and served with distinction and bravery. "He was not merely strong and active as a hoplite on military service, but capable of bearing fatigue or hardship, and indifferent to heat or cold, in a measure which astonished all his companions."[480]

Although Athens was defeated at the battle of Delium in 424 B.C., Socrates distinguished himself. His "bravery, both in the battle and the retreat, was much extolled by his friends ... his patience under hardship, and endurance of heat and cold, being not less remarkable than his personal courage. He [Socrates] and his friend Laches were among those hoplites who in the retreat from Delium, instead of flinging away their arms and taking to flight, kept their ranks, their arms, and their firmness of countenance; insomuch that the pursuing cavalry found it dangerous to meddle with them, and turned to an easier prey in the disarmed fugitives."[481]

Socrates' actions at Delium were also noted by Strabo. "The Athenians, after their defeat in battle, fled in disorder. In the flight, Socrates the philosopher (who having lost his horse, was serving on foot) observed Xenophon, the son of Gryllus, upon the ground, fallen from his horse; he raised him upon his shoulders and carried him away in safety, a distance of many stadia, until the rout was at an end."[482]

Athenaeus (fl. A.D. 2nd–3rd century), however, was skeptical of Socrates' military achievements. He noted that the primary historian of the Peloponnesian War, Thucydides, never mentioned Socrates, and questioned, "what is there in common between a shield and a philosopher's staff?"[483]

Socrates was a homely man. "His friends, who communicate to us his great bodily strength and endurance, are at the same time full of jests upon his ugly physiognomy — his flat nose, thick lips, and prominent eyes like a satyr."[484] In Plato's *Symposium*, Alcibiades said "you yourself will not deny, Socrates, that your face is like that of a satyr."[485]

Socrates believed that an inner voice, "divine sign," or "demon" guided his actions. However, this voice was peculiar in that it never prompted him to do anything in particular, but warned him when it thought he was on the verge of making a mistake. Socrates described the inner demon in Plato's *Apology*. "You have heard me speak at sundry times and in diverse places of an oracle or sign which comes to me ... this sign, which is a kind of voice, first began to come to me when I was a child; it always forbids but never commands me to do anything which I am going to do."[486]

Socrates was temperate, frugal and self-disciplined. He eschewed wealth, dressed in rags, and devoted his life to intellectual pursuits. "Socrates ... was not only the most rigid of all men in the government of his passions and appetites, but also most able to withstand cold, heat, and every kind of labor; and, besides, so inured to frugality, that, though he possessed very little, he very easily made it a sufficiency."[487] Xenophon quoted Socrates as stating, "I ... am constantly preparing my body by exercise to endure whatever may happen to it."[488] Socrates justified his devotion to physical culture by noting that "those who have their bodies in a good state are healthy and strong."[489]

Socrates' impoverished lifestyle must have been a stress on family life. In Xenophon's *Banquet*, Socrates' wife, Xanthippe, is characterized as "the most ill-conditioned of all women that are in existence, and ... of all that ever were and ever will be."[490] Diogenes Laërtius related that Xanthippe once "threw water at him [Socrates]," and "attacked him in the market-place, and tore his cloak off."[491] Socrates bore this abuse because "Xanthippe brings me children [three sons]."[492] As an additional reason, Socrates offered the explanation that living with Xanthippe was sort of an extreme training for him. "I, wishing to converse and associate with mankind, have chosen this wife, well knowing that if I shall be able to endure her, I shall easily bear the society of all other people."[493] Cato the Censor (234–149 B.C.)

reportedly stated that "he admired the ancient Socrates for nothing so much as for having lived a temperate and contented life with a wife who was a scold, and children who were half-witted."[494]

Socrates was a well-known figure in Athens, and spent his day in public discourse. Xenophon said "He [Socrates] was constantly in public, for he went in the morning to the places for walking and the gymnasia; at the time when the market was full he was to be seen there; and the rest of the day he was where he was likely to meet the greatest number of people; he was generally engaged in discourse, and all who pleased were at liberty to hear him."[495]

## *Moral Philosophy*

According to Diogenes Laërtius, Socrates was "a disciple of Archelaus, the natural philosopher."[496] Archelaus was "the first person who imported the study of natural philosophy from Ionia to Athens."[497] However Socrates evidently became frustrated by the inability of natural philosophers to discover anything of practical use. "He would ask ... whether they will be able to carry into effect what they have learned, either for themselves, or for any one else."[498]

Socrates became convinced that natural philosophers were engaged in idle, useless speculation. As evidence for this, he noted that all of the explanations proposed by the natural philosophers disagreed with each other. "He [Socrates] did not dispute about the nature of things as most other philosophers disputed ... but endeavored to show that those who chose such subjects of contemplation were foolish ... it is impossible for man to satisfy himself on such points, since even those who pride themselves most on discussing them, do not hold the same opinions with one another."[499] In Plato's *Apology*, Socrates explicitly disavowed any interest in natural philosophy. "I [do not] mean to speak disparagingly of any one who is a student of natural philosophy ... but I have nothing to do with physical speculations."[500]

Instead of natural philosophy, Socrates chose to devote his intellectual energies to ethical considerations and moral philosophy. "He was the first who turned his thoughts and discussions distinctly to the subjects of ethics."[501] Cicero noted that "Socrates was the first who brought philosophy down from the heavens, placed it in cities, introduced it to families, and obliged it to examine into life and morals, and good and evil."[502]

In a letter to a friend, Xenophon said "When ... did any one ever hear Socrates discoursing about the heavenly bodies, or exhorting men to learn geometry in order to improve their morals? ... He was constantly discussing with his friends what propriety was, or fortitude, or justice, or other virtues ... these he called the important concerns of mankind."[503] Socrates would "hold discourse, from time to time, on what concerned mankind, considering what was pious, what impious; what was becoming, what unbecoming; what was just, what unjust; what was sanity, what insanity; what was fortitude, what cowardice; what a state was, and what the character of a statesman; what was the nature of government over men, and the qualities of one skilled in governing them."[504]

George Grote explained that at this time "physical or astronomical science was narrow in amount, known only to a few; and even with those few it did not admit of being expanded, enlivened, or turned to much profitable account in discussion. But the moral and political phenomena, on which Socrates turned the light of speculation, were abundant, varied, familiar, and interesting to everyone."[505]

Socrates' condemnation of the natural philosophers of his day was not altogether unwarranted. For two hundred years, the natural philosophers had done little more than speculate and theorize. There was no recognition of the need for a systematic experimental method, or any intimation that science should be applied to the purposes of life. No philosopher had discovered anything that made any person's life better. Ionian science had undergone an auspicious beginning, but was now mired in futility.[506]

Although he eschewed natural philosophy in favor of moral philosophy, Socrates made several contributions to the development of scientific reasoning. First, he insisted upon clear definitions. Second, he advocated logical reasoning along with skepticism of authority and dogma. Thirdly, he taught that individuals must be truthful and moral; "the bad citizen cannot be a good scientist."[507] Fourth, he taught that knowledge begins with skepticism. Fifth, and perhaps most importantly, he fostered the inductive method of reasoning that eventually would become the central tenet of modern science. In the pursuit of meaningful definitions of abstract moral quantities such as justice, Socrates would invariably insist on generalization from particulars. "His conversations exhibit the main features of a genuine inductive method."[508] In *Metaphysics*, Aristotle said "Socrates ... was seeking the universal in these ethical matters, and fixed thought for the first time on definitions."[509]

## *The Elenchos*

Sometime in middle life, Socrates received what he viewed as his mission in life from the oracle at Delphi. The oracle was located at the city of Delphi on the slopes of Mt. Parnassos. "The foundation of the temple of Delphi itself reaches far beyond all historical knowledge, forming one of the aboriginal institutions of Hellas."[510] According to mythology, Delphi was "the center of the world,"[511] and "whatever the priestess at Delphi said would happen infallibly came to pass."[512]

The temple of the oracle was constructed directly on top of a chasm from which vapors arose from the underworld. The oracle was a priestess possessed of occult powers of divination. She sat directly above the chasm where she could inhale the intoxicating gasses. Presumably her predictions were delivered in response to generous offerings. The Delphic Oracle was a central tenet of Greek culture, and even the educated believed in its veracity.

The story was related by Socrates himself in Plato's *Apology*. "Chaerephon; he was early a friend of mine ... asked the oracle to tell him whether any one was wiser than I was, and the Pythian prophetess answered, that there was no man wiser."[513] Upon hearing the declaration that he was the wisest man in the world, Socrates set out to prove the oracle wrong. "I thought of a method of trying the question. I reflected that if I could only find a man wiser than myself, then I might go to the god with a refutation in my hand."[514]

Socrates traveled everywhere speaking to the wisest men about the nature of abstract moral qualities such as justice, goodness, virtue, and truth. The result was always the same. Upon the initial questioning, the expert Socrates spoke to would always claim that they knew exactly what they were talking about. But Socrates would then proceed to systematically expose their ignorance and conceit through the *elenchos*. Today, the elenchos is more commonly referred to as the *Socratic Method*, a method of questioning a person until they are logically led to a conclusion that contradicts their stated premises. The elenchos is thus a method of demonstration or logical proof; it is a variation of the method of *reductio ad absurdum* employed in Greek geometry.

In the elenchos, the questioner does the thinking for the interlocutor (the person being questioned). The purpose of the elenchos is self examination of our most closely held beliefs. The procedure sharpens our minds and mitigates our ignorance. It is "painful mental surgery, in which, indeed, the temporary pain experienced is one of the conditions almost indispensable to the future beneficial results."[515] We undergo this procedure because "the unexamined life," Socrates maintained, "is not worth living."[516] "The elenchos, as Socrates used it, was animated by the truest spirit of positive science, and formed an indispensable precursor to its attainment."[517]

An example of Socrates in action is found in Plato's *Euthyphro*. Euthyphro is a young man who has charged his own father with murder. The circumstances are as follows. The father had two servants, one who became drunk and violent. The violent drunk slew another servant. The father thereupon had the killer bound and thrown into a ditch for safekeeping while he summoned the authorities. While the killer was in the ditch, he himself died from "the effect of cold and hunger and chains upon him."[518] Whereupon Euthyphro declared that his father was guilty of murder.

When Socrates heard that Euthyphro had his own father arrested for murder, he was astonished and professed his belief that Euthyphro was being overzealous. However, Euthyphro explained that Socrates should not be concerned, because he [Euthyphro] was an expert on the subject of piety and his action was truly a pious one. Socrates asked Euthyphro to enlighten him by defining piety. Euthyphro replied, as Socrates' interlocutors invariably did, not by offering a general definition of piety, but a specific example. He said "piety is doing as I am doing; that is to say, prosecuting any one who is guilty of murder."[519]

Socrates pressed Euthyphro for a true definition. "Remember that I did not ask you to give me two or three examples of piety, but to explain the general idea which makes all pious things to be pious."[520] Forced to define piety, Euthyphro said "piety ... is that which is dear to the gods, and impiety is that which is not dear to them."[521] But upon questioning by Socrates, Euthyphro was forced to admit "that the gods fought with one another, and had dire quarrels, battles, and the like,"[522] because "they have differences of opinion ... about good and evil, just and unjust, honorable and dishonorable."[523] Therefore, the god themselves had no unanimity of agreement concerning what is pious or impious. It followed irresistibly that Euthyphro's definition of piety as "that which is dear to the gods" was both wrong and meaningless.

Euthyphro not only failed to thank Socrates for exposing his ignorance, but even refused to admit his own contradiction. In Socrates' elenchos, the typical response of the interlocutor is to maintain that Socrates has merely confused him, or that Socrates is playing semantic tricks.

When Socrates demonstrated to his friend Meno, a professor of virtue, that Meno could not define virtue, Meno refused to admit his ignorance but rather insisted that Socrates had merely obfuscated the dialogue by confusing him.

> O Socrates, I used to be told, before I knew you, that you were always doubting yourself and making others doubt; and now you are casting your spells over me, and I am simply getting bewitched and enchanted, and am at my wit's end. And if I may venture to make a jest upon you, you seem to me both in your appearance and in your power over others to be very like the flat torpedo fish, who torpifies those who come near him and touch him, as you have now torpified me, I think. For my soul and my tongue are really torpid, and I do not know how to answer you; and though I have been delivered of an infinite variety of speeches about virtue before now, and to many persons-and very good ones they were, as I thought-at this moment I cannot even say what virtue is. And I think that you are very wise in not voyaging and going away from home, for if you did in other places as do in Athens, you would be cast into prison as a magician.[524]

Meno was kind. Others, upon having their ignorance exposed by Socrates, would become angry. "Very often, while arguing and discussing points that arose, he [Socrates] was treated with great violence and beaten, and pulled about, and laughed at and ridiculed by the multitude."[525]

According to Socrates, the result of his interrogations was always the same. "I found that the men most in repute were all but the most foolish; and that others less esteemed were really wiser and better."[526] Socrates' dialogues finally convinced him that the Oracle was correct. Socrates was the wisest man in the world, not because he understood the true meaning of justice, goodness, or any other abstract moral concept. Socrates was the wisest man because he was the only one who realized that he did not know. Only after admitting his own ignorance is the philosopher ready to begin the voyage toward knowledge. A human being is capable of obtaining knowledge, but this knowledge is never final or complete.

Socrates believed that no one was ever voluntarily bad or evil in nature. He "resolved all virtue into knowledge or wisdom; all vice into ignorance or folly."[527] If a person understood what is good, they would not choose what is bad. Therefore all incorrect behavior was due to ignorance. "The worst of all ignorance ... was when a man was ignorant of himself, fancying that he knew what he did not really know, and that he could do, or avoid, or endure, what was quite beyond his capacity."[528] Socrates believed it was his sacred mission to expose ignorance, for only by exposing ignorance could he lead a person to good.

## *Trial and Death*

Although Socrates was famous for the contributions he made to philosophy during his life, it was his death that immortalized him. Following the defeat and expulsion of the Persians in 479 B.C., Greece became the most progressive and civilized country in the world. The attainment was to be short-lived, because Greece tore herself to pieces in the disastrous Peloponnesian War (431–404 B.C.).

The *Peloponnese* is essentially southern Greece, and the Peloponnesian War was primarily a battle between Sparta and Athens. "The Peloponnesian War was a protracted struggle, and attended by calamities such as Hellas had never known."[529] Thucydides (c. 460–395 B.C.), "whose judgment in politics is never at fault,"[530] attributed the true cause of the conflict "to have been the growth of the Athenian power, which terrified the Lacedaemonians."[531]

The Peloponnesian War lasted for twenty-seven years, ending with the defeat of Athens by Sparta and her allies in 404 B.C.[532] When Athens was forced to surrender, the Corinthians and Thebans urged the total destruction of the city. But the Spartans acknowledged the greater role Athens had played in the Persian Wars. They said "they would never reduce to slavery a city which was itself an integral portion of Hellas, and had performed a great and noble service to Hellas in the most perilous of emergencies."[533]

Although Athens had governed itself democratically since 507 B.C., the victorious Spartans replaced the Athenian democracy with an oligarchy. The oligarchy consisted of thirty aristocrats who were hostile to democracy; they became known as the Thirty Tyrants. The function of the Thirty was to "maintain the city [Athens] in a state of humiliation and dependence upon Lacedaemon."[534] The Thirty "embarked on a course of wholesale butchery, in which many were sacrificed to the merest hatred, many to the accident of possessing riches."[535] "Multiplied cases of execution and spoliation naturally filled the city [Athens] with surprise, indignation, and terror."[536]

The reign of the Thirty Tyrants was short-lived; after only a year of rule, they were deposed and democracy restored in 403 B.C.[537] By the time Socrates was brought to trial in 399 B.C., a democratic form of government had been in power for four years. Socrates had gone out of his way to not serve the Thirty Tyrants, and had, in fact, put himself at risk at one point by offending them. The philosopher's resistance to the Thirty Tyrants was "inflexible."[538] However, it was well known that Socrates was hostile to democracy. He thought it ridiculous to place people in positions requiring the greatest skill and intelligence on the basis of popularity rather than merit. Socrates said "it was absurd to believe that men could not become skilled in the lowest mechanical arts without competent instructors, and to imagine that ability to govern a state, the most important of all arts, might spring up in men by the unassisted efforts of nature."[539]

In *The Republic*, Socrates' student, Plato, excoriated democracy as the wellspring of tyranny.

> And does not tyranny spring from democracy[?] ... the father grows accustomed to descend to the level of his sons and to fear them, and the son is on a level with his father, he having no respect or reverence for either of his parents.... In such a state of society the master fears and flatters his scholars, and the scholars despise their masters and tutors; young and old are all alike; and the young man is on a level with the old, and is ready to compete with him in word or deed; and old men condescend to the young and are full of pleasantry and gaiety; they are loath to be thought morose and authoritative, and therefore they adopt the manners of the young.... The last extreme of popular liberty is when the slave bought with money, whether male or female, is just as free as his or her purchaser; nor must I forget to tell of the liberty and equality of the two sexes in relation to each other ... the horses and asses have a way of marching along with all the rights and dignities of freemen; and they will run at anybody who comes in their way if he does not leave the road clear for them: and all things are just ready to burst with liberty ... see how sensitive the citizens become; they chafe impatiently at the least touch of authority and at length, as you know, they cease to care even for the laws, written or unwritten; they will have no one over them. Yes, he said, I know it too well. Such, my friend, I said, is the fair and glorious beginning out of which springs tyranny ... so tyranny naturally arises out of democracy, and the most aggravated form of tyranny and slavery out of the most extreme form of liberty.[540]

Two thousand years later, Plato's opinion was echoed by James Madison (1751–1836). Madison, the chief author of the Constitution of the United States, was an advocate of republican government. He disparaged pure democracies, contending "democracies have ever been spectacles of turbulence and contention; have ever been found incompatible with personal security or the rights of property; and have in general been as short in their lives as they have been violent in their deaths."[541]

In 399 B.C., Socrates was accused of not worshipping the gods of the state, of introducing new divinities, and of corrupting the youth of Athens. The charges were brought by Anytus, a leader of the democrats, in concert with two obscure fellows named Meletus and Lykon.

In the case of Anytus, the motivation was partly personal. His son had become a follower of Socrates, and had laughed at and mocked his father's beliefs. The charges were insincere, politically motivated, and seemed designed to force Socrates out of Athens into exile. For example, the charge of "not worshipping the gods of the state" should be viewed in the understanding that in Socrates' time, no educated man believed in the existence of the Greek gods. Furthermore, the uneducated likely knew little about them, and therefore cared little for them. There was no church, no priesthood, and therefore no state religion or religious orthodoxy.[542] The accusers demanded the death penalty; in this tactic they most likely sought to force Socrates to flee even before the trial.

The account of the trial is Plato's narrative, *The Apology*. Socrates' "apology" was not an apology as we understand the term, but rather his defense to the charges brought against him. The lead prosecutor was Meletus. In *The Apology*, Socrates proceeded to question Meletus. Because he was charged with corrupting the youth of Athens, Socrates asked Meletus to name the improvers of youth. Meletus was led by Socrates into admitting that he considered every resident of Athens save Socrates to be an improver of youth instead of a corrupter. Socrates pointed out that Meletus' assertion was improbable. "Happy indeed would be the condition of youth if they had one corrupter only, and all the rest of the world were their improvers.... Am I, at my age, in such darkness and ignorance as not to know that if a man with whom I have to live is corrupted by me, I am very likely to be harmed by him; and yet I corrupt him, and intentionally, too—so you say, although neither I nor any other human being is every likely to be convinced by you."[543]

During the questioning by Socrates, Meletus conclusively declared that in his opinion Socrates was "a complete atheist."[544] This allowed Socrates to show that his accusers had contradicted themselves, for their own indictment of Socrates had charged him with "introducing new divinities," surely a hard task for an atheist. Socrates showed his prosecutors to be complete fools, but this did not help him: it only aggravated his true crime of exposing men's ignorance and foolishness.

Socrates was unapologetic and uncompromising. He explained that if acquitted he would not cease the activities that caused him to be charged. "Men of Athens, I honor and love you; but I shall obey God rather than you, and while I have life and strength I shall never cease from the practice and teaching of philosophy."[545]

Socrates asked for acquittal, but not for his own sake. He warned the assembled Athenians that his conviction and death would harm not so much him, as it would the state of Athens.

> I would have you know, that if you kill such an one as I am, you will injure yourselves more than you will injure me.... For if you kill me you will not easily find a successor to me, who, if I may use such a ludicrous figure of speech, am a sort of gadfly, given to the state by God; and the state is a great and noble steed who is tardy in his motions owing to his very size, and requires to be stirred into life. I am that gadfly which God has attached to the state, and all day long and in all places am always fastening upon you, arousing and persuading and reproaching you. You will not easily find another like me, and therefore I would advise you to spare me.[546]

Socrates said that he would not bring his family to court in order to arouse the sympathy of the judges, because it would be dishonorable.

> My friend, I am a man, and like other men, a creature of flesh and blood, and not "of wood or stone," as Homer says; and I have a family, yes, and sons, O Athenians, three in number, one almost a man, and two others who are still young; and yet I will not bring any of them hither in order to petition you for an acquittal. And why not? Not from any self-assertion or want of respect for you. Whether I am or am not afraid of death is another question, of which I will not now speak. But, having regard to public opinion, I feel that such conduct would be discreditable to myself, and to you, and to the whole state.[547]

The jury of Athenians narrowly voted to find Socrates guilty "by two hundred and eighty-one votes, being six more than were given in his favor."[548] In ancient Athens, it was the custom for the convicted to propose to the jury an alternative punishment to that demanded by the prosecutors. The alternative punishment suggested by the convicted was, of course, always less severe than that demanded by the prosecution. However, in order for the jury to accept the alternative, it had to be realistic. In Socrates' case, this would have entailed his proposing exile or a stiff fine. If Socrates had done this he would have, in effect, changed his plea to guilty. He would have dishonored himself, implicitly admitted his entire defense to be a lie, and his philosophy false.

Socrates proceeded honestly to the end. Explaining that he still viewed his work to be beneficial and in the public interest, he proposed that his punishment consist of being served free meals in a public building. This was an honor usually reserved for Olympic champions. Unrepentant, Socrates explained "As I am convinced that I never wronged another, I will assuredly not wrong myself. I will not say of myself that I deserve any evil, or propose any penalty. Why should I?"[549] Outraged by his apparent impudence, the jury voted for Socrates to be put to death. The margin was greater than that which had voted for his conviction.

The method of execution in Athens was for the convicted man to drink a cup of hemlock, a poison extracted from a plant of the same name. Ordinarily, the execution immediately followed the conviction. However, it was the festival of the Delian Apollo. Every year at this time, the Athenians sent a garlanded ship to Delos. The law of Athens forbade executions during the absence of the ship. This period lasted thirty days. During this time, Socrates friends and admirers came from all over Greece and its colonies to visit with him. Socrates' supporters urged him to escape; he might have done so easily. But if Socrates had fled, he would have been a lawbreaker and a fugitive. His flight would have handed a moral victory to his prosecutors. The only honorable choice was for Socrates to stay and drink the poison cup of hemlock.

Socrates' last hours in prison were recorded by Plato in the dialogue *Phaedo*.

> [Socrates]: "I do not think that I should gain anything by drinking the poison a little later; I should only be ridiculous in my own eyes for sparing and saving a life which is already forfeit. Please then to do as I say, and not to refuse me." Crito made a sign to the servant, who was standing by; and he went out, and having been absent for some time, returned with the jailer carrying the cup of poison ... raising the cup to his lips, quite readily and cheerfully he [Socrates] drank off the poison. And hitherto most of us had been able to control our sorrow; but now when we saw him drinking, and saw too that he had finished the draught, we could no longer forebear, and in spite of myself my own tears were flowing fast; so that I covered my face and wept, not for him, but at the thought of my own calamity in having to part from such a friend. Nor was I the first; for Crito, when he found himself unable to restrain his tears, had got up, and I followed; and at that moment, Apollodorus, who had been weeping all the time, broke out in a loud and passionate cry which made cowards of us all. Socrates alone retained his calmness: "What is this strange outcry?" he said. "I sent away the women mainly in order that they might not misbehave in this way, for I have been told that a man should die in peace. Be quiet, then, and have patience." When we heard his words we were ashamed, and refrained our tears; and he walked about until, as he said, his legs began to fail, and then he lay on his back, according to the directions, and the man who gave him the poison now and then looked at his feet and legs; and after a while he pressed his foot hard, and asked him if he could feel; and he said, "no;" and then his leg, and so upwards and upwards, and showed us that he was cold and stiff. And he felt them himself, and said: "When the poison reaches the heart, that will be the end." He was beginning to grow cold about the groin, when he uncovered his face, for he had covered himself up, and said — they were his last words — he said: "Crito, I owe a cock to Asclepius; will you remember to pay the debt?" "The debt shall be paid," said Crito; "is there anything else?" There was no answer to this question; but in a minute or two a movement was heard, and the attendants uncovered him; his eyes were set, and Crito closed his eyes and mouth. Such was the end, Echecrates, of our friend; concerning whom I may truly say, that of all the men of his time whom I have known, he was the wisest and justest and best.[550]

The execution of Socrates was a sin against philosophy, and a stain on the glory of Athens. In defense of Athens, George Grote argued that the persecution of Socrates was essentially the exception that proves the rule. "Intolerance is the natural weed of the human bosom ... [at Athens] liberty of speech was consecrated, in every man's estimation, among the first of privileges; every man was accustomed to hear opinions opposite to his own constantly expressed, and to believe that others had a right to their opinions as well as himself ... it is certain that there was at Athens both a keener intellectual stimulus, and greater freedom of thought as of speech, than in any other city of Greece."[551]

A few decades earlier, in his funeral oration (c. 431 B.C.), Pericles (c. 495–429 B.C.) had boasted of Athens' toleration for individual differences, noting "in our private intercourse we are not suspicious of one another, nor angry with our neighbor if he does what he likes."[552]

Thus our wonder should be excited, not that Athens put Socrates to death, but that this city could have tolerated the philosopher for so long, finally moving against him only when he had reached the age of seventy.

Diogenes Laërtius stated "the Athenians immediately repented of their action ... and the banished his [Socrates'] accusers, and condemned Meletus to death." But George Grote was skeptical, concluding "I know not upon what authority this statement is made, and I disbelieve it altogether."[553] As evidence, Grove cited the fact that Socrates' student, Plato, left Athens after the trial and "remained absent for some time."[554]

Socrates' death ended the golden age of Greece. However Hellas was yet to produce two great philosophers: Plato and Aristotle. Their contrasting views have divided the intellectual world for more than two thousand years.

## *Plato (428–348 B.C.)*

### *Early Life*

With the exception of Aristotle, Plato is generally regarded as the most influential and famous philosopher in all of European history. According to Aristotle, Plato had such a noble nature that he was "a man whom the wicked have no place to praise."[555] The British philosopher and mathematician Alfred North Whitehead (1861–1947) summarized Plato's importance succinctly by noting "the safest general characterization of the European philosophical tradition is that it consists of a series of footnotes to Plato."[556]

Plato was born into an aristocratic Athenian family around 428 B.C., twenty-eight years before Socrates was tried and put to death. His father was Ariston, and his mother, Perictone, "traced her family back to Solon [c. 640–559 B.C.]."[557] According to Diogenes Laërtius, "Perictone was very beautiful."[558] Ariston died when Plato was very young, and Perictone married her uncle, Pyrilampes.[559] Perictone's brother, Charmides, was a close friend of Socrates. In *Memorabilia*, Xenophon records that Socrates perceived Charmides to be a man of considerable gifts, and chastised him for not entering public life.[560] Accordingly, Plato must have been acquainted with Socrates from the time of his boyhood.

Plato's birth name was Aristocles, but he acquired the nickname of Plato (meaning "broad") "on account of the breadth either of his forehead or of his shoulders."[561] As a young man, Plato excelled in numerous fields of endeavor. "He learnt gymnastic exercises under the wrestler Ariston of Argos ... he applied himself to the study of painting ... and he wrote poems."[562] "He [Plato] is said to have displayed both diligence and remarkable quickness of apprehension, combined too with the utmost gravity and modesty."[563] Plato must have entered compulsory military duty in 409 B.C., around the age of 18.[564] Diogenes

Laërtius said "he was three times engaged in military expeditions ... in the battle of Delium he obtained the prize of pre-eminent in valor."[565] But this is impossible. The battle of Delium was fought in 424 B.C. when Plato was about four years old.[566]

At the age of twenty, Plato "became a pupil of Socrates."[567] In the *Apology*, it was recorded that "Plato is present,"[568] during the trial of Socrates. But in the narrative *Phaedo*, which relates Socrates' last days in prison, it is stated that Plato was absent because of illness.[569] In *Metaphysics*, Aristotle stated that Plato "in his youth first became familiar with Cratylus* and with the Heraclitean doctrines ... Socrates ... was busying himself about ethical matters ... [and] Plato accepted his teaching."[570]

In *Epistle 7*, Plato wrote that "when I was a young man ... I thought ... to betake myself immediately to the public affairs of the state."[571] But the trial and execution of Socrates disillusioned the young Plato. He concluded "that all states existing at present were badly governed ... the human race will never cease from ills, until the race of those, who philosophize correctly and truthfully, shall come to political power, or persons in power in states, shall, by a certain divine allotment, philosophize really."[572] More succinctly stated in *The Republic*, Plato believed that the affairs of men would not be justly governed until "philosophers are kings,"[573] or kings become philosophers.

Following the death of Socrates in 399 B.C., Plato traveled for the next twelve years, although he may have returned periodically to Athens.[574] "When he [Plato] was eight and twenty years of age ... he withdrew to Megara [Greece] to Euclid ... and subsequently, he went to Cyrene [North Africa] to Theodorus the mathematician; and from thence he proceeded to Italy to the Pythagoreans ... and from thence he went to Eurytus [Egypt] to the priests there."[575] "There was apparently nothing that he did not study."[576]

In 387 B.C., at age 40, Plato visited Syracuse in Sicily where he ran afoul of the tyrant, Dionysius I. Dionysius sold Plato into slavery. "The tyrant became very indignant, and at first was inclined to put him [Plato] to death ... but gave him to Pollis, the Lacedaemonian ... to sell as a slave."[577] Plato was rescued by Anniceris, "who ransomed him for twenty minae ... and sent him to Athens, to his companions, and they immediately sent Anniceris his money: but he refused to receive it, saying that they were not the only people in the world who were entitled to have a regard for Plato."[578]

## *The Academy and the Dialogues*

Upon his return to Athens, Plato purchased a tract of land that had belonged to a man named Academus. "In this precinct there were both walks, shaded by trees, and a gymnasium for bodily exercise."[579] Here Plato set up a school that became known as the *Academy*; it was essentially the world's first known institute for higher education.

The study of mathematics was considered a prerequisite for admission to the Academy. According to tradition, the requirement was inscribed on the portal above the entrance to the Academy, "let no one unversed in geometry enter my doors."[580] In effect, this meant that students should be acquainted with the concept of demonstrative proof. Students at the *Academy* paid no tuition, but we may surmise that their families were obligated to support the institution with philanthropic gifts.[581]

The *Academy* of Plato was not a trade school. There was no formal instruction in mathematics or the natural sciences; no classes, no examinations, and no degrees. The aim

*Athenian philosopher, 5th century B.C., student of Heraclitus

of the School was to make philosophers of the students, to inculcate in them a love of knowledge and wisdom, and to teach them the principles of education, ethics, and politics. Neither Plato or his students had any interest in practical matters; the pursuit of wisdom was for them the Pythagorean quest for spiritual purification. The *Academy*, in one form or another, was in existence for some nine hundred years, finally being shut down in A.D. 529 by the Byzantine Emperor Justinian as an undesirable remnant of pagan philosophy.

Plato transmitted his philosophy orally through teaching and dialogues in the *Academy*. He wrote a narrative of Socrates trial and death titled *The Apology*, and thirteen epistles, some of which may be spurious. However nearly all of what we know concerning Plato's philosophy is contained in a series of twenty-six books known as the *Dialogues*. In these dramas, Plato himself never appears, but issues such as the nature of justice, law, or the ideal state are explored in a series of conversations.

The chief figure in the *Dialogues* is nearly always Socrates. The *Dialogues* are not only great philosophy, but great literature as well. They are the "dramatic manifestation of Hellenic philosophy,"[582] and Plato's "style is the perfection of Attic prose of the golden age."[583] With some exceptions, Plato's personal philosophy is hard to pin down. In good part, this is because he believed philosophy to be a personal process instead of a set of dogmatic beliefs. By never introducing himself in the *Dialogues*, he is not committed to any views. This approach is consistent with Socrates' demonstration that any attempt to define abstract moral quantities such as justice, truth, or courage was doomed from the outset.

The process of philosophy nominally seeks truth as it can be ascertained by human reason, but "it is more proper that philosophy should be reasoned, than that it should be true; because, while truth may perhaps be unattainable by man, to reason is certainly his province, and within his power."[584]

The purpose of science is to discover reliable knowledge. But philosophy is more concerned with process than ends. D. R. Khashaba explained, "the whole aim of the Socratic dialectic is not to establish demonstrable truth (something which, I maintain, is foreign to philosophical thinking) but to build up intelligibility; not to argue demonstratively, but to make our ideas harmonize in an intelligible whole. The end of philosophical thinking is not to attain truth, analytic or synthetic, but to affirm the intellectual integrity of the thinker."[585]

The process of philosophy aims at "becoming reasoned truth: an aggregate of matters believed or disbelieved after conscious process of examination gone through by the mind, and capable of being explained to others."[586] This differs from unexamined truths derived from tradition or authority and believed by the ordinary person. In the later case, "there has been no conscious examination — there is no capacity of explaining to others— there is no distinct setting out of primary truths [axioms] assumed — nor have any pains been taken to look out for the relevant reasons on both sides, and weigh them impartially."[587]

In *Epistle 7*, Plato complained that some people had published interpretations of one of his doctrines. He maintained that he had no intention of putting his views in writing, stating "there is no writing of mine on this subject, nor ever shall be."[588] John Burnet explained that philosophy was an individual process, not a doctrine to be memorized. "[This] is simply a statement of the true theory of all higher education. To be of any use, philosophy must be a man's very own; it ceases to be philosophy if it is merely an echo of another's thought."[589]

## *Doctrine of Forms*

Despite the fact that Plato tended to be reluctant to align himself with any particular view, there are two aspects of his philosophy that can be discerned. One of these is his *Doctrine of Forms* or *Doctrine of Ideas*.

The Pythagoreans had observed the natural world and inductively inferred that "all things are numbers."[590] Objects in nature contained crude similarities to perfect mathematical forms. Tree trunks were round, and sometimes stones were. However, the perfect circle was never found in nature. It was a mathematical ideal, a *noumenon*—an object of intellectual intuition, rather than a sensual perception. Noumena originate in our minds, and the Pythagoreans had never divorced their study from the external world of phenomena. It was Parmenides, the "dissident Pythagorean,"[591] who concluded that the world of the senses was totally illusory, and that truth could only be obtained by reasoned cognition.

Socrates had carried the inductive process into the moral realm by searching for the definition of moral abstractions such as truth and justice. Whenever Socrates would ask someone "what is true?" or "what is good?" they would invariably respond by citing examples of what was "true" or "good." Whereupon, Socrates would patiently explain that he had not asked for an example, but rather a definition. Pushed a little, Socrates interlocutor would then offer up a definition which Socrates would destroy. According to the Socratic Method, wisdom was obtained by dialectic, a mental process of reasoned discussion.

Plato's *Doctrine of Forms* was a synthesis of Pythagorean induction with the Socratic method of seeking truth through pure intellectual inquiry. Postulating the existence of ideal forms inductively is a necessary part of science. The laws of nature are idealized generalizations. However, Plato turned this on its head. Scientists observe the natural world and inductively infer the existence of forms. They reason from the particular to the general. Plato assumed the existence of ideal forms, and from this reasoned as to the behavior of both man and nature.

Scientists study nature, but Plato argued that such study was a waste of time. Heraclitus taught that the world revealed to us by our senses is in a state of perpetual flux. Perception is also subjective. Any two people studying the same phenomenon will always have different accounts of it. Indeed, even the same person seeing the same phenomenon at different times will perceive it differently. In *Theaetetus*, Socrates says "no one knows whether what appears to him is the same as what appears to another, and everyone knows that what appears to himself in one way at one time appears to him differently at another."[592] Nothing related to a specific sense perception or dealing with observation could be an object of scientific knowledge. "Whether a man gapes at the heavens or blinks on the ground, seeking to learn some particular of sense, I would deny that he can learn, for nothing of that sort is [a] matter of science."[593]

True knowledge dealt only with pure abstract forms. The philosopher was the person "who recognizes the existence of absolute beauty [and] is able to distinguish the idea from the objects which participate in the idea."[594] Because specific, sensible things were always changing, there could be no true knowledge concerning them, only opinion. "Particulars are the object of opination, as distinguished from universal entities, forms, or ideas, which are the object of cognition."[595] If truth is that which never changes, then truth can never be found from an examination of the external world. Another reason that study of the external world can never provide us with true knowledge is that forms reside in our souls, not in sense perceptions.

The inherent limitations of empiricism were eventually overcome by utilizing systematic and controlled experimentation validated by a criterion of repeatability. In modern science, anecdotal information is given little weight and observations are required to be repeatedly confirmed by independent observers. Repeatability is a realistic methodology, but it depends on printing presses, academic journals, and an established scientific community. In Plato's time, none of these existed.

In Plato's *Doctrine of Forms*, the only things that truly exist are ideal mental forms, perfect, unchanging, and eternal. The world of ideas or forms is the mind of God; the greatest of the forms is the idea of "good." To know good is to know God. In the *Republic*, "the idea of good" is said to be "the highest knowledge" through which "all other things become useful and advantageous."[596] Plato's good is an ineffable abstract spiritual quality. Beauty is good, but good itself is more than just beauty. "The good may be said to be not only the author of knowledge to all things known, but of their being and essence, and yet the good is not essence, but far exceeds essence in dignity and power."[597]

Like Parmenides, Plato concluded that certain knowledge could only be obtained by reasoned cogitation. In *Metaphysics*, Aristotle explained that in his youth Socrates became acquainted with the doctrine of Heraclitus "that all sensible things are ever in a state of flux and there is no knowledge about them."[598] This evolved into Plato's *Doctrine of Forms*. "Plato accepted his [Socrates'] teaching, but held that the problem applied not to sensible things but to entities of another kind—for this reason, that the common definition could not be a definition of any sensible thing, as they were always changing. Things of this other sort, then, he called *Ideas*, and sensible things, he said, were all named after these."[599]

Because Plato emphasized that truth was to be obtained from pure cognition, and that the evidence of the senses was unreliable, Benjamin Farrington (1891–1974) concluded "Plato not only made no direct contribution to positive science, but did much to discourage it."[600]

Evidence of Plato's hostility toward technology is shown by his treatment of Archytas (c. 428–350 B.C.) and Archytas' student Eudoxus (c. 400–347 B.C.), natural philosophers in the tradition of Thales.

Archytas was a Pythagorean philosopher who lived in Italy. Plato had made his acquaintance during a sojourn there and they had become close friends. Archytas was an adept mathematician. Because of his mechanical inventions, he is also sometimes called the founder of the science of mechanics. Archytas was the "alleged inventor of the screw."[601] Diogenes Laërtius described Archytas as "the first person who applied mathematical principles to mechanics."[602] In his eclectic book *Attic Nights*, the Latin author Aulus Gellius (c. A.D. 125–180) related that Archytas had constructed a mechanical bird driven by compressed air. "The model of a pigeon formed in wood by Archytas, was so contrived, as by a certain mechanical art and power to fly: so nicely was it balanced by weights, and put in motion by hidden and enclosed air."[603] Archytas was also the inventor of the rattle, for which he was praised by Aristotle. "Children should have something to do, and the rattle of Archytas, which people give to their children in order to amuse them and prevent them from breaking anything in the house, was a capital invention, for a young thing cannot be quiet."[604]

Plutarch (c. A.D. 50–120) said that Archytas and Eudoxus had been criticized by Plato for employing mechanical aids in mathematical calculations.

> Eudoxus and Archytas had been the first originators of this far-famed and highly-prized art of mechanics, which they employed as an elegant illustration of geometrical truths, and as means

of sustaining experimentally, to the satisfaction of the senses, conclusions too intricate for proof by words and diagrams. As, for example, to solve the problem, so often required in constructing geometrical figures, given the two extremes, to find the two mean lines of a proportion, both these mathematicians had recourse to the aid of instruments, adapting to their purpose certain curves and sections of lines. But what with Plato's indignation at it, and his invectives against it as the mere corruption and annihilation of the one good of geometry, which was thus shamefully turning its back upon the unembodied objects of pure intelligence to recur to sensation, and to ask help (not to be obtained without base supervisions and depravation) from matter; so it was that mechanics came to be separated from geometry, and, repudiated and neglected by philosophers, took its place as a military art.[605]

The classic illustration of the *Doctrine of Forms* is the metaphor of *Plato's Cave*. Plato likened the human condition to that of troglodytes chained to the floor of a cave. "Behold! human beings living in an underground den, which has a mouth open towards the light and reaching all along the den; hence they have been from their childhood, and have their legs and necks chained so that they cannot move, and can only see before them, being prevented by the chains from turning round their heads."[606]

On the cave wall in front of them, the unfortunate cave dwellers could see shadows cast by a brilliant source of illumination. But they were restricted by their chains from directly turning and observing the source of the light. They could not directly observe reality, only vague images in the form of shadows on the wall in front of them. These shadows represented the perceptible world, a world consisting of information obtained through the human senses. The true reality of mental forms or ideas that lay behind the shadows, the intelligible world, could only be understood by intellectual means.

Therefore, it was largely a waste of time to study the phenomenological world — the proper way to find truth was by pure thought and contemplation. If the chains that prevent man from perceiving reality directly are the limits of the senses, it evidently never occurred to Plato that the chains could be broken by aid of mechanical devices, systematic experimentation, or consensus through repeatability and the dissemination of information.

## *The Timaeus*

The *Timaeus* is unique among the platonic Dialogues, because it deals chiefly with natural philosophy instead of moral philosophy. Chalcidius (4th century A.D.) translated the first half of the *Timaeus* into Latin, and it was the only one of Plato's *Dialogues* known to the Latin West until the middle of the twelfth century.[607] Plato's natural philosophy was never strongly influential in medieval Europe, and after Aristotle's works became widely available in translation during the twelfth and thirteenth centuries, he was supplanted.[608] The significance of the *Timaeus* is that it is the earliest account of a cosmological system "that has come to us in the words of the author himself."[609] The works of the presocratic natural philosophers mentioned by Plato, Aristotle, and others now only exist in fragments. The *Timaeus* also reflects a strong Pythagorean influence, and thus provides us with some insight into the doctrines of this philosophic sect.

The *Timaeus* is the original source for the fable of the lost island of Atlantis, and the inclusion of the story at the beginning of the *Dialogue* is incongruous with the cosmological exposition that follows. The speaker was Critias, and he attributed the origin of the Atlantean myth to a tale told to the Athenian Solon (c. 640–559 B.C.) by an Egyptian priest. The priest claimed that it was a rendition of events that had occurred 9,000 years earlier.[610]

There was at this time "a mighty power which unprovoked made an expedition against the whole of Europe and Asia, and to which your city [Athens] put an end. This power came forth out of

> the Atlantic Ocean, for in those days the Atlantic was navigable; and there was an island situated in front of the straits which are by you called the pillars of Hercules; the island was larger than Libya and Asia put together, and was the way to other islands.... Now in this island of Atlantis there was a great and wonderful empire which had rule over the whole island and several others, and over parts of the continent, and, furthermore, the men of Atlantis had subjected the parts of Libya within the columns of Hercules as far as Egypt, and of Europe as far as Tyrrhenia [central Italy] ... [Athens] defeated and triumphed over the invaders ... but afterwards there occurred violent earthquakes and floods; and in a single day and night of misfortune all your [Athen's] warlike men in a body sank into the earth, and the island of Atlantis in like manner disappeared in the depths of the sea.[611]

After Critias related the Atlantean fable, he introduced Timaeus, "who is the most of an astronomer amongst us, and has made the nature of the universe his special study." Cicero (106–43 B.C.) described Timaeus as a Pythagorean: "he dedicated his time to the disciples of Pythagoras and to their opinions."[612]

Timaeus began by relating Plato's Doctrine of Forms. Forms or universals were eternal and their study could lead to true knowledge through demonstration and abstract ratiocination. But the world of the senses was in a continual state of flux, and therefore any knowledge obtained through the senses could never be more than opinion. "That which is apprehended by intelligence and reason [universals or forms] is always in the same state; but that which is conceived by opinion with the help of sensation and without reason, is always in a process of becoming and perishing and never really is."[613]

The reader was warned that the cosmological system to be described was not a certainty, but only a probable tale. "We ought to accept the tale which is probable, and enquire no further."[614] This is a concomitant corollary to the Doctrine of Forms. John Burnet explained, "The world of experience is only, after all, an image, and it belongs to the region of becoming, and we can therefore do no more than tell 'likely tales' about it. Cosmology is not, and cannot be science, any more than theology or psychology. It is only a form of 'play.' Science, in the strict sense, must be mathematical."[615]

Like Hesiod ("verily at the first Chaos came to be"[616]), Plato began with chaos or "disorder."[617] The ordered universe, or cosmos, is formed out of chaos by a deity that can be described as a creator, demiurge, or architect. Plato's creator is a god, but not the omnipotent God of Christianity or Islam. There is no creation *ex nihilo*. The creator of the *Timaeus* must work with matter that is pre-existing. Furthermore, the demiurge is constrained by an ambiguous concept translated as "necessity." "The creation is mixed, being made up of necessity and mind. Mind, the ruling power, persuaded necessity to bring the greater part of created things to perfection, and thus and after this manner in the beginning, when the influence of reason got the better of necessity, the universe was created."[618]

Plato's conception of "necessity to which we are here introduced is not by any means an easy one."[619] But by this term it seems clear that he is referring to "random, indeterminate, chaotic, pre-existent, spontaneity of movement or force."[620]

By relying upon a creator, Plato was invoking supernaturalism. But Plato's supernaturalism was intermediate between the materialism of the atomists and the earlier polytheistic beliefs found in Homer and Hesiod. Gregory Vlastos (1907–1991) explained, "if you cannot expunge the supernatural, you can rationalize it, turning it paradoxically into the very source of the natural order, restricting its operation to a single primordial creative act which insures that the physical world would be not chaos but cosmos forever after."[621]

The creator was "good," and his motive for creation was "he desired that all things should be as like himself as they could be."[622] "Good is something which Plato in other works often talks about, but never determines ... but so far as we can understand him, it means order, regularity, symmetry, proportion — by consequence, what is ascertainable and predictable."[623] Plato's creator is the personification of "*nous*, or reason, or artistic skill ... he is the regularizing agent by whom order, method, and symmetry are copied from the [intangible] ideas and partially realized among the intractable data of necessity."[624]

The first thing created was the Earth, or the world. The world was formed as a sphere, because "having its extremes in every direction equidistant from the center, [a sphere is] the most perfect and the most like itself of all figures."[625] After creating the physical world, the creator endowed it with a soul possessing intelligence. The Earth became literally "a living creature truly endowed with soul and intelligence."[626] The world is "a blessed god,"[627] and Plato's entire cosmos is in fact a rational animal.

Heaven is created after the world. Each of the celestial bodies has both a soul and an intelligence. The stars [including the Sun, Moon, and planets] became "living creatures."[628] The Sun, Moon, and five planets were created for the purpose of marking time. In fact, "time ... and the heaven came into being at the same instant."[629] The "Moon [is] in the orbit nearest the earth, and next the Sun, in the second orbit above the earth; then came the morning star [Venus] and the star sacred to Hermes [Mercury], moving in orbits which have an equal swiftness with the Sun."[630]

Plato's student, Eudoxus (c. 400–347 B.C.), developed the first mathematical model of planetary motions. Eudoxus' geocentric system consisted of rotating concentric spheres.[631] But even in Plato's time, it must have been apparent that simple circular motion would not adequately explain the motion of the celestial bodies. Plato assigned his students the task of finding "the uniform and ordered movements by the assumption of which the apparent movements of the planets can be accounted for."[632]

In the *Timaeus*, celestial motions are explained by two circular motions, "the motion of the same," and "the motion of the other."[633] The circular motion of the "same" is motion about the equator, while the circular motion of the "other" is motion about the ecliptic.[634] This is the simplest device to explain the fact that the motion of the planets differs from that of the fixed stars. Circular motion is chosen because it is perfect, regular, and beautiful. This is the only type of motion compatible with the eternal and ordered nature of the heavens.

Plato and his contemporaries evidently recognized that this simple system was inadequate, for Plato apparently noted the existence of retrograde motion when he mentioned that there are "stars [planets] which reverse their motion."[635] But the conception that complex celestial motions could be represented by compounding circular motions became the central axiom of Greek astronomy, and reached its highest development in the Ptolemaic System, a model that was not superseded until the Polish astronomer Nicolaus Copernicus (1473–1543) published *De Revolutionibus Orbium Coelestium* in A.D. 1543.[636] The problems of celestial motions are passed over in the *Timaeus* with the explanation that further discussion is impractical without resort to a physical model.[637]

Next in the order of creation was the traditional panoply of Greek polytheistic gods, including Oceanus, Tethys, Phorcys, Cronos, Rhea, Zeus, and Hera. Plato appeared to be acknowledging convention when he confessed that his basis here was not logic, but nothing more than tradition. "We must accept the traditions of the men of old time who affirm themselves to be the offspring of the gods."[638] In this affirmation, Plato appeared to be doing nothing more than making a token gesture of piety for the sake of avoiding political complications and controversies. He "could not have forgotten the fate of his master Socrates,"[639] who was charged with impiety and put to death.

The lesser gods are subsequently charged by the creator with the task of creating animals, including man. These creations are to be mortal, but possess immortal souls. The mortal bodies are fashioned by the "younger gods," and to these the creator attaches an immortal soul.[640] The total number of animal souls is equal to the number of stars in the sky. A man who "lived well during his appointed time was to return and dwell in his native star."[641] But if he lived an evil or immoral life, he was reincarnated as a lesser being. "At the second birth he would pass into a woman," and from there downward into more brutish beings.[642]

The Pythagorean concept of transmigration is associated with a hierarchical ladder of creation. Men "who were cowards or led unrighteous lives ... [are] changed into the nature of women in the second generation."[643] The next lower step is occupied by birds. "The race of birds was created out of innocent light-minded men."[644] Below the birds are "pedestrian animals," reincarnations of men "who had no philosophy in any of their thoughts."[645] The lowest class consists of "the inhabitants of the water."[646] "Fishes and oysters" have been assigned "the most remote habitations as a punishment of their outlandish ignorance."[647] Plato's entire conception is remarkably anthropomorphic in its value judgments. It must have been difficult to avoid anthropomorphism when it appeared that Earth, the home of man, occupied a privileged place at the center of the cosmos.

The creation and composition of the four elements was explained. The elements of fire, air, earth, and water were not regarded as elementary or ultimately indivisible. "Water, by condensation, becomes stone and earth; and this same element, when melted and dispersed, passes into vapor and air."[648]

The Pythagorean influence is manifest again in Plato's reduction of the four elements into geometrical forms, the five regular solids. Fire is composed of pyramids, air of octahedrons, water of icosahedrons, and earth of cubes.[649] The last of the five regular solids, the dodecahedron, "was assigned as the basis of structure for the spherical cosmos itself."[650] Like atoms, these geometrical bodies are too small to be visible.[651]

But the five regular solids are not themselves truly elementary, and can be further decomposed into two-dimensional geometrical forms, triangles. The entire model is based on the principle of parsimony, which is itself a form of beauty and therefore good. "Of the infinite forms we must select the most beautiful."[652]

Continuing the strong anthropomorphic trend, Plato described the construction of the human body in excruciating detail, even going so far as to explain why men were equipped with "skin, hair, and [finger] nails."[653] Plato implicitly endorsed the practice of divination when he observed that the gods "placed in the liver the seat of divination."[654]

Vision is explained by a confluence of both external light and a stream of light that issues from the eyes. The awkward fact that no light can be seen to come from human eyes in the dark is explained by the rationale that the stream of light from the eyes is extinguished in the absence of external light. "When night comes on and the external and kindred fire departs, then the stream of vision is cut off."[655]

The concept of soul as intelligence becomes confused when Plato describes that there is a second, or mortal, soul located "in the breast, and in what is termed the thorax."[656] There is even a third type of soul that "lives and does not differ from a living being, but is fixed and rooted in the same spot, having no power of self-motion."[657] This is the type of soul that is found in plant life, but animal life also "partakes of the third kind of soul, which is said to be seated between the midriff and the navel."[658]

The Greek physician Galen (c. A.D. 129–200) accepted Plato's theory of the three souls, and characterized them as the "rational, the energetic, and the appetitive soul."[659] Galen located these physically, and respectively, in the brain, the heart, and the liver.[660]

The causes of human disease are discussed in the *Timaeus*, and Plato adopted the Hippocratic theory of the four humors.[661] Following Socrates, Plato declares that "no man is voluntarily bad."[662] Badness or evil arises from either illness or ignorance. Plato offers the aphorism, if a man "neglects education he walks lame to the end of his life."[663]

There is no solution to the problem of evil offered in the *Timaeus*. At the end, the world is described as a living god, "the greatest, best, fairest, [and] most perfect."[664] But there is no explanation as to why the "most perfect" creation contains imperfections. "If we inquire why the wise constructors put together their materials in so faulty a manner, the only reply to be made is, that the counteracting hand of necessity was too strong for them."[665]

## *Political Philosophy*

Plato's political philosophies were set forth in the Dialogues *Republic* and *Laws*. The lengthy *Republic*, consisting of ten books, has been called "the grandest" of Plato's writings.[666] The *Republic* in fact has little to do with the subject of republican government. The professed theme is a discussion of the nature and definition of justice.

At the beginning of the *Republic*, Thrasymachus puts forth the cynical opinion that justice is best defined as what is in the interest of the powerful. "I say that in all states there is the same principle of justice, which is the interest of the government; and as the government must be supposed to have power, the only reasonable conclusion is, that everywhere there is one principle of justice, which is the interest of the stronger."[667]

There follows a heated debate, but at the end of this preliminary discussion there is no consensus other than everyone perceives justice to be what is in his personal self-interest. "He only blames injustice who, owing to cowardice or age or some weakness, has not the power of being unjust. And this is proved by the fact that when he obtains the power, he immediately becomes unjust as far as he can be."[668]

Adeimantus is unhappy with this, as the traditional understanding is that justice is good and injustice evil. Therefore he calls upon Socrates to "prove to us that justice is better than injustice," and explain what "makes the one to be a good and the other an evil."[669]

Socrates undertakes the challenge, and the latter half of Book Two, as well as the remaining eight books of the *Republic* are devoted to his discussion and definition of justice. Socrates explains that as either a state or an individual human being may be considered just, he will find it easier to begin by describing the ideal or just state. "I propose, therefore that we enquire into the nature of justice and injustice, first as they appear in the state, and secondly in the individual."[670]

Socrates begins by arguing that a state arises from the mutual benefits that people derive from living cooperatively. The state, or city, makes it easier to produce the necessities of life, food, shelter, and clothing, because the state (meaning civilization) makes possible the existence of specialized occupations. There include carpenters, blacksmiths, shepherds, shoemakers, and weavers.[671]

Having established that specialization is "reciprocally advantageous to all,"[672] Socrates characterizes military service as the most vital of all occupations. Soldiers are the all-important guardians of the state that preserve and protect it from both internal and external threats. Ideally, these guardians have the temperament of guard dogs. They are "dangerous to their enemies, and gentle to their friends."[673]

The central argument that follows is that the guardians and rulers of the state must be both warriors and philosophers.[674] Only philosophers acquainted with Plato's Doctrine of Forms can rule, for only philosophers trained in dialectic can appreciate the abstraction of the *good*, and "the idea of good is the highest knowledge ... [by which] all other things become useful and advantageous."[675] The point is made emphatically: "until philosophers are kings, or the kings and princes of this world have the spirit and power of philosophy ... then only will our state have a possibility of life and behold the light of day."[676]

An objection is raised that philosophers are not suited to rule because "the greater number of them are arrant rogues, and the best are useless."[677] Surprisingly, Socrates, who presumably is the mouthpiece of Plato, agrees. He characterizes the great number of people who currently call themselves philosophers as "unworthy of education."[678] Their "natures are imperfect," and their "souls maimed and disfigured by their meannesses, as their bodies are by their trades and crafts."[679]

Socrates explains that what he had in mind was an ideal sort of philosophy that had not yet been realized. "I declare that states should pursue philosophy, not as they do now, but in a different spirit."[680]

In the discussion that follows, Socrates lays out a utopian scheme for a perfect society that is stratified into different classes. In this idealized state, the guardians, or philosopher-kings, are systematically selected through a rigorous program of testing and education.

The state begins with a unifying myth, an "audacious fiction,"[681] and the division of society into three rigid classes.

> [People of the state] "are to be told that their youth was a dream, and the education and training which they received from us, an appearance only; in reality during all that time they were being formed and fed in the womb of the earth, where they themselves and their arms and appurtenances were manufactured; when they were completed, the earth, their mother, sent them up; and so, their country being their mother and also their nurse, they are bound to advise for her good, and to defend her against attacks, and her citizens they are to regard as children of the earth and their own brothers.[682]

Society is now divided into three classes. "Citizens, we shall say to them in our tale, you are brothers, yet God has framed you differently. Some of you have the power of command, and in the composition of these he has mingled gold, wherefore also they have the greatest honor; others he has made of silver, to be auxiliaries; others again who are to be husbandmen and craftsmen he has composed of brass and iron."[683]

It is of the utmost importance that the class distinctions, or the "purity of the race,"[684] be preserved, because if "a man of brass or iron guards the state, it will be destroyed."[685] However, classes are not strictly determined by birth, but by merit. "If the son of a golden or silver parent has an admixture of brass and iron, then nature orders a transposition of ranks."[686]

The topmost class of state guardians to be trained as philosophers are chosen as youths by being subjected to a series of trials. "Toils and pains and conflicts [are] prescribed for them."[687] "He who at every age, as boy and youth and in mature life, has come out of the trial victorious and pure, shall be appointed a ruler and guardian of the state."[688]

The education of the youths chosen to be guardians is of the utmost importance. A bad education can ruin the best of persons. "Great crimes and the spirit of pure evil spring out of a fullness of nature ruined by [bad] education."[689] A youth "is like a plant which, having proper nurture, must necessarily grow and mature into all virtue, but, if sown and planted in an alien soil, becomes the most noxious of all weeds."[690]

The education of the young is to be both physical and intellectual. They are to be trained in both gymnastics and music. "Music includes all training by means of words or sounds: speech and song, recital and repetition, reading and writing."[691] So that youth will not be exposed to corrupting or degrading influences, children's books must be censored. "The first thing will be to establish a censorship of the writers of fiction, and let the censors receive any tale of fiction which is good, and reject the bad; and we will desire mothers and nurses to tell their children the authorized ones only."[692]

Specific authors to be censored include both Homer and Hesiod.[693] The primary criterion to judge whether or not a literary work is to be prohibited is whether or not they contain "the fault of telling a lie."[694] But the rulers of the state, the guardians, are "to have the privilege of lying."[695] They are "allowed to lie for the public good."[696] And if "the ruler catches anybody beside himself lying in the state, he will punish him for introducing a practice which is equally subversive and destructive."[697]

Education in music is especially important, because "rhythm and harmony find their way into the inward places of the soul."[698] Artists in the Platonic state must "be those who are gifted to discern the true nature of the beautiful and graceful."[699] "Poets, painters, architects, and artisans, must be prohibited from embodying in their works any ungraceful or unseemly type."[700] It follows that the state must regulate the fine arts. Bad music is to be banned. Only the "Dorian and Phrygian harmonies"[701] are to be allowed. Even musical instruments are outlawed. "Only the lyre and the harp [are permitted] for use in the city, and the shepherds may have a pipe in the country."[702]

Surprisingly, Plato admits women to the role of guardian. "The general inferiority of the female sex"[703] is maintained, but this is qualified by the existence of individual variation amongst the sexes. "Many women are in many things superior to many men."[704] Plato concludes that "men and women alike possess the qualities which make a guardian; they differ only in their comparative strength or weakness."[705]

The formal education of the guardians includes only those disciplines which are abstract, or might have some practical military application. They are to study arithmetic, geometry, solid geometry, and astronomy.[706] Astronomy is admitted grudgingly, because it deals with sense perceptions and involves observation. Even though a person may be gazing at the heavens, "his soul is looking downwards."[707]

The testing of the potential guardians culminates in two exams administered at the ages of twenty and thirty. Presumably, the percentage of young adults who end up passing the final exam is very small. Only after overcoming this last hurdle at the age of thirty are the guardians to be admitted to the highest study, the study of dialectic.[708] "Dialectic, and dialectic alone, goes directly to the first principle and is the only science which does away with hypotheses ... [it] is the coping-stone of the sciences, and is set over them; no other science can be placed higher."[709] Dialectic is not to be taught at an earlier age, because youth will not have the discipline and maturity to utilize it dispassionately as a tool in the search for truth. Instead, they will engage in disputation for the sake of disputation, arguing for "the sake of amusement."[710]

The training in dialectic and platonic philosophy lasts for five years. At the age of thirty-five, the guardians take their place in society and assume "any military or other office which young men are qualified to hold."[711] The guardians continue in their practical occupations for fifteen years. Only at the age of fifty are they finally ready to assume the duties of the philosopher-kings and rule the state.[712]

The state is to regulate the pairing of men and women and their sexual reproduction.

The argument is made that if care is not taken in the breeding of dogs, birds, horses, and other animals, it is widely acknowledged that the species and varieties would deteriorate. Human beings are more important than animals, therefore the state must practice eugenics.[713]

It is not entirely clear if Plato's eugenics is to be applied only to the guardian class, or throughout the entire state. But he wrote that the state will take "the offspring of the inferior," and "put [them] away in some mysterious, unknown place." Thus it is clearly implied that the state will be involved in regulating the reproductive activity of all classes.[714]

The guardians must reproduce, but they are not allowed to have families or own private property. "The wives of our guardians are to be common, and their children are to be common, and no parent is to know his own child, nor any child his parent."[715] "The guardians were not to have houses or lands or any other property; their pay was to be their food, which they were to receive from the other citizens, and they were to have no private expenses."[716]

After a woman gives birth to a child, the child is taken from her. She is allowed to nurse, but "the greatest possible care [is taken so] that no mother recognizes her own child."[717] Incest is forbidden. This presents a practical problem, in that the identity of parents and children is to be concealed from each other. The solution proposed is that older adults are forbidden from having intercourse with any younger person born at the same time their child was.[718] The ages for reproduction are regulated. Women are allowed to bear children only between the ages of twenty and forty. Men are permitted to father children beginning at the age of twenty-five and desisting at fifty-five.[719]

Communal ownership is designed to promote harmony and prevent quarrels and disputes. "The community of property and the community of families ... [will] tend to make them more truly guardians; they will not tear the city in pieces by differing about 'mine' and 'not mine.'"[720]

Lest anyone claim that Plato's utopian vision is unrealistic, in book eight of the *Republic* he acknowledges that his perfect state is doomed to failure because "everything which has a beginning has also an end."[721] Five different forms of government are described and compared. Ranked from best to worst, these are aristocracy, timocracy, oligarchy, democracy, and tyranny. The Platonic state is a aristocracy, because it is "government of [by] the best."[722] Aristocracy inevitably degrades into the second-best form of government, timocracy. *Timocracy* is defined as "government of honor."[723] Crete and Sparta are offered as examples of timocracies existing in Plato's Greece.[724]

The process of degradation continues, with timocracy degenerating into oligarchy, oligarchy into democracy, and democracy finally sinking into tyranny. Plato reserved some of his most bitter and biting rhetoric for democracy.[725] "Tyranny naturally arises out of democracy, and the most aggravated form of tyranny and slavery out of the most extreme form of liberty."[726]

The *Republic* is poorly organized, and the dialogue between the speakers rambles haphazardly. A topic is introduced, then the conversation digresses, later returning to the same subject. The all-important education of the guardians is first discussed in Book 2.[727] But a discussion of their systematic training is deferred to Book 7.[728]

The *Republic* also contains some implicit contradictions. Plato elaborates a plan for a totalitarian state so extreme that even possession of the wrong musical instruments is a crime.[729] Yet he states that the "guardians, setting aside every other business, are to dedicate themselves wholly to the maintenance of freedom in the state."[730] Philosophers, as "lovers of wisdom," are described as "those who love the truth."[731] But the guardians, or philosopher-kings, are to "have the privilege of lying."[732]

The ideal state described in Plato's last work, *Laws*, is based on law rather than personal rule by philosopher-kings. In the *Republic*, the rule of the philosopher-kings was to be absolute. But in *Laws*, it is admitted that "no human nature invested with supreme power is able to order human affairs and not overflow with insolence and wrong."[733] Therefore, men must be governed by laws, not other men. "For that state in which the law is ... above the rulers, and the rulers are the inferiors of the law, has salvation, and every blessing which the gods can confer."[734]

The ideal city envisaged in *Laws* is based on communism. "The first and highest form of the state and of the government and of the law in that in which ... 'friends have all things in common ...' [and] in which the private and the individual is altogether banished from life."[735] It is conceded this perhaps is an unobtainable ideal, but if communism could be instituted, there would never be "a state which will be truer or better or more exalted in virtue."[736]

The model city begins with 5,040 citizens, because this is "a convenient number."[737] Each citizen receives an equal allotment of land. Stagnation is an ideal that permeates *Laws*. The number of citizens or "families should be always retained, and neither decreased nor diminished."[738] Usury is forbidden.[739] No one is allowed to possess either silver or gold. The city or state is to have its own base metal currency, because this is "necessary in dealing with artisans, and for payment of hirelings."[740] Some inequality of wealth is permitted to exist, but no one is allowed to become so poor that they lose their basic allotment of land. A man is permitted to acquire wealth up to an amount four times the value of his land allotment, but any amount in excess of that is confiscated by the state.[741]

Marriage and reproduction are overseen by the state. The state is to define the age periods during which marriage may take place, but the numbers given in *Laws* are ambiguous.[742] To maintain equality among the classes, "the rich man shall not marry into the rich family, nor the powerful into the family of the powerful."[743] This contrasts with the *Republic*, where the maintenance of class distinctions is of paramount importance. However, the perfect state depicted in the *Republic* is a meritocracy, where societal stratifications are based on ability, not wealth or power.

The primary vehicle by which the state is to be organized is by means of a state religion. The state defines and maintains orthodoxy in religion, and therefore is a theocracy. No one is "allowed to practice religious rites contrary to law," nor is anyone to "have sacred rites" or "possess shrines of the gods" in private residences.[744] Heresy and impiety are crimes, and these can take many forms. "There are many kinds of unbelievers."[745] If someone is impious or irreligious "not from malice or an evil nature," he is to be imprisoned for "not less than five years."[746] The purpose of imprisonment in this case is not so much punishment, as re-education. The prisoner is held "with a view to the improvement of their soul's health."[747] The view is taken that anyone who questions the state religion must be insane, because if they have not acquired a "sound mind" at the end of five years imprisonment, they are to be "punished with death."[748]

*Laws* presents military discipline as an ideal to be maintained in civilian life.

> For expeditions of war ... the great principle of all is that no one of either sex should be without a commander; nor should the mind of any one be accustomed to do anything, either in jest or in earnest, of his own motion, but in war and in peace he should look to and follow his leader, even in the least things being under his guidance; for example, he should stand or move, or exer-

> cise, or wash, or take his meals ... when he is bidden ... and we ought in time of peace from youth upwards to practice this habit of commanding others, and of being commanded by others.[749]

As in the *Republic*, the state regulates the fine arts by establishing standard models for artistic expression. "No one in singing or dancing shall offend against public and consecrated models."[750] Youth are to be prevented, "in every possible way," "from even desiring to imitate new modes either in dance or song."[751] "Change," is characterized as "the most dangerous of all things."[752] It is "quite unreasonable," to suppose that "poets are to be allowed to teach in the dance anything which they themselves like."[753] The stagnant art of Egypt is held up as an ideal worthy of emulation. "Music and dancing in Egypt ... [are] fixed ... and no painter or artist is allowed to innovate ... or to leave the traditional forms and invent new ones. To this day, no alteration is allowed either in these arts, or in music at all. And you will find that their works of art are painted or molded in the same forms which they had ten thousand years ago—this is literally true and no exaggeration."[754]

No criticism or even questioning of the state's laws is to be permitted, especially public questioning by the young. "One of the best of them [the state's laws] will be the law forbidding any young men to enquire which of them are right or wrong; but with one mouth and one voice they must all agree that the laws are all good."[755] However, an older man who notes any defect in the state's laws is allowed to discreetly "communicate his observation to a ruler."[756] And lawgivers may employ "useful lies" in their management of the state.[757]

With criticism disallowed, the *elenchos* of Socrates most certainly would have been illegal in Plato's city. George Grote concluded, "during the interval of forty-five years between the trial of Socrates and the composition of the *Laws*, Plato had passed from sympathy with the freespoken dissenter to an opposite feeling—hatred of all dissent, and an unsparing employment of penalties for upholding orthodoxy."[758]

Order in the state is to be maintained by magistrates obedient to the dictates of the law. But above these is the Nocturnal Council. The Nocturnal Council is composed of the "ten oldest guardians of the law," together with ten young men chosen by the older members.[759] The ten young men are to be "not less than thirty years of age."[760] The Nocturnal Council is "the anchor of the state."[761] It is "the conservative organ of the Platonic city ... [it] serve[s] as the perpetual embodiment of the original lawgiver."[762]

Like the *Republic, Laws* is poorly organized and contains apparent internal contradictions. Although there is little freedom of any type in the state that Plato describes, in Book 3 of *Laws* he states that "the lawgiver ought to have three things in view: first, that the city for which he legislates should be free."[763]

For apparently endorsing totalitarianism, Plato has been excoriated by modern writers. George Sarton dismissed the *Republic* as "the work of a disgruntled fanatic," and concluded that Plato "would have forbidden the very things that we have in mind when we speak of the glory of Greece."[764]

The most pointed criticisms were made by Karl Popper in the first volume of *The Open Society*.[765] Plato "hated the individual and his freedom,"[766] and he betrayed his teacher, Socrates, who would never have been tolerated in Plato's totalitarian state.[767] Popper concluded that Plato was mistaken in identifying "altruism with collectivism, and all individualism with egoism."[768] In Popper's view, altruism and individualism are not mutually exclusive. Their union is the basis of Western Civilization.[769]

As always, it is difficult to discern Plato's personal views from any of the *Dialogues*. In the case of at least the *Republic*, it is possible that the state depicted is merely a metaphor for an individual human being. In Book 2, it is implied that a person is a state in microcosm.[770] Thus the process by which the guardians are educated may merely be a means of describing the ideal training for the human mind and soul so that an individual can realize the potential good inherent in them.

## *Philip (382–336 B.C.) and Alexander (356–323 B.C.)*

### *Philip Unifies Greece*

At the end of the Peloponnesian War and the dawn of the 4th century B.C., Greece found itself exhausted. The Greek city-states had fought among themselves almost unceasingly, never able to unify and produce a lasting peace and stability. Both Sparta and Athens were in decline. Athens lost the Peloponnesian War to Sparta after overextending itself in an ill-considered offensive expedition against Syracuse in 413 B.C. The Athenians were utterly defeated in Syracuse, losing thousands of hoplites and hundreds of triremes. "Athens ... [was left] utterly defenseless. Her treasury was empty, her docks nearly destitute of triremes, the flower of her hoplites, as well as her seamen, had perished in Sicily."[771]

Sparta was victorious, but became spoiled by success. The old Lycurgean reforms that had instilled the legendary martial discipline of Sparta were corrupted. The Spartan middle class was destroyed by the accumulation of wealth and land into the hands of a relatively few citizens. "Many new sources of corruption were sufficient to operate most unfavorably on the Spartan character."[772] Xenophon, writing c. 394–371 B.C.,[773] noted that the traditional Lycurgean discipline was in decline. "If any one should ask me, whether the laws of Lycurgus appear to me to continue even at the present time intact, I could certainly no longer reply with confidence in the affirmative."[774] With a diminished middle class, Sparta had fewer citizens to draw upon for its formidable armies. In 371 B.C., Spartan power was broken by Thebes at the Battle of Leuctra. "Such was the overwhelming force of the Theban charge ... that even the Spartans, with all their courage, obstinacy, and discipline, were unable to stand up against it."[775] In *Politics*, Aristotle concluded "the city [Sparta] sank under a single defeat; the want of men was their ruin."[776]

In 359 B.C., Philip (382–336 B.C.) became King of Macedonia. He began to unify and strengthen his country. Philip had genius, and was both tireless and unscrupulous in his methods. He bought friendships with bribery and suppressed rivals with cunning. One of Philip's first moves was to invade Athenian-controlled Thrace to the east of Macedonia and seize control of its rich gold and silver mines. Athens was too weak to do anything about the loss. The income from the mines, "not less than 1,000 talents per annum,"[777] gave Philip a tool for increasing his power.

Philip ruthlessly exploited the Greek's own enmities and divisions. If one Greek city-state began to move against him, he would bribe the politicians in another city-state to attack the first, deflecting attention from Macedonia. In personal bravery and valor, Philip was unexcelled. His injuries in battle included the loss of one eye, a broken shoulder, and the loss of an arm and leg to paralysis. Philip had no qualms about resorting to lies and bribery, as he considered these to be humane alternatives to war. When he was victorious in battle, Philip usually gave his defeated foes fair and generous terms. He was well liked and highly respected by most of his contemporaries.

Philip's Macedonian army was the most advanced of its time. He replaced the hoplite's traditional eight-foot-long spear with a spear twenty-one feet long. The spear was weighted in the end, so that it could be held near the back and project fifteen feet forward. Philip's columns were sixteen men deep, and the longer spears enabled the first five rows of spears to project in front of the phalanx.[778]

Infantry were supplemented with armored cavalry that could swiftly maneuver to take advantage of any weakness that might develop in the enemy's position. Behind the infantry were archers and siege machines such as catapults and battering rams. Philip assembled an army of 10,000 men and drilled and disciplined them relentlessly until they were the most efficient fighting force the world had ever seen.

From the time of his early youth, Alexander was trained to succeed his father, as ruler and military commander. "While he was yet very young, he entertained the ambassadors from the King of Persia, in the absence of his father.... Whenever he heard Philip had taken any town of importance, or won any signal victory, instead of rejoicing at it altogether, he would tell his companions that his father would anticipate everything, and leave him and them no opportunities of performing great and illustrious actions."[779]

One of Alexander's first instructors was the soldier Leonidas. Leonidas "directed all his efforts to the production of a Spartan endurance of hardship and contempt of danger. He was accustomed to ransack his pupil's trunks for the purpose of discovering any luxurious dress or other means of indulgence which might have been sent to him by his mother."[780]

The young Alexander astonished his father by taming the wild horse, Bucephalus, that others found to be "vicious and unmanageable."[781] When Alexander was thirteen years old,[782] Philip "sent for Aristotle, the most learned and most celebrated philosopher of his time.... Alexander received from him not only his doctrines of morals and of politics, but also something of those more abstruse and profound theories which these philosophers ... professed to reserve for oral communications to the initiated."[783] Plutarch described Alexander as "a great lover of all kinds of learning and reading ... he constantly laid Homer's *Iliad* ... under his pillow, declaring that he esteemed it a perfect portable treasure of all military virtue and knowledge."[784]

In 338 B.C., Philip seized control of Greece by defeating the allied forces of Thebes and Athens at the Battle of Chaeronea. Although only eighteen years old, Alexander was entrusted with a command by his father and acquitted himself well. "Alexander, earnest to give an indication of his valor to his father, charged with a more than ordinary heat and vigor, and, being assisted by many stout and brave men, was the first that broke through the main body of the enemy."[785]

Philip was harsh in his treatment of Thebes. "He sold the Theban captives into slavery ... he put to death several of the leading citizens, banished others, and confiscated the property of both ... [and] a council of three hundred, composed of philippizing Thebans, for the most part recalled from exile, was invested with the government of the city."[786]

In contrast, Athens received favorable terms of surrender, calculated to assist in Philip's securing his control of Greece. Two thousand Athenians taken prisoner were returned without ransom. But the Athenians were required to serve Philip.

> The Athenians were required to acknowledge the exaltation of Philip to the headship of the Grecian world, and to promote the like acknowledgment by all other Greeks, in a congress to be speedily convened. They were to renounce all pretensions to headship, not only for themselves, but for every Grecian state ... [and] to acquiesce in the transition of Greece from the position of a free, self-determining, political aggregate, into a provincial dependency.... The recognition of his

> [Philip's] Hellenic supremacy by Athens was the capital step for the prosecution of his objects. It insured him against dissentients among the remaining Grecian states.[787]

Following his victory at Chaeronea, Philip marched into Lacedaemonia to consolidate his power. But there is no evidence that he directly attacked Sparta. "The Spartans could not resist; yet would they neither submit, nor ask for peace."[788] Philip's dominion over Greece was formalized at a conference at Corinth. "At Corinth ... he [Philip] declared his design to make war upon the Persians ... and therefore desired the council to join with him as confederates in the war. At length he was created general of all Greece, with absolute power, and thereupon he made mighty preparations for that expedition; and, having ordered what quota of men every city should send forth; he returned into Macedonia."[789]

Philip's downfall was caused by his own appetites. Although disciplined in statesmanship and battle, he was prone to drink heavily. He humiliated and angered Alexander's mother, Olympias, by marrying other women. Olympias was not of a forgiving character. She had a violent temper and was "jealous, cruel, and vindictive."[790]

Things came to a head when Philip married Cleopatra, niece of his general, Attalus.

> Amidst the intoxication of the marriage banquet, Attalus proposed a toast and prayer, that there might speedily appear a legitimate son, from Philip and Cleopatra, to succeed to the Macedonian throne. Upon which Alexander exclaimed in wrath—"Do you then proclaim me as a bastard?"—at the same time hurling a goblet at him. Incensed at this proceeding, Philip started up, drew his sword, and made furiously at his son, but fell to the ground from passion and intoxication. This accident alone preserved the life of Alexander, who retorted—"Here is a man, preparing to cross from Europe into Asia, who yet cannot step surely from one couch to another."[791]

Shortly thereafter, Philip was assassinated by one of his own bodyguards, Pausanias. Pausanias took revenge upon Philip because Philip would not give him the justice he felt he deserved. According to Diodorus Siculus (fl. 1st century B.C.), "Attalus, one of [Philip's] courtiers ... invited Pausanias to a feast, and, after he had made him drunk, exposed his body ... to be abused by the filthy lusts of a company of base, sordid fellows."[792]

As Attalus was the uncle of Philip's new bride, Cleopatra, he [Attalus] was too powerful to be held accountable for the crime. So Philip tried to pacify Pausanias. "He [Philip] bestowed upon him [Pausanias] many rich gifts, and advanced him to a more honorable post in his guards. But Pausanias' anger was implacable."[793]

Pausanias resolved to revenge himself, not upon Attalus, who was unapproachable, but upon Philip. "Perceiving that the king was alone, [Pausanias] ran up to him, and, stabbing him in the side, through the body, laid him dead at his feet, and forthwith fled ... he would certainly have escaped, but that a branch of a vine caught hold of the heel of his shoe, and so entangled him that he fell: upon which Perdiccas, with the others, came upon him as he was endeavoring to rise, and, after many wounds given him, there slew him.... And thus Philip ... came to his end [336 B.C.]."[794]

## *Succession and Consolidation*

Alexander "succeeded to a kingdom, beset on all sides with great dangers and rancorous enemies."[795] In a situation where most men would have hesitated and watched their kingdoms crumble, twenty-year-old Alexander acted swiftly and decisively to consolidate his power. He sent one of his friends, Hecateus, into Asia, to assassinate the disloyal general Attalus, who was conspiring with Athens to displace Alexander.[796]

The Greek cities were on the verge of revolt. "The Athenians rejoiced at the news of Philip's death ... and stirred up many of the [Greek] cities to assert their liberties."[797] Alexander "marched into Peloponnesus as soon as he had secured the regal power,"[798] and called a council at Corinth in the autumn of 336 B.C.[799] Upon this show of military power, the Greeks "created him [Alexander] general of all Greece, and decreed him aid and assistance against the Persians."[800] Alexander organized the Greek city-states as a confederacy, with himself as the head. Each Greek city was allowed to be "free and autonomous," and govern itself, so long as they acknowledged Alexander's supremacy.[801] In Asia, Hecateus eventually caught up with Attalus and killed him, "upon which the Macedonian army in Asia laid aside all thoughts of defection."[802]

Having consolidated his power both internally and in Greece, Alexander "marched into Thrace ... then he invaded Poeonia and Illyria, and the people bordering upon them, and having subdued those that had revolted, he likewise brought under his dominion the barbarians next adjoining."[803] While still embarked upon these wars, news reached Alexander in September of 335 B.C. that Thebes was in revolt.

Thebes was betrayed by Athens. "The Athenians, though they had decreed aid to the Thebans, yet they sent no forces thither, intending first to observe how matters were likely to go."[804] The Thebans soon found Alexander at their city gates, along with thirty thousand infantry and three thousand cavalry, all veteran soldiers.[805] Still, the Thebans refused to yield. "Pushed forward by the heat of their spirits, [they] encouraged one another with the remembrance of their famous victory at the battle of Leuctra ... [and] ran headlong to the ruin of themselves and their country."[806] So Alexander "resolved to destroy the city, and by that means to terrify all others who for the future should dare to rebel."[807]

The Thebans fought with great resolve, inflamed by "the love of liberty."[808] But Alexander used his superior numbers to great advantage. As soon as he perceived that the Macedonian fighters were fatigued, they were replaced by fresh troops. But the Thebans had no reserves to call upon. "The Macedonians coming with a fierce and sudden charge upon the Thebans, now even tired out, bore them down and killed multitudes of them."[809]

During the heat of the battle, Alexander happened to spy an unguarded entrance into Thebes. His troops entered the city, opening a second front. As the Thebans retreated behind their city walls, the battle turned into a rout. "While they [the Thebans] made into the city in this trepidation and confusion, many were killed by running upon their own weapons in the strait and narrow passages ... [the Macedonians] issued forth like a rapid torrent upon the backs of the Thebans, and fell upon them as they were in this disorder and confusion, and cut them down in heaps."[810]

Outraged by the Thebans' stiff resistance, the Macedonians showed them no mercy. "By indiscriminate slaughter [the Macedonians] vented their rage against them [the Thebans] ... not even the women and children being spared."[811] "Multitudes of all sorts of cruelties were acted within the walls ... without any pity or compassion [the Macedonians] put all to the sword that were in their way."[812]

Thebes was totally destroyed. Six thousand Thebans were killed, and the remaining population of three thousand were sold into slavery. The city itself was razed to the ground. Alexander's intention to make an example of Thebes was successful, because his actions "struck a terror into all the Grecians that had revolted."[813] Macedonian casualties amounted to five hundred.

After the fall of Thebes, in the autumn of 335 B.C., Alexander traveled to Corinth where he "presided at a meeting of the Grecian synod," and was met with "universal deference and submission."[814] The singular exception was Alexander's encounter with the cynical philosopher, Diogenes of Sinope (c. 412–323 B.C.). According to Diogenes Laërtius, the philosopher "was very violent in expressing his haughty disdain of others," maintaining "there was not a more foolish animal than man."[815]

The story of Alexander's meeting with Diogenes was related by Plutarch.

> Diogenes of Sinope, who then was living at Corinth, thought so little of him [Alexander], that instead of coming to compliment him, he never so much as stirred out of the suburb called the Cranium, where Alexander found him lying alone in the sun. When he [Diogenes] saw so much company near him, he raised himself a little, and vouchsafed to look upon Alexander, and when he [Alexander] kindly asked him [Diogenes] whether he wanted anything, "Yes," said he, "I would have you stand from between me and the sun." Alexander was so struck at this answer, and surprised at the greatness of the man, who had taken so little notice of him, that as he went away he told his followers, who were laughing at the moroseness of the philosopher, that if he were not Alexander, he would choose to be Diogenes.[816]

Having secured Greece, Alexander "returned into Macedonia."[817] Alexander's counselors advised him to forego his father's planned conquest of Asia, to settle into his new role as ruler of Macedonia and Greece, and to marry and sire heirs. But Alexander was ambitious. He replied that "it was a mean and unworthy thing for him who was created general of all Greece, and had the command of an army that never knew what it was to be conquered, to stay at home merely to marry and beget children."[818]

## *Invasion of Asia*

Alexander began the invasion of Asia in spring of 334 B.C. In his *Anabasis of Alexander*, Arrian gave the number of Alexander's troops as 30,000 infantry and "more than 5,000 cavalry."[819] *The History* of Diodorus Siculus is in concurrence on the size of Alexander's infantry, but estimates the cavalry as amounting to 4,500.[820] To secure Europe, Alexander left 12,000 infantry and 1,500 cavalry under command of the general Antipater. To ensure Antipater's faithfulness, Alexander took three of Antipater's sons with him on the Asian expedition.[821]

Upon arriving in Asia in the spring of 334 B.C.,[822] Alexander visited the tomb and monument of Achilles, legendary hero of the Trojan War (c. 1250 B.C.). Alexander "went through the customary ceremony of anointing himself with oil and running naked up to it [the tomb]."[823] The ritual helped to excite wonder in his troops; they regarded Alexander as the new Achilles. Armed with the power of myth and the belief in their own destiny, Alexander's army was now unstoppable.

## *Battle of Granicus*

In the months preceding Alexander's invasion, the Persian commanders "met together to consult how to manage the war."[824] They decided to let Alexander enter Asia unopposed, but to meet him at the river Granicus, where the stream itself afforded them the virtue of a natural defense.

In most confrontations, the Persians outnumbered the Macedonians and their Greek allies. Alexander relied upon the superior quality, discipline, and organization of his troops.[825] At the river Granicus, the opposing forces met for the first time. Alexander's chief general, Parmenio, advised caution, "for it is clear that many parts of the stream are deep, and you see that these banks are very steep and in some places abrupt."[826] The Persians waited for the Macedonians to cross the river. "Both armies stood a long time at the margin of the river, keeping quiet from dread of the result; and profound silence was observed on both sides."[827]

Eventually, Alexander decided to launch the assault, with himself in the front of the ranks. "The Macedonians ... suffered severely at the first outset ... [because] the Persians were fighting from the top of the bank, which gave them an advantage."[828] But the Macedonians began to cross the river in increasing numbers, beating back the defenders on the opposite bank. The two forces engaged their cavalry, and "Alexander's men began to gain the advantage ... through their superior strength and military discipline."[829] "Alexander ... was the first that charged, and rushing into the thickest of his enemies, made great slaughter amongst them."[830]

The defense was led by Spithrobates, Darius' son-in-law, and governor of Ionia. Alexander "fought with him hand to hand."[831] They initially wounded each other, then Alexander "having got another lance, threw it directly into his [Spithrobates'] face, and pierced him through the head."[832] In the intensity of the battle, Alexander "received three strokes through his breast-plate, and one cut upon his helmet, and had his buckler ... thrice pierced through,"[833] but received no serious or incapacitating wounds.

After Spithrobates fell, the Persian cavalry was routed. "They fled in all directions, pursued by the Macedonians."[834] Without the protection of their cavalry, the Persian infantry was "cut to pieces on the field."[835] At the end of the battle, the Macedonians had lost 60 cavalry and 30 infantry.[836] The Persians lost 1,000 cavalry and their entire infantry was wiped out, with the exception of 2,000 men taken prisoner.[837]

## *Consolidation of Ionia*

Following the victory at Granicus, Alexander marched southward "towards Sardis, the chief town of Lydia, and the main station of the Persians in Asia Minor ... when he arrived within eight miles of Sardis ... he met ... a deputation of the chief citizens ... [and] the Persian governor ... who delivered [Sardis] up to him without a blow."[838] Alexander "granted freedom to the Sardians and to the other Lydians generally, with the use of their own Lydian laws ... [but] did not exonerate them from paying the usual tribute."[839]

Alexander now marched toward the Mediterranean coast. At Ephesus, he met no resistance. Upon occupying the city, Alexander overthrew the Persian oligarchy and established a democratic government.[840] He sent a deputy to the other Ionic cities of Greek heritage with orders to "to break up the oligarchies everywhere, to set up the democratical form of government, to restore their own laws to each of the cities, and to remit the tribute which they were accustomed to pay to the foreigners."[841]

At Miletus, Alexander was resisted. Miletus was the sanctuary for those Persians who had survived the battle of Granicus.[842] The Persian commander at Miletus, Hegesistratus, was reassured by "the fact that the Persian fleet was not far off," with about 400 ships.[843] But the Grecian fleet arrived first, and anchored nearby with 160 ships.[844] Parmenio advised Alexander to engage the Persians in a sea-battle, but Alexander demurred, reasoning "he should get the mastery over the Persian fleet by defeating their army on land."[845]

Alexander's decision to engage the Persians by land, and not by sea, was based in part on the ambiguous interpretation of a superstitious omen. One of the reasons that Parmenio advised Alexander to fight the Persians at sea was that "an eagle had been seen sitting upon the shore, opposite the sterns of Alexander's ships ... [Alexander] admitted that the eagle was in his favor; but as it was seen sitting on the land, it seemed to him rather to be a sign that he should get the mastery over the Persian fleet by defeating their army on land."[846]

The histories of Alexander by Arrian (c. A.D. 86–186) and Diodorus Siculus (fl. 1st century B.C.) are replete with repeated instances of serious people drawing serious inferences from superstitious omens. The ancient Greeks and Romans had a sincere conviction in the authenticity of divination. George Sarton characterized their "firm belief in divination" as "the outstanding superstition of classical antiquity."[847] This superstition could have devastating consequences. On the night of August 27, 413 B.C., an eclipse of the moon prevented the Athenian navy from fleeing Syracuse. Subsequently, the Athenians suffered a complete defeat at the hands of the Syracusans, and Athenian power was broken forever.[848]

As Alexander began to besiege Miletus, the Greek navy "shut off the Persian fleet from the port," making it impossible for the Milesians to receive reinforcements from the sea.[849] The Macedonians soon "forced their way through a breach of the walls," and Alexander took Miletus.[850] Having occupied Miletus, Alexander now "dismissed his fleet."[851] Arrian explained the strategy. "He [Alexander] considered, that now he was occupying Asia with his land force, he would be able to break up that of the Persians, if he captured the maritime cities; since they would neither have any ports from which they could recruit their crews, nor any harbor in Asia to which they could bring their ships."[852]

By this time, "the whole Asiatic coast northward of Miletus ... had either accepted willingly the dominion of Alexander, or had been reduced by his detachments. Accordingly, Alexander now directed his march southward."[853] The march southward turned east, and Alexander met occasional resistance as he progressed into Asia. After consolidating his control over the southern coast of Asia Minor, Alexander turned north into Phrygia. In the spring of 333 B.C., he arrived at the town of Gordium where he cut the Gordian Knot.[854]

> There was preserved in the citadel [of Gordium] an ancient wagon of rude structure, said by the legend to have once belonged to the ... primitive rustic kings of Phrygia ... the cord attaching the yoke of this wagon to the pole, was so twisted and entangled as to form a knot of singular complexity, which no one had ever been able to untie. An oracle had pronounced that to the person who should untie it the empire of Asia was destined.... Alexander, on inspecting the knot, was as much perplexed as others had been before him, until at length, in a fit of impatience, he drew his sword and severed the cord in two. By every one this was accepted as a solution of the problem, thus making good his title to the empire of Asia.[855]

Hence the term "cutting the Gordian Knot" became a metaphor for solving a difficult problem by means of a bold or imaginative stroke.

## *Battle of Issus*

The Persian king, Darius, gathered his vast forces for an inevitable showdown with Alexander. "Their numbers became greater and greater, amounting at length to a vast and multitudinous host, the total of which is given by some as 600,000 men, by others as 400,000 infantry and 100,000 cavalry ... there were among them between 20,000 and 30,000 Grecian mercenaries."[856]

Among the Greeks in the service of Darius was Charidemus, an Athenian exile. He had been forced to flee Greece after the destruction of Thebes. Charidemus advised Darius that a large number of inferior troops was unlikely to be of much use. "He advised Darius to place no reliance on Asiatics, but to employ his immense treasuries in subsidizing an increased army of Grecian mercenaries."[857] When Darius' Persian generals heard this advice, they flew into a rage, and denounced Charidemus "as a traitor who wished to acquire the king's confidence in order to betray him to Alexander."[858] Darius rashly accepted this judgment, and had Charidemus executed, but regretted the decision almost instantly.

Darius assumed the personal command of his forces and marched out to meet Alexander. In addition to his force of 600,000 men, "his mother, his wife, his harem, his children, his personal attendants of every description, accompanied him, to witness what was anticipated as a certain triumph."[859] At first, Darius judiciously waited for Alexander to march eastward and meet him on a battle plain where the larger size of the Persian forces could be used to the Persian's advantage. Darius was advised "not to abandon" his favorable position on "a plain in the land of Assyria which stretched out in every direction, suitable for the immense size of his army and convenient for the evolutions of cavalry."[860]

But as time passed, and Alexander held back, Darius became overconfident. "He came to the conclusion that Alexander was no longer desirous of advancing further, but was shrinking from an encounter on learning that Darius himself was marching against him ... the worse advice, because at the immediate time it was more pleasant to hear, prevailed."[861] Darius' Grecian advisors counseled against advancing into the mountain passes and the "narrow defiles ... where his numbers and especially his vast cavalry, would be useless ... their advice, however, was not only disregarded by Darius, but denounced by the Persian counselors as traitorous."[862]

As battle became imminent, Alexander called together his commanders and generals and delivered an address designed to fortify their resolve. He told them that they were facing "a foe who had already been beaten," an enemy "similar to them neither in strength nor in courage."[863] The "Persians and Medes [were] men who had become enervated by a long course of luxurious ease ... the most robust and warlike of men in Europe, were about to be arrayed against the most sluggish and effeminate races of Asia."[864]

The two armies arrayed before each other, Alexander led the Macedonians toward the waiting Persians slowly at first, "with measured step."[865] Due to the narrow width of the flat terrain, "the great numerical mass of his [Darius'] vast host [had] ... no room to act; accordingly, they remained useless in the rear of his Greek and Asiatic hoplites."[866] Polybius said that "the width of the ground from the foot of the mountain to the sea was not more than fourteen stades [2.59 kilometers]."[867]

Plutarch observed, "[Darius] began to perceive his error in engaging himself too far in a country in which the sea, the mountains, and the river Pinarus running through the midst of it, would necessitate him to divide his forces, render his horse [cavalry] almost unserviceable, and only cover and support the weakness of his enemy. Fortune was not kinder to Alexander in the choice of the ground."[868]

Alexander's strategy was to aim his attack directly at Darius' position with his best cavalry, reasoning that if he could put the king to flight, the Persian army would be in disarray. According to Arrian, "When they [the Macedonians] came within range of darts, Alexander himself and those around him being posted on the right wing, dashed first into the river with a run, in order to alarm the Persians by the rapidity of their onset, and by coming sooner to close conflict to avoid being much injured by the archers. And it turned out just as Alexander had conjectured; for as soon as the battle became a hand-to-hand one, the part of the Persian army stationed on the left wing was put to rout; and here Alexander and his men won a brilliant victory."[869]

But the other sections of the Macedonian front encountered stiff resistance from the Persians and their Greek mercenaries. "The struggle was desperate; the Grecian mercenaries of Darius fighting in order to push the Macedonians back into the river ... a fierce cavalry battle ensued; for the Persians did not give way until they perceived that Darius had fled and the Grecian mercenaries had been cut up by the [Macedonian] phalanx and severed from them."[870]

The battlefront reached Darius' position, and "the dead bodies rose up in heaps round the chariot of Darius.... Darius himself was in a great terror and consternation; which, when some of the Persians discerned, they began first to fly."[871] Soon, Darius "began to flee in his chariot," but when he encountered uneven terrain, Darius "left the chariot there ... and mounting a horse continued his flight."[872]

As soon as it became apparent to the Persian troops that their king had fled the battlefield, the conflict turned into a rout. "So great was the slaughter that ... the men who were ... pursuing Darius, coming in the pursuit to a ravine, passed over it upon the corpses."[873] Persian casualties amounted to 100,000 infantry, and 10,000 cavalry. The Macedonians lost 300 infantry and 150 cavalry.[874] Traditional wisdom is that Darius' defeat was due to engaging the enemy on unsuitable terrain. However George Grote maintained that "the result would have been the same had the battle been fought in the plains to the east ... superior numbers are of little avail on any ground unless there be a general who knows how to make use of them ... the faith of Darius in simple multitude was altogether blind and childish."[875]

## *Sieges of Tyre and Gaza*

Following the victory at Issus, Alexander did not press eastward, but decided to further consolidate his position by marching south, along the Phoenician coast of the eastern Mediterranean. He received a letter from Darius, asking for the return of his "wife, mother, and children,"[876] whom had been captured by Alexander at Issus. Darius offered friendship, and the prospect of forming an alliance. But Alexander was unremitting. In his reply, he demanded Darius' submission: "Come to me therefore, since I am lord of all Asia ... and ask for your mother, wife, and children, and anything else you wish.... But for the future, whenever you send to me, send to me as the king of Asia, and do not address to me your wishes as to an equal.... If you act otherwise ... I intend to march against you."[877]

Tyre, the capital of Phoenecia, refused to submit to Alexander, feigning neutrality. "The people passed a decree ... not to admit into the city any Persian or Macedonian."[878] They had no desire to "admit within their walls an irresistible military force."[879] Alexander resolved to take Tyre by siege. He brought his generals together, and explained his strategy.

> An expedition to Egypt will not be safe for us, so long as the Persians retain the sovereignty of the sea; nor is it a safe course ... for us to pursue Darius, leaving in our rear the city of Tyre itself in doubtful allegiance, and Egypt and Cyprus in the occupation of the Persians.... But if Tyre were captured, the whole of Phoenicia would be in our possession, and the fleet of the Phoenicians, which is the most numerous and the best in the Persian navy, would in all probability come over to us ... [and] we shall acquire the absolute sovereignty of the sea, and at the same time an expedition into Egypt will become an easy matter for us. After we have brought Egypt into subjection, no anxiety about Greece and our own land will any longer remain, and we shall be able to undertake the expedition to Babylon with safety in regard to affairs at home.[880]

To take Tyre was no easy matter for Alexander. The city was located entirely on an island, half a mile (805 meters) from shore. The intervening water reached a maximum depth of 18 feet (5.5 meters). The city of Tyre itself was surrounded by walls, the least of which was 150 feet (46 meters) high. The inhabitants were well prepared to withstand a siege, having provisions, armaments and "other things essential to defense."[881]

Alexander undertook the siege of Tyre by constructing a massive causeway, or mole, from the mainland to the island. The structure was 200 feet (61 meters) in breadth. Thousands of local residents were pressed into service for the project.[882] "Whole trees, with their arms and branches, were cast into the sea, upon which they flung great heaps of stones; these were covered with a new course of trees, which they covered again with earth, till by successive lays of trees, stones and earth, the whole work became one solid body."[883]

At first, the Tyrians scoffed at the prospect of Alexander's forces being able to successfully breach the gap between their city and the land. They "came out in their boats," and taunted the Macedonians, asking "whether Alexander was greater than Neptune?"[884] But the Tyrians laughter turned to concern as the causeway grew. They began to harass the workers, "attacking with their darts those that guarded it," and "the Macedonians were forced to interrupt the work for some time to defend themselves."[885]

"The Tyrians ... omitted nothing that ingenuity could invent to render the Macedonians' labor ineffectual. The greatest help they received was from their divers, who entering the water out of the enemies sight, swam down unperceived to the very pier, and with hooks dragged after them the branches that stuck out of the stones, which drew along with them the other materials into the deep ... so that the foundation falling, the whole superstructure followed."[886]

It became apparent to Alexander that to mount a successful attack on Tyre, he would have to take control of the sea surrounding the island. So he put an end to the Tyrian harassment by assembling a fleet of 200 ships and positioning them around, and adjacent to, Tyre.[887] The Tyrians "were now pressed on all sides; for the pier was advanced within the cast of a dart, and the fleet surrounded the walls; so that they were annoyed both from the sea and land at the same time."[888]

When the causeway was complete, and the sea was controlled by the Macedonians, Alexander began the assault upon Tyre. He "collected many engineers," and constructed "many engines of war."[889] Eventually, the Macedonians found a weakness on Tyre's south wall. "He [Alexander] shook down a large piece of the wall; and when the breach appeared to be sufficiently wide, he ... brought up two [ships] ... which carried the bridges, which he intended to throw upon the breach in the wall."[890] "In the mean time, the rams battered down a great part of the wall in another place. And now the Macedonians entered through the breach on one side, Alexander with his party passed over the wall in another, so the city was now taken ... yet the Tyrians ... fought it out to the last man, insomuch that above seven thousand were cut in pieces."[891] Two thousand male Tyrians who were not killed in the assault were hung by Alexander on the seashore. A total of 30,000 women, children, and slaves were sold "to the slave-merchant."[892]

While Alexander was still engaged in the siege of Tyre, a remarkable incident occurred that is perfectly illustrative of Alexander's character and self-confidence.

> Ambassadors came to him [Alexander] from Darius, announcing that he [Darius] would give him [Alexander] ten thousand talents in exchange for his mother, wife, and children; that all the territory west of the river Euphrates, as far as the Grecian Sea, should be Alexander's; and proposing that he should marry the daughter of Darius, and become his friend and ally. When these proposals were announced in a conference.... Parmenio is said to have told Alexander, that if he were Alexander, he would be glad to put an end to the war on these terms, and incur no further hazard of success. Alexander is said to have replied, so would he also do, if he were Parmenio, but as he was Alexander he replied to Darius as he did. For he said that he was neither in want of money from Darius, nor would he receive a part of his territory instead of the whole; for that all his money and territory were his; and that if he wished to marry the daughter of Darius, he would marry her, even though Darius refused her to him.[893]

"When Darius received Alexander's answer, he was out of all hopes of putting an end to the war by letters and messages, and therefore he trained his soldiers every day."[894]

On the way to Egypt, Alexander met stiff resistance at the city of Gaza in Palestine. The governor, a eunuch named Batis, refuse to surrender the city to Alexander. Alexander's "engineers expressed the opinion that it was not possible to capture the wall [of Gaza] by force, on account of the height of the ground. However, the more impracticable it seemed to be, the more resolutely Alexander determined that it must be captured. For he said that the action would strike the enemy with great alarm from its being contrary to their expectation; whereas his failure to capture the place would redound to his disgrace."[895]

Alexander ordered a "mound to be constructed quite round the city on all sides, two stades [370 meters] in breadth and 250 feet [76 meters] in height."[896] The siege engines that had been employed at Tyre were sent for, and arrived by sea. The mound, or ramp, to the walls of Gaza being complete, Alexander began his assault. While the siege engines tried to bring down the city walls, Alexander's forces simultaneously weakened the walls of Gaza by tunneling beneath them. At length, the walls were breached, "[falling] down in many parts."[897]

The Gazans fought with great resolve. Four attacks were necessary to take the city. "The Gazans ... stood together and fought; so that they were all slain fighting there."[898] In *The Histories*, Polybius praised the Gazans for their courage. "It seems to me to be at once just and proper to give the people of Gaza the praise which they deserve ... on the invasion of Alexander, when not only did the other cities surrender, but even Tyre was stormed and its inhabitants sold into slavery; and when it seemed all but hopeless for any to escape destruction, who resisted the fierce and violent attack of Alexander, they alone of all the Syrians withstood him, and tested their powers of defense to the uttermost."[899]

Although normally he would be respectful of courage shown by an enemy, Alexander was exasperated at having his plans for conquest delayed by the obstinacy of Tyre and Gaza. He had personally received a "severe wound" during the assault on Gaza.[900] Furthermore, there was no honor in the victory. The honor, rather, belonged to the defeated, due to their courage against overwhelming odds.

All the male Gazans were slain,[901] and "Alexander sold their wives and children into slavery." The governor of Gaza, the eunuch Batis, was captured alive and brought before Alexander, who "resolved to inflict a punishment as novel as it was cruel. He directed the feet of Batis to be bored, and brazen rings to be passed through them; after which the naked body of this brave man, yet surviving, was tied with cords to the tail of a chariot driven by Alexander himself, and dragged at full speed amidst the triumphant jeers and shouts of the army. Herein Alexander, emulous even from childhood of the exploits of his legendary ancestor Achilles, copies the ignominious treatment described in the *Iliad* as inflicted on the dead body of Hector."[902]

## *Founding of Alexandria*

In the autumn of 332 B.C., "Alexander now led his army into Egypt,"[903] where he was unopposed. In Egypt, Alexander visited the ancient temple at Ammon to consult with the oracle. One of the priests told Alexander "that his wonderful successes and prosperous achievement, were evidences of his divine birth: for, as he was never yet overcome by any, so he should be ever victorious for the time to come."[904] The priest's declaration marked Alexander's "increasing self-adoration and inflation above the limits of humanity."[905]

Near the island of Pharos, Alexander founded the city named after him, Alexandria. It was probably intended to be "a place from which he could conveniently rule Egypt,"[906] but it would evolve into the center of the intellectual world in the centuries following Alexander's passing. Alexander's deliberate and methodical conquest would remake the Mediterranean world, and Western Civilization itself by spreading and cementing the influence of Greek culture. "The advance of the Macedonian leader was no less deliberate than rapid: at every step the Greek power took root, and the language and the civilization of Greece were planted from the shores of the Aegean to the banks of the Indus, from the Caspian and the great Hyrcanian plain to the cataracts of the Nile; to exist actually for nearly a thousand years, and in their effects to endure forever."[907]

## *Battle of Arbela*

Having secured all the lands bordering on the Mediterranean Sea, Alexander now marched east for the ultimate showdown with the Persians. Since Alexander had rejected Darius's offers of peace, the Persian king "employed himself wholly in making preparations for war."[908] "The whole army of Darius was said to contain 40,000 cavalry, 1,000,000 infantry, and 200 scythe-bearing chariots."[909]

Darius encamped his forces near "a village called Arbela,"[910] on a "spacious and level plain,"[911] and waited for Alexander to arrive. Alexander had crossed the Tigris river and was advancing. The Macedonian force consisted of 7,000 cavalry and 40,000 infantry,[912] a force that Edward Creasy assessed as "wholly composed of veteran troops in the highest possible state of equipment and discipline, enthusiastically devoted to their leader, and full of confidence in his military genius and his victorious destiny."[913] After learning the size and position of the Persian force, Alexander rested his men for four days before moving for a confrontation with Darius' army.

Parmenio advised Alexander to launch a night attack, but Alexander demurred. The Persians expected an assault after dark, and their troops were weakened by staying up the entire night before the battle. But Alexander slept. "He laid himself down in his tent and slept the rest of the night more soundly than was usual with him, to the astonishment of the commanders."[914]

On the morning of October 1, 331 B.C., the Macedonians marched forward to battle.[915] Alexander divided his troops into two lines, "so that his phalanx might be a double one."[916] The Macedonian second line, or rear guard, could respond if their line was flanked by the superior number of the enemy. Alexander's strategy was to make a concentrated attack on one specific point of the Persian front with the hope of breaking the cohesiveness of their force and perhaps panicking the troops.

According to Creasy, "Alexander found that the front of his whole line barely equaled the front of the Persian center, so that he was outflanked on his right by the entire left wing of the enemy, and by their entire right wing on his left. His tactics were to assail some one point of the hostile army, and gain a decisive advantage; while he refused, as far as possible, the encounter along the rest of the line. He therefore inclined his order of march to the right, so as to enable his right wing and center to come into collision with the enemy on as favorable terms as possible though the maneuver might in some respect compromise his left."[917]

Darius launched his chariots against the center of the Macedonian forces with the intent of throwing them into disarray and confusion. But the Macedonians were prepared for the chariot attack. Instead of directly resisting it, they ran off to the sides and from there assailed the horses and their drivers. "The men stood apart and opened their ranks, as they had been instructed, in the places where the chariots assaulted them ... [they] hurled their javelins at some of the horses; others they seized by the reins and pulled the drivers off, and standing round the horses killed them."[918]

After a while, Alexander perceived a weakness or gap in the Persian front, and he launched a direct assault on the Persian center, aiming for Darius. Alexander "led them with a quick charge and loud battle-cry towards Darius himself. For a short time there ensued a hand-to-hand fight; but when the Macedonian cavalry, commanded by Alexander himself, pressed on vigorously, thrusting themselves against the Persians and striking their faces with their spears, and when the Macedonian phalanx in dense array and bristling with long pikes had also made an attack upon them, all things together appeared full of terror to Darius, who had already long been in a state of fear, so that he was the first to turn and flee."[919] After Darius "himself set the example of flight ... from this moment the battle, though it had lasted so short a time, was irreparably lost [by the Persians]."[920]

But Alexander's charge had left the rest of the Macedonian line weakened. The Macedonians were "in a state of confusion from being attacked on all sides," "[and] Parmenio sent a messenger to Alexander in haste, to tell him that their side was in a critical position and that he must send him aid.... Alexander ... turned back ... and then ensued the most obstinately contested cavalry fight in the whole engagement."[921] "There was here a close hand-to-hand fight, which lasted some time ... [but] at length the Macedonian discipline and valor again prevailed."[922]

Casualty estimates are uncertain, with the ancient historians offering different numbers. "According to Arrian, 300,000 Persians were slain, and many more taken prisoners. Diodorus puts the slain at 90,000, Curtius at 40,000. The Macedonian killed were, according to Arrian, not more than 100 — according to Curtius, 300 ... Diodorus states the slain at 500, besides a great number of wounded."[923]

Having decisively defeated the Persian army, Alexander moved to occupy and plunder Persia. According to Diodorus, at the city of Susa Alexander "gained possession of ... the most beautiful palace in the universe ... [and] found there above forty thousand talents of uncoined gold and silver."[924] From Susa, Alexander marched to Persepolis, "the heart of Persian nationality," a city that had "more wealth, public and private, than any place within the range of Grecian or Macedonian knowledge."[925]

At Persepolis, Alexander "met ... certain Greeks, whom the former kings of Persia had made captives and slaves, and fell down at his feet; they were near eight hundred, most of them old men, and all maimed, some having their hands, others their feet, some their ears, and others their noses cut off.... Alexander so pitied their sad condition, that he could not refrain from weeping."[926]

Alexander allowed his soldiers to "plunder and spoil" Persepolis, reserving the king's palace for himself.[927] The Macedonians "put all the men to the sword ... [and] they first ravished the women as they were in their jewels and rich attire, and then sold them for slaves."[928] In the public treasury, Alexander recovered "a hundred and twenty thousand talents."[929]

At Persepolis, "Alexander made a sumptuous feast for the entertainment of his friends in commemoration of his victory, and offered magnificent sacrifices to the gods. At this feast were entertained whores, who prostituted their bodies for hire, where the cups went so high, and the reins so let loose to drunkenness and debauchery, that many were both drunk and mad."[930] The mistress of Alexander's general, Ptolemy, Thais, said "it would please her much better if, while the king looked on, she might in sport, with her own hands, set fire to the court."[931] Alexander himself, torch in hand, led the way, and the palace at Persepolis was burnt.

Alexander's most trusted and experienced general, Parmenio, was essentially second-in-command of the Macedonian forces. Parmenio had three sons, two of whom had died in battle fighting for Alexander. The third, Philotas, was one of Alexander's commanders. One day, inebriated, Philotas boasted to his mistress that "all the great actions were performed by him and his father, the glory and benefit of which, he said, together with the title of king, the boy Alexander reaped and enjoyed by their means."[932] The indiscretion was reported to Alexander, and Philotas came under suspicion. Alexander's suspicion was heightened when he learned that Philotas had obtained advanced information on a plot to assassinate him, but had not informed Alexander.

Philotas was arrested and tortured. His interrogators "made him pass through the severest torments ... [and] most miserably tore his body ... they made use of both fire and scourges, rather by the way of punishment than examination."[933] "The tortures inflicted, cruel in the extreme and long continued, wrung from him at last a confession implicating his father along with himself."[934]

Philotas was executed, and Alexander immediately ordered the assassination of Parmenio, now age seventy, "because Alexander deemed it incredible that Philotas should conspire against him and Parmenio not participate in his son's plan; or perhaps, he thought that even if he had had no share in it, he would now be a dangerous man if he survived."[935] Alexander commissioned reliable men to kill Parmenio, and when the deed was accomplished, the head of his old friend was delivered to him as proof.[936]

## *Death of Alexander*

Alexander spent the next seven years consolidating his control over the eastern part of the Persian empire.[937] He pursued Darius, but found only a body. Darius had been assassinated by one of his own satraps, Bessus. This angered Alexander, because a live Darius could have helped him govern the empire. Eventually, Alexander captured Bessus alive, and "he ordered his nose and ears to be cut off."[938] The mutilation was only the beginning of Bessus' punishment. Alexander "gave him up into the hands of Darius' brother ... to punish him in such manner as they thought fit, who, after they had put him to all manner of torments, and used him with all the despite and disgrace imaginable, cut his body into small pieces, and hurled every part here and there out of their slings."[939]

As time passed, Alexander began to believe the adulation that was heaped upon him. He "became gradually corrupted by unexampled success and Asiatic influences."[940] After visiting the temple of Ammon in Egypt (c. 332 B.C.), Alexander left "with a full and sincere faith that he really was the son of Zeus Ammon ... though he did not directly order himself to be addressed as the son of Zeus, he was pleased with those who volunteered such a recognition, and angry with skeptics or scoffers."[941]

Plutarch disagreed, arguing that Alexander's godhood was a political device employed to help him govern the superstitious. "Alexander in himself was not foolishly affected, or had the vanity to think himself really a god, but merely used his claims to divinity as a means of maintaining among other people the sense of his superiority."[942]

Whatever Alexander's true intentions may have been, within a few years "he could no longer be satisfied without obtaining prostration or worship from Greeks and Macedonians as well as from Persians, a public and unanimous recognition of his divine origin and superhuman dignity."[943] In Greece, Alexander's godhood was treated with disinterested amusement. Unimpressed, the Spartans said "we are content that Alexander be called a god."[944]

As Alexander began to "indulge in the soft and effeminate manners of the Persians, and to imitate the luxury of the Asiatic kings,"[945] resentment began to build among his Greek and Macedonian compatriots. Things came to a head at a banquet. Clitus was one of Alexander's most loyal commanders. At the battle of Granicus, he had saved Alexander's life by cutting off the arm of the Persian general, Spithradates, when he was about to strike Alexander from behind.[946] But "Clitus had long been vexed at Alexander for the change in his style of living in excessive imitation of foreign customs, and at those who flattered him with their speech." His tongue loosened by too much wine, Clitus "affirmed Alexander's deeds were neither in fact at all so great or marvelous as they represented in their laudation; nor had he achieved them by himself, but for the most part they were the deeds of the Macedonians.... Clitus ... then began to put Philip's achievements in the first rank, and to depreciate Alexander and his performances."[947]

Alexander, as inebriated as Clitus, flew into a rage, and "looked about for his sword. But Aristophanes, one of his life-guard, had hid that out of the way.... Clitus still refusing to yield, was with much trouble forced by his friends out of the room. But he came in again immediately at another door ... [and] at last, Alexander, snatching a spear from one of his soldiers ... ran him [Clitus] through the body."[948] Alexander immediately sobered up, and went into a deep depression. "He did not cease calling himself the murderer of his friends; and for three days rigidly abstained from food and drink, and paid no attention whatever to his personal appearance."[949]

Another person who vexed Alexander and unjustly incurred his wrath was Aristotle's nephew, Callisthenes. Callisthenes fell under suspicion of participating in an assassination plot, a supposition that was completely without factual support. Alexander had the nephew of his teacher "first put to the torture and then hanged."[950] It was a sin against philosophy, for which Alexander was excoriated by later generations of writers. In *Natural Questions*, Seneca (c. 4 B.C.–A.D. 65), concluded "Callisthenes" was endowed with a lofty intellect, and he dared to brave the wrath of a king. His death is an eternal blot on the memory of Alexander, which no valor and no success in war can ever remove ... granted that he surpassed all former precedents of generals and kings, yet of all that he did, nothing will match his guilt in slaying Callisthenes.[951]

In 324 B.C., Alexander's friend from boyhood, Hephaestion, became ill. Hephaestion was described by Arrian as the person "who was the dearest to [Alexander] in the world."[952] "When Alexander was informed that Hephaestion was in a critical state, he went to him without delay, but found him no longer alive ... his [Alexander's] grief was great."[953] Alexander "cast himself on the ground near the dead body, and remained there wailing for several hours; he refused all care, and even food, for two days; he cut his hair close, and commanded that all the horses and mules in the camp should have their manes cut close also ... [and] he interdicted all music and every sign of joy in the camp."[954]

Alexander's extreme grief at the death of Hephaestion has led to the supposition that their relationship was sexual. There is no direct indication of this in the ancient biographies. If that was the case, Alexander would have been bisexual, not homosexual. Diodorus reported that Alexander had a passion for intercourse with the female sex. Alexander "began likewise to carry his concubines along with him from place to place ... and these were the greatest beauties that could be found throughout all Asia. These stood round the king's bed every night, that he might take his choice of whom he pleased to lie with him."[955] Plutarch reported that Alexander was highly offended at the suggestion of having a sexual relationship with young boys. "When Philoxenus, his lieutenant ... wrote to him [Alexander] to know if he would buy two young boys of great beauty ... [Alexander] was so offended that he often expostulated with friends what baseness Philoxenus had ever observed in him that he should presume to make him such a reproachful offer."[956]

After Hephaestion's death, Alexander became increasingly morose and bad-tempered. "Cassander, who had lately arrived, and had been bred up in Greek manners, the first time he saw some of the barbarians adore the king could not forbear laughing at it aloud, which so incensed Alexander that he took him by the hair with both hands and dashed his head against the wall."[957] Alexander's "temper became so much more irascible and furious, that no one approached him without fear, and he was propitiated by the most extravagant flatteries."[958]

In June of 323 B.C., Alexander fell ill with a fever after drinking for two days. His fever festered for several days, and he became weaker and weaker. "Calling to his friends to draw nearer (for his voice began already to fail him) he took his ring off his finger, and gave it to Perdiccas, enjoining him to convey his body to Hammon; and as they asked him, 'to whom he bequeathed his kingdom' he answered, 'to the most worthy....' These were the king's last words, a little after which he expired."[959]

At the time of his death, Alexander was 32 years old. He had reigned as Macedonian king for 12 years. His "achievements ... during his twelve years of reign, throwing [his father] Philip into the shade, had been on a scale so much grander and vaster, and so completely without serious reverse or even interruption, as to transcend the measure, not only of human expectation, but almost of human belief."[960]

Although Alexander passed away in early middle age, his conquest created a cultural unity that would persist for nearly a thousand years. Greek science, art, literature, and culture now dominated the Mediterranean world. Their influence would define Western Civilization as we know it. By breaking down the wall between Asia and Europe, Alexander prepared the way for the Roman Empire and the spread of Christianity. As Alexander united the political world; his teacher Aristotle would unite the intellectual.

## *Aristotle (384–322 B.C.)*

### *Early Life*

Aristotle "unquestionably ranks among the greatest* thinkers that the world has known. He surveyed the whole field of human knowledge as it was in his day, and his encyclopedic writings dominated men's thinkings in many fields for better or worse for nearly 2,000 years."[961] Fifteen-hundred years after his death, Aristotle was referred to simply as "The Philosopher,"[962] and his scientific works were considered state-of-the-art. George Sarton noted the encyclopedic nature of Aristotle's work by observing that "one cannot deal with any science, or with any branch of science, without having to drag him [Aristotle] in."[963] English poet Samuel Taylor Coleridge (1772–1834) referred to Aristotle as "the sovereign lord of the understanding ... the greatest that ever animated the human form."[964]

*It is not hagiographic to describe Aristotle as "great." Adulation is a continual temptation, but it is also possible to go too far the other way.

Dante Alighieri's (A.D. 1265–1321) epic masterpiece is the allegorical narrative, *The Divine Comedy*. In the first part, "The Inferno," he described his descent into hell where he met Socrates, Plato, and an entire host of Greek philosophers from antiquity. Above all of these was Aristotle, the "master of those who know."

> I saw the Master of those who know,
> seated amid the philosophic family;
> all regard him, all do him honor
> Here I saw Socrates and Plato,
> who in front of the others stand nearest to him....[965]

Aristotle was born in the city of Stagira[966] in the province of Chalcidice in 384 B.C. Located on the northern shore of the Aegean Sea, Stagira resembled "both in soil and appearance ... the southern part of the bay of Naples."[967] It had been founded in 654 B.C. by emigrants from the island of Andros, itself a colony of the Greek city of Eretria on the west coast of Euboea.[968] "Stagira was in the fullest sense a Greek town."[969]

Aristotle's father, Nicomachus, was court physician to Philip's father, Amyntas II.[970] We are informed by the physician Galen (c. A.D. 129–200), that the sons of the Greek physicians "were from their childhood exercised by their parents in dissecting just as familiarly as in writing and reading; so that there was no more fear of their forgetting their anatomy than of their forgetting their alphabet."[971] Other than this inference, we know nothing of Aristotle's elementary education. As a youth, it is likely that Aristotle accompanied his father when he attended Amyntas II, and thus made the acquaintance and friendship of Philip from an early age.[972]

At the age of seventeen or eighteen, Aristotle's father died, and he traveled to Athens to attend Plato's Academy. Aristotle spent nineteen or twenty years in Athens, departing only after the death of Plato.[973] From the fact that his father was a physician, it seems likely that Aristotle's early interests were physiology and natural philosophy.[974] But from the time of his arrival at the Academy, it is more likely that his studies would have been concerned with the subjects discussed in Plato's *Dialogues*, dialectics and moral philosophy.

The superiority of Aristotle's intellect was recognized by Plato, who referred to him as "the reader" and "the mind" of the Academy.[975] Aristotle necessarily and eventually asserted his intellectual independence from Plato. Diogenes Laërtius related that Plato complained, "Aristotle has kicked us off just as chickens do their mother after they have been hatched."[976] In Aristotle's writings, Plato's "opinions are often controverted, but always with fairness, and never with discourtesy."[977] In the *Nicomachean Ethics*, Aristotle explained that although he loved Plato, he loved the truth that much more. "It would perhaps be thought to be better, indeed to be our duty, for the sake of maintaining the truth even to destroy what touches us closely, especially as we are philosophers or lovers of wisdom; for, while both are dear, piety requires us to honor truth above our friends."[978]

In 347 or 348 B.C., Plato died. Plato's nephew and heir, Speusippus, was chosen to be the new head of the Academy. Around this time, Philip of Macedonia was extending and consolidating his empire. He attacked the Chalkidic peninsula,[979] and Aristotle's native city of Stagira was "captured and ruined" by Philip "during the Olynthian war of 349–347 B.C."[980]

In the company of fellow student Xenocrates, Aristotle left Athens and the Academy to take "residence at Atarneus or Assos, in Mysia,"[981] a province in northwest Asia Minor. He had been invited there by another student of the Academy, Hermias. According to Strabo,

"Hermias was a eunuch, servant of a money-changer. When he was at Athens he was the hearer both of Plato and of Aristotle. On his return he became the associate in the tyranny of his master, who attacked the places near Atarneus and Assos. He afterwards succeeded his master, sent for both Aristotle and Xenocrates, and treated them with kindness. He even gave his niece in marriage to Aristotle."[982] The account given by Diogenes Laërtius is slightly different. "Aristotle was enamored of the concubine of Hermias, and that, as Hermias gave his consent, he married her."[983]

Hermias was betrayed by Mentor, a "general in the service of the Persians."[984] Strabo stated that Mentor "invited to his house Hermias, under the mask of friendship, and on pretense of business. He seized Hermias, and sent him to the king, who ordered him to be hanged. The philosophers, avoiding places in possession of the Persians, escaped by flight."[985]

According to Diodorus Siculus, Mentor captured the cities under Hermias' rule without a fight. "Possessing himself [Mentor] of his [Hermias'] seal-ring, he wrote letters in his [Hermias] name, to the several cities, signifying that through the means of Mentor he had been restored to the king's favor.... [The] governors of the cities, giving credit to the letters, and being, likewise, very desirous of peace, delivered up all the towns and forts to the king in every place."[986]

Following the demise of Hermias, Aristotle and his wife fled to Mitylene, a city on the island of Lesbos.[987] After two or three years in Mitylene, Aristotle was invited by Philip to join the Macedonian court and become the teacher of Alexander, who was then thirteen years old.[988] Aristotle's price was the restoration of his native city, Stagira, which had been destroyed by Philip during the Olynthian war of 349–347 B.C. "He [Philip] repeopled his [Aristotle's] native city Stagira, which he had caused to be demolished a little before, and restored all the citizens, who were in exile or slavery, to their habitations."[989] Plutarch (A.D. 46–120) stated that in his day he visited Stagira and was shown "Aristotle's stone seats, and the shady walks which he was wont to frequent."[990]

Aristotle remained in Macedonia as Alexander's teacher for eight years,[991] albeit with "occasional visits" to Athens.[992] When Alexander departed on his conquest of Asia in 335 B.C., Aristotle returned to Athens and opened a new school, the Lyceum. "Much of his instruction is said to have been given while walking in the garden, from when the students and the sect derived the title of Peripatetics."[993]

In *Attic Nights*, Aulus Gellius (c. A.D. 125–180), stated that Aristotle had two courses of teaching at the Lyceum. Routine subjects were covered in evening lectures, but morning lectures were devoted to instruction in more abstruse and difficult topics.

> The philosopher Aristotle ... is said to have had two kinds of lectures ... one of which he called *exoteric* [ordinary], the other *acroatic* [abstruse]. Those were called exoteric which involved the study of rhetoric, logical subtleties, and a knowledge of politics; those were called acroatic, which had concern with a more profound and recondite philosophy ... to the cultivation of this science which I have called the acroatic, he gave up the morning in the Lyceum; nor did he admit any person to this lecture till he had previously made examination concerning his talents, his elementary knowledge, and his zeal and industry in the pursuit of learning. The exoteric lectures he delivered in the same place, in the evening.[994]

We are told by Diogenes Laërtius that Aristotle was somewhat effeminate and weak. "He [Aristotle] had a lisping voice ... he had also very thin legs, they say, and small eyes ... [and] he used to indulge in very conspicuous dress, and rings, and used to dress his hair carefully."[995] But there is no question that Aristotle was heterosexual. He was "deeply attached" to his first wife, Pythias, by whom he had a daughter of the same name.[996] After

Pythias died, he married "a woman of Stagira, name Herpyllis, who bore him a son called Nicomachus."[997]

The scope of Aristotle's knowledge and teaching was encyclopedic. He developed and taught a curriculum that included biology, astronomy, physics, metaphysics, logic, ethics, politics, and poetry. It would be unreasonable to conclude that all of Aristotle's knowledge was original. He must have been acquainted with a significant body of presocratic works on natural philosophy that are now completely lost or exist only in fragments.

In the twelve or thirteen years that he taught at the Lyceum, Aristotle wrote extensively. Diogenes Laërtius numbered Aristotle's works as "nearly four hundred, the genuineness of which is undoubted."[998] But only thirty-five of Aristotle's complete books are extant. From this, we might surmise that ninety percent or more of the philosopher's work has been lost. However it is likely that later editors and compilers condensed shorter works into single volumes. It is probable that "we only possess less than a third of Aristotle's total work."[999]

The manner in which these writings have come down to us is instructive, for it illustrates the fragile nature of knowledge before the advent of the printing press. According to Strabo, Aristotle's scrolls were inherited by his friend and successor at the Lyceum, Theophrastus. Theophrastus in turn willed them to his nephew, Neleus.

> Neleus carried it [Aristotle's library] to [the city of] Scepsis, and bequeathed it to some ignorant person who kept the books locked up, lying in disorder. When the Scepsians understood that the Attalic kings, on whom the city was dependent, were in eager search for books, with which to furnish the library at Pergamus, they hid theirs in an excavation under-ground; at length, but not before they had been injured by damp and worms, the descendants of Neleus sold the books of Aristotle and Theophrastus for a large sum of money to Apellicon of Teos [1st century B.C.]. Apellicon was rather a lover of books than a philosopher; when therefore he attempted to restore the parts which had been eaten and corroded by worms, he made alterations in the original text and introduced them into new copies; he moreover supplied the defective parts unskillfully, and published the books full of errors.[1000]

Henceforth, the story of Aristotle's manuscripts was related by Plutarch in his *Life of Sulla*. The Roman General, Sulla (138–78 B.C.), after capturing Athens in 86 B.C., "seized for his use the library of Apellicon ... in which were most of the works of Theophrastus and Aristotle, then not in general circulation. When the whole was afterwards conveyed to Rome, there, it is said, the greater part of the collection passed through the hands of Tyrannion, the grammarian, and that Andronicus, the Rhodian, having through his means the command of numerous copies, made the treatises public, and drew up the catalogues that are now current."[1001]

## *Biology*

Aristotle's greatest scientific contributions were in the field of biology. He assembled facts, organized them, and inductively reasoned to generalized conclusions. The biologic investigations were assisted by aid from Alexander, who reportedly donated "eight hundred talents,"[1002] and placed the services "thousands of men"[1003] at Aristotle's disposal. Aristotle's methodology was essentially modern, and the resources provided by Alexander enabled his biological investigations to be large in scope.

Aristotle's writings contain descriptions of more than 500 animals, many of these in great detail. He dissected about 50 different species of animals, and described their structure, habitat, and method of reproduction.[1004] Aristotle commented on comparative anatomy and physiology, embryology, and the geographic distribution of animals. He also invented ecology, the science that deals with the relationships between organisms and their environments.

In 1882, William Ogle (1827–1912) sent Charles Darwin (1809–1882) a copy of his translation of Aristotle's book, *Parts of Animals.* In his reply to Ogle, dated February 22, 1882, the master biologist of the nineteenth century indicated his high esteem for Aristotle by confessing, "Linnaeus and Cuvier have been my two gods, though in different ways, but they were mere schoolboys to old Aristotle."[1005]

In his book, *On the Parts of Animals*, Aristotle contrasted biology with astronomy, underlined the importance of data collection and direct observation, and correctly concluded that his knowledge of biology was much greater than his knowledge of astronomy.

> Some members of the universe are ungenerated, imperishable, and eternal, while others are subject to generation and decay. The former are excellent beyond compare and divine, but less accessible to knowledge. The evidence that might throw light on them, and on the problems which we long to solve respecting them, is furnished but scantily by sensation; whereas respecting perishable plants and animals we have abundant information, living as we do in their midst, and ample data may be collected concerning all their various kinds, if only we are willing to take sufficient pains. Both departments, however, have their special charm. The scanty conceptions to which we can attain of celestial things give us, from their excellence, more pleasure than all our knowledge of the world in which we live; just as a half glimpse of persons that we love is more delightful than a leisurely view of other things, whatever their number and dimensions. On the other hand, in certitude and in completeness our knowledge of terrestrial things has the advantage. Moreover, their greater nearness and affinity to us balances somewhat the loftier interest of the heavenly things that are the objects of the higher philosophy. Having already treated of the celestial world, as far as our conjectures could reach, we proceed to treat of animals, without omitting, to the best of our ability, any member of the kingdom, however ignoble. For if some have no graces to charm the sense, yet even these, by disclosing to intellectual perception the artistic spirit that designed them, give immense pleasure to all who can trace links of causation, and are inclined to philosophy ... every realm of nature is marvelous.[1006]

In some cases, Aristotle's conclusions were regarded as being false for hundreds of years until they were verified by modern investigators. In describing the breeding habits of the catfish, Aristotle noted that once the female had laid the eggs, the male guarded them until they hatched. "The male watches over the place where the greatest number of ova are deposited, and ... remains by them for forty or fifty days, in order that they may not be devoured by fish chancing to come that way."[1007] This story was disbelieved for centuries, because catfish native to Western Europe do not exhibit this behavior. The first verification of Aristotle's observations was done by the nineteenth-century geologist Louis Agassiz. In 1856, Agassiz discovered that American catfish behave exactly as Aristotle wrote. Subsequently, Agassiz obtained access to some catfish from the Achelous River in Greece and was able to corroborate Aristotle's claims. The authenticity of Aristotle's remarks on the breeding habits of catfish became well established by the year 1906.[1008]

It is difficult for us today to appreciate the conditions under which Aristotle worked. He lacked instrumentation, even a tool as basic as a magnifying glass or wristwatch. He had no reference books, not even a dictionary, and had to invent terms and words as necessary to describe what he saw.

In *History of Animals*, Aristotle painstakingly described the embryological development of a chicken. The description was thorough, perspicuous, and must have been based on a long and careful period of observation. "About the twentieth day, if the hatching has been delayed beyond this period, the young bird is able to chirp when moved externally, and if the shell is taken off, by this time also it is downy. The head is placed over the right leg upon the side, and the wing is over head."[1009]

Aristotle also made observations of the life of bees that were detailed and thorough. His attention to minutiae is recorded in *History of Animals*. "The bees collect the wax by climbing actively on the flowers with their fore feet. They cleanse these upon the middle pair of legs, and their middle legs again on the curved part of their hind legs, and thus loaded they fly away. They are evidently heavily loaded. During each flight the bee does not settle upon flowers of different kinds, but as it were from violet to violet, and touches no other species till it returns to the hive. There they are unloaded, and two or three bees follow every one on its return to the hive."[1010]

In Aristotle's book, *On the Generation of Animals*, there is a passage that is remarkably prescient of the modern experimental method, and which seems to contradict the prevailing Greek view that theory was superior to practice. After discussing the habits of bees, Aristotle remarked that precedence must be given to empirical knowledge, and that theories must be discarded if they are not in accordance with observed facts. "Such appears to be the truth about the generation of bees, judging from theory and from what are believed to be the facts about them; the facts, however, have not yet been sufficiently grasped; if ever they are, then credit must be given rather to observation than to theories, and to theories only if what they affirm agrees with the observed facts."[1011]

Not all of Aristotle's conclusions were correct. One of his more glaring errors was that he did not understand the true function of the human heart and brain. He considered the heart to be the seat of intelligence, and thought that the primary function of the brain was to cool the blood.[1012] There were other errors as well. In *History of Animals* Aristotle said that certain fish come into being through spontaneous generation. "It is evident from the following considerations that some of them [species of fish] are of spontaneous growth, and do not originate either in ova or semen."[1013]

In consideration of his role in originating the biologic sciences, it was inevitable that Aristotle would make a number of mistakes. But it would be a mistake to assess him by enumerating his correct conclusions and comparing them to the incorrect ones. Aristotle's emphasis on direct observation and systematic classification distinguish him from his predecessors.

Although he did not emphasize it, Aristotle recognized the value of observation and experiment. In his book *On the Parts of Animals*, he first described the sponge as a form of sea life that resembles a plant because it spends its entire life in one place, attached to a rock. He then described other sea plants that live in the water unattached to any substrate, and land plants that "are independent of the soil."[1014] In an apparent effort to convince his readers that such a thing is possible, he made reference to experiment. "Even among land-plants there are some that are independent of the soil, and that spring up and grow, either parasitically upon other plants, or even entirely free. Such for example is the plant which is found on Parnassus, and which some call the Epipetrum. This you may take up and hang from the rafters, and it will yet live for a considerable time."[1015]

Plato believed he could find truth through pure mentality. But Aristotle confronted nature on her own terms. When Plato was absorbed in dialectic or contemplation, Aristotle was out in the field ripping up plants by the roots, in the meadows watching bees dart from flower to flower, or in the hen house cracking open chicken eggs. Criticizing Plato's theory of matter, Aristotle noted that any theory which dealt with the observable side of nature must be constrained by observation, because "the principles of sensible things are sensible."[1016] Therefore, "practical knowledge culminates in the work produced, natural philosophy in the facts as presented consistently and indubitably to sense-perception."[1017]

A teleological vein ran through all of Aristotle's biology, not just in an incidental sense, but as a unifying and central theme. Aristotle recognized a natural order in living things. He classified them from the lowest (nonliving) to the highest (man). Aristotle's "ladder of nature" did not necessarily imply evolution; for him it "was a means of illustrating the fundamental unity and order of nature."[1018]

In Aristotelean biology, each organism and each organ within every living thing had a purpose. In *On the Generation of Animals*, Aristotle noted "we must observe how rightly nature orders generation in regular gradation."[1019] In *History of Animals*, he described a continuous variation in living things, placing plants lower on the scale and animals higher. "Nature proceeds little by little from things lifeless to animal life in such a way that it is impossible to determine the exact line of demarcation ... after lifeless things in the upward scale comes the plant, and of plants one will differ from another as to its amount of apparent vitality ... there is observed in plants a continuous scale of ascent towards the animal."[1020]

In *On the Parts of Animals*, Aristotle noted that everything in nature was ordered, and the resultant harmony was beautiful. "Absence of haphazard and conduciveness of everything to an end are to be found in nature's works in the highest degree, and the resultant end of her generations and combinations is a form of the beautiful."[1021]

## *Chemistry*

Aristotle and others apparently adopted Empedocles' theory of the four elements as the best extant at the time. However the Greeks were aware that the four elements could not be ultimately elementary, for they observed them change and transform into each other. The most obvious example was provided by the evaporation and condensation of water. In *Timaeus*, Plato noted "water ... when melted and dispersed, passes into vapor and air ... [and] air when collected and condensed, produces cloud and mist; and from these, when still more compressed, comes flowing water."[1022]

In *De Caelo* (*On the Heavens*), Aristotle noted "all [the elements] change into one another."[1023] In *Meteorology*, he described the four elements as interchangeable. "Fire, air, water, earth, we assert, originate from one another, and each of them exists potentially in each, as all things do that can be resolved into a common and ultimate substrate."[1024] The four elements or "bodies" owed their existence to "the four principles" of hot, cold, moist, and dry.[1025] "Fire is hot and dry, whereas air is hot and moist ... water is cold and moist, while earth is cold and dry."[1026]

In discussing any important subject, it was Aristotle's habit to begin with a dispassionate review of other people's ideas and contributions. Every idea was examined critically, and its strengths and weaknesses considered. In *Metaphysics*, Aristotle explained "he who has heard all the contending arguments, as if they were the parties to a case, must be in a better position for judging."[1027] In *On Generation and Corruption*, Aristotle discussed the question of the fundamental nature of matter. He mentioned concepts introduced by Empedocles, Anaxagoras, Leucippus, and Democritus. After examining the atomic theory of Leucippus and Democritus,[1028] Aristotle rejected atomism on the grounds that matter is a continuum.[1029]

## *Geology*

Aristotle made perspicacious observations of the Earth and had acute insights into geological processes that were well ahead of his time. In *Meteorologica (Meteorology)*, Aristotle showed that he had a nearly complete understanding of the hydrologic cycle. He understood that water could turn into air [water vapor], and then back into water when cooled, and that this process was cyclic.

"We get a circular process that follows the course of the Sun. For according as the Sun moves to this side or that, the moisture rises or falls. We must think of it as a river flowing up and down in a circle and made up partly of air, partly of water."[1030] Aristotle believed that rivers were recharged by moisture condensing in cool, underground caverns. But he almost hit upon the modern conception of the Earth as a porous medium when he wrote, "mountains and high ground, suspended over the country like a saturated sponge, make the water ooze out and trickle together in minute quantities but in many places."[1031]

Aristotle had an appreciation for the vastness of geologic time. "The whole vital process of the Earth takes place so gradually and in periods of time which are so immense compared with the length of our life, that these changes are not observed, and before their course can be recorded from beginning to end whole nations perish and are destroyed ... there will be no end to time and the world is eternal."[1032] Europeans would not begin to appreciate the great age of the Earth until the end of the eighteenth century. John Playfair (1748–1819), recounting his visit to Siccar Point with James Hutton (1726–1797), wrote "the mind seemed to grow giddy by looking so far into the abyss of time."[1033]

Aristotle thought that earthquakes were caused by the action of wind inside the Earth. "A great wind is compressed into a smaller space and so gets the upper hand, and then breaks out and beats against the Earth and shakes it violently."[1034] This is the wrong theory, but the discussion in *Meteorology* shows that Aristotle recognized the importance of observation and experiment. Justifying his theory of earthquakes, he wrote "our theory has been verified by actual observation in many places."[1035] In his discussion of evaporation, Aristotle recounted that he had reached his conclusions by performing experiments. "Salt water when it turns into vapor becomes sweet, and the vapor does not form salt water when it condenses again. This I know by experiment."[1036]

In his short treatise, *Mechanics*, Aristotle mentioned the uniformity of nature, a key precept of modern geological thinking, and, indeed, the indispensable corollary to naturalism itself. "Nature ... always follows the same course without deviation."[1037] The author of *Mechanics* may have been Archytas (c. 428–350 B.C.), but the statement nevertheless indicates that the principle of uniformity was known and stated explicitly at this time.[1038] The principle of uniformity was also invoked by Aristotle in *De Generatione et Corruptione* (*On Generation and Corruption*). "It is a law of nature that the same cause, provided it remain in the same condition, always produces the same effect."[1039]

## *Cosmology*

By the time of Aristotle in the later part of the fourth century B.C., the Greeks knew considerably more about astronomy than the presocratic philosophers of the sixth and fifth centuries. They knew that the Earth was spherical, recognized five planets that moved against the fixed background of stars, and understood that eclipses were occultations. In *De Caelo* (*On the Heavens*), Aristotle stated that Anaximenes (c. 570–500 B.C.), Anaxagoras (c. 500–428), and Democritus (460–370 B.C.) had maintained that the Earth was flat.[1040] But in *Phaedo*, Plato wrote "the Earth is a round body in the center of the heavens."[1041]

The Greeks also had some grasp of relative distances. In *De Caelo* (*On the Heavens*),

Aristotle noted that the Moon had been observed to obscure the planet Mars by moving in front of it, and therefore had to be closer to the Earth than Mars.[1042] Aristotle explained that Greek astronomers had been able to increase their knowledge at least in part by utilizing observational data provided to them by "the Egyptians and the Babylonians."[1043]

Aristotle's cosmological model was geocentric. The Earth was located at the center of universe, and did not move. In *On the Heavens*, Aristotle concluded "[the Earth] has its center at the center of the universe ... it is clear that the Earth does not move, neither does it lie anywhere but at the center."[1044]

Aristotle's explanation for the movements of the Sun, Moon, planets, and stars was derived from that of the Greek astronomer Eudoxus (c. 400–347 B.C.), a pupil of Plato. Eudoxus' model explained the motions of the heavenly bodies by invoking a series of concentric spheres, "it was the beginning of scientific astronomy."[1045]

> All the spheres were concentric, the common center being the center of the Earth ... each planet was fixed at a point in the equator of the sphere which carried it, the sphere revolving at uniform speed about the diameter joining the corresponding poles ... but one such circular motion was not enough; in order to explain the changes in the speed of the planets' motion, their stations and retrogradations, as well as their deviations in latitude, Eudoxus had to assume a number of such circular motions working on each planet and producing by their combination that single apparently irregular motion which can be deduced from mere observation.[1046]

To account for the motions of the Sun and the Moon required three spheres of rotation; the planets required four each. Finally, one sphere accounted for the daily rotation of the fixed stars, thus the entire system consisted of twenty-seven concentric spheres.[1047]

The Eudoxian system was "purely geometrical and theoretical; there was nothing mechanical about it."[1048] Eudoxus said nothing concerning "the causes of these rotational motions or the way in which they were transmitted from one sphere to another; nor did he inquire about the material of which they were made, their sizes and mutual distances."[1049] The significance of Eudoxus' model was a demonstration that natural phenomena could be explained by a mathematical model. "It seemed to show how extremely complicated phenomena might be reduced to simple regular movements."[1050]

Aristotle "transformed the purely abstract and geometrical theory [of Eudoxus] into a mechanical system of spheres, i.e. spherical shells, in actual contact with one another."[1051] This necessitated the introduction of additional spheres. In *Metaphysics*, Aristotle stated "the number of all the spheres ... will be fifty-five."[1052]

A central tenet of Aristotelian cosmology was the "doctrine of the incorruptibility of the heavens."[1053] Aristotle divided the universe into two realms, the terrestrial and the heavenly, each characterized by distinctly different natures and chemistries. The lower, terrestrial realm, was characterized by corruption, decay, generation, and change. All things on Earth were composed of the four elements: earth, water, air, and fire.

In contrast, the heavenly realm was incorruptible and unchanging. This was clearly demonstrated to Aristotle and his contemporaries by observation. In *On the Heavens*, Aristotle explained "the truth of it [the eternal and unchanging nature of the heavens] is also clear from the evidence of the senses, enough at least to warrant the assent of human faith; for throughout all past time, according to the records handed down from generation to generation, we find no trace of change either in the whole of the outermost heaven or in any one of its proper parts."[1054]

Because the heavens were not subject to corruption or decay, it was clearly implied that they were composed of a fifth element, ether.[1055] Aristotle explained that this was an old idea. "The ancients ... believing that the primary body was something different from earth and fire and air and water, gave the name *ether** to the uppermost region."[1056]

The conception of the universe as a series of neatly arranged concentric spheres led to an easy accommodation of the idea that the Earth, too, was spherical. But Aristotle also had sound physical reasons for concluding that the shape of the Earth was spherical. In *On the Heavens*, he argued that the Earth's "shape must be spherical ... if the Earth were not spherical, eclipses of the Moon would not exhibit segments of the shape which they do ... the boundary is always convex."[1057]

A second reason was that stars would disappear below the horizon as an observed traveled from north to south. "Observation of the stars also shows not only that the Earth is spherical but that it is of no great size, since a small change of our position on our part southward or northward visibly alters the circle of the horizon, so that the stars above our heads change their position considerably."[1058] Aristotle cited a value for the Earth's circumference of 400,000 stades [74,000 Kilometers], but did not explain how the number was calculated.[1059] The modern estimate is 40,029 Kilometers.[1060]

Aristotle believed that both meteors ("shooting stars") and comets were sublunary phenomena. In Aristotelean chemistry, fire was a hot and dry element, that, while not actually burning, was "the most inflammable of all bodies."[1061] When the sun warmed the Earth, this fire element evaporated and rose to fill the outermost concentric layer of the Earth's atmosphere, adjacent to the motion of the heavenly spheres. Occasionally, the friction or motion of the spheres would cause the fire element to ignite, and this was the cause of both meteors and comets.

> The dry and warm exhalation is the outermost part of the terrestrial world which falls below the circular motion. It ... is carried round the Earth by the motion of the circular revolution. In the course of this motion it often ignites wherever it may happen to be of the right consistency, and this we maintain to be the cause of the "shooting" of scattered "stars" ... a comet is formed when the upper motion introduces into a gathering of this kind a fiery principle not of such excessive strength as to burn up much of the material quickly.[1062]

## *Physics*

Like his biology, Aristotle's physics was teleological. In *Physics*, he observed "it is plain then that nature is a cause, a cause that operates for a purpose."[1063] In Aristotelean physics all things have four causes: material, formal, efficient, and final.[1064] These are best illustrated by example.

Consider the construction of a house. The *material cause* consists of the wood, pipe, insulation, wiring, and other materials that make up the house. The *formal cause* is the form or shape that the house is to have; it is recorded as an architect's blueprint. The *efficient cause* is the activity of the contractors who construct the house. The *final cause* is the desire of the homeowner to construct a house.

In Aristotelean physics it is the final cause that is the most important, because it is the final cause that brings the other causes into existence. According to Aristotle, to understand a thing, we must understand all four causes of its existence. Aristotle maintained that natural objects including living things have a final cause. However, this doctrine did not necessarily imply the existence of a supernatural intelligence (God) as designer. Instead, the final cause of every object was embedded within that object. It is the purpose or final

*or "aether," "aither," etc.; spellings vary.

end of each thing to realize its own form and perform the function for which it is best suited.

In Aristotelean physics, there are two types of motion: natural and unnatural (or violent). A *natural* motion is the motion that an object will take when left to its own devices. An *unnatural* motion is a motion imposed by an external agent. A stone falling naturally from a cliff exhibits natural motion. However, if a human picks a stone up and throws it, the motion is unnatural.[1065]

The Earth and the heavens are governed by different physics. In the terrestrial realm, below the orbit of the moon, the natural movements of objects are linear and directed toward their natural location. Heavy elements such as earth move downward in a straight line; lighter elements such as fire move upward. "All motion is either natural or unnatural, and that motion which is unnatural to one body is natural to another, as the [linear] motions up and down are natural or unnatural to fire and earth respectively."[1066]

In the heavens, the natural direction of movement is not linear, but circular. The Sun, Moon, and planets move in circles at constant speed. "If circular motion is natural to anything, it will clearly be one of the simple and primary bodies ... thus the reasoning from all of our premises goes to make us believe that there is some body separate from those around us here, and of a higher nature in proportion as it is removed from the sublunary world."[1067]

Aristotle's characterization of the heavens having a "higher nature," meaning a moral superiority, was one of the roots of the Medieval concept of the *Chain of Being*. The *Chain of Being* "is the idea of the organic constitution of the universe as a series of links or gradations ordered in a hierarchy of creatures, from the lowest and most insignificant to the highest."[1068] In Medieval European thought, this idea would be reconciled with Christian theology that maintained man was a fallen spirit contaminated by original sin. The abode of man was the corrupt and changing Earth, while the home of God was the incorruptible and unchanging heavens. Thus Medieval thinkers obtained perfect consilience and correspondence between the physical and moral worlds.

Aristotle's law of motion was that the speed of an imposed or unnatural motion was proportional to the motive force. "If a given force move[s] a given weight a certain distance in a certain time and half the distance in half the time, half the motive power will move half the weight the same distance in the same time."[1069] It is thus implied that a constant motive force will result in a constant velocity, and that a motive force of zero would result in a velocity of zero.

Although wrong, this law is a natural induction that follows directly from observation. Consider, for example, a horse pulling a cart. It follows that vacuums do not exist in nature. A vacuum would offer zero resistance to motion and accordingly the speed of an object propelled by an outside force — no matter how small — would be infinite. "It moves through the void with a speed beyond any ratio."[1070] Infinite velocities are never observed. Therefore, a vacuum cannot exist.

An obvious problem with Aristotle's law of motion is that a thrown object continues to move once it has left the thrower's hand and the impelling force has been removed. Aristotle explained this by arguing that the impelling force was transmitted to the air that surrounded the moving object. "The original force transmits the motion by, so to speak, impressing it upon the air. That is the reason why an object set in motion by compulsion continues in motion though the mover does not follow it up."[1071]

Aristotle also maintained that the speed at which an object fell was proportional to its weight. "If a certain weight move a certain distance in a certain time, a greater weight will move the same distance in a shorter time, and the proportion which the weights bear to one another, the times too will bear to one another."[1072] In other words, in Aristotelean physics heavy objects are predicted to fall faster than light ones, for the impelling force is greater for a heavier object.

Aristotelean physics is partly consonant with observations. For example, stones obviously fall faster than feathers. But there were problems. If an unnatural motion required the presence of an impelling force, why did an arrow continue to move after it left the bowstring? Aristotle had argued that the motive force was somehow transferred to the air surrounding the arrow. The argument was made that during the arrow's flight, air was pushed out of the way. The deflected air traveled to the rear of the arrow and exerted a push. But this argument can be defeated by a simple experiment. If air were pushing at the back of an arrow, a thread tied to the back of the arrow ought to point forward — but it does not. The thread trails the arrow, indicating that the air motion at the back of the arrow is not forward, but backward.[1073]

There was another common observation that was not adequately explained by Aristotelian physics. If a constant force produces uniform motion, why does a falling body accelerate with passing time? If the weight of an object and the resistance to motion offered by air are constant, the velocity of a falling object should not change. But it does change — a falling body moves faster as it approaches the ground. Once again, an argument was found and offered by way of explanation. Aristoteleans said that as a falling body approached closer to the ground, the column of air below it became shorter and thus the air resistance diminished and the object accelerated.[1074]

It is astonishing to us today, not that people should have once thought that heavy objects fall faster than light ones, but that this view should have persisted for nearly two thousand years among the world's greatest thinkers. Herbert Butterfield (1900–1979) observed that these simple problems baffled the greatest human thinkers for centuries. "Things which would strike us as the ordinary natural way of looking at the universe, the obvious way of regarding the behavior of falling bodies, for example — defeated the greatest intellects for centuries, defeated Leonardo da Vinci and at the marginal point even Galileo, when their minds were wrestling on the very frontiers of human thought with these very problems."[1075]

If we start off with the view that Aristotelian physics was bad science, we will get nowhere in our understanding. Aristotelian physics was good science. It was progressive, simple, and derived from observation. Although of necessity it separated the terrestrial and celestial realms, it unified the observations pertaining to each.

For a more accurate physics to develop, men had to learn to think of motion in the absence of resistance. The friction and resistance that is universal on Earth is absent in the vacuum of space. Only when Galileo began rolling spheres down smooth inclined planes was the role of resistance minimized to the point that its effects on motion could be neglected and the true laws of motion discovered. In the seventeenth century, Newton showed that it was possible to derive a physics that was not only more accurate than Aristotle's, but also more unifying, because it simultaneously applied to both heaven and Earth.[1076]

In fact, well before Galileo and Newton, some writers had noted the failure of Aristotelean physics to conform with observation. Strato of Lampsacus (c. 340–270 B.C.) noted that objects accelerated as they fell, contrary to Aristotle's assertion of constant velocity.[1077] John Philoponus (c. A.D. 490–570) rejected the idea that thrown objects (such as an arrow) continued in motion because they were pushed by the air. He suggested that "some incorporeal kinetic power is imparted by the thrower to the objects thrown."[1078]

Philoponus' arguments adumbrated the exposition of the theory of impetus by Jean Burridan (c. 1295–1360) in the 14th century. The impetus theory is "the doctrine that all motion depends on the transmission of an exhaustible moving force which passes from a moving cause to a movable object and acts on it instantaneously."[1079] Some have argued that the concept of impetus may have originated with Hipparchus in the 2nd century B.C., but "the evidence is scanty."[1080] Buridan explained that an impetus "would endure forever if it were not diminished and corrupted by an opposed resistance or by something tending to an opposed motion."[1081] Philoponus also foreshadowed Galileo by noting "that two unequal weights dropped from a given height strike the ground at almost the same time."[1082]

If there were striking and obvious failures to conform to observation, why did Aristotelian physics persist for so long? The answer to this question goes to the heart of science as well as to the core of human psychology. An answer was not provided until a twentieth-century scholar, Thomas Kuhn (1922–1996), identified the existence of what he called *paradigms*.[1083] Aristotelian physics was a paradigm, an internally consistent and unified way of looking at the world. Paradigms are only abandoned when better ones become available. It is not enough to show that an existing theory has some problems or inconsistencies. A replacement must be offered, for problematic explanations are better than no explanations at all. Aristotelian physics was retained for two thousand years because it had a "highly integrated structure,"[1084] and there was nothing better available to take its place.

Because science is a progressive and continuous process of accumulating knowledge, it is the destiny of all scientific theories to eventually be replaced. In this sense, the correctness of any scientific view, theory, or conjecture, is only temporary. Butterfield warned against impugning Aristotle for incorrect physics. "The conflicts of the later medieval and early modern centuries ought not to be allowed to diminish our impression of the greatness of this ancient teacher, who provoked so much thought and controversy, and who kept the presiding position for so long."[1085]

## *Metaphysics and Theology*

Metaphysics is the branch of philosophy that deals with the nature of ultimate reality and existence; it is also the title of one of Aristotle's works. Aristotle defined metaphysics as "a science which investigates being as being."[1086] It is a field that concerns itself with all things that exist. Unlike the specialized sciences of physics, biology, and astronomy, metaphysics is concerned with study of the "truths [that] hold good for everything that is, and not for some special genus apart from others."[1087] In *Metaphysics*, Aristotle sought "to develop a first philosophy, or science of being, in complete abstraction from matter."[1088] He discussed "concepts such as being, substance, causality, potency, actuality, unity, and pure thought."[1089]

In the beginning of *Metaphysics*, Aristotle contrasted the relative superiority of practice and theory. He embraced experience as an asset to theory. "Men of experience succeed even better than those who have theory without experience."[1090] But Aristotle concluded that theoretical knowledge was superior to the purely empirical. "We think that knowledge and understanding belong to art rather than to experience.... For men of experience know that the thing is so, but do not know why, while the others know the 'why' and the cause."[1091]

Aristotle disavowed Plato's Doctrine of Forms. It was a rejection of the most fundamental part of Plato's philosophy. Aristotle's heresy was also symbolic of a fundamental epistemological division in philosophy. For two thousand years, philosophers argued as to whether knowledge was best obtained through cogitation and reason, or through experience and observation. Coleridge observed, "every man is born an Aristotelian, or a Platonist."[1092]

Aristotle conceded that objects consist of both matter and form, but maintained that form does not and cannot exist without matter. Only physical objects that are directly perceptible have a real existence. In *Metaphysics*, Aristotle explained that universals (forms) have no substance. "It is plain that no universal attribute is a substance."[1093]

Aristotle did not deny that Plato's forms or universals existed. However, he did deny that they had an existence independent of physical objects. "Beauty exists only insofar as particular beautiful things do."[1094] The corollary is that the true path to knowledge is the study of nature, not mental introspection. The forms of Plato are only mental abstractions. "A universal [form] is not a substance, but the work of the understanding: it is the intellect which gives universality to forms. The purpose of science is not to know universal realities; it is to know particular things in a universal way, by abstracting [through induction] from singular things the natures that their matters make individual."[1095]

In Aristotle's teleological universe there had to be a highest spiritual principle, God. Although the popular Olympian religion in Greece was polytheistic, monotheism was very old, even in Aristotle's time. "Before 1500 B.C. a generalized monotheism was already the possession of the Babylonian and Egyptian priesthoods ... [although] the vulgar simply accepted the polytheistic myths."[1096]

Xenophanes of Colophon (c. 570–475 B.C.), a presocratic philosopher of the Eleatic School, "denied the anthropomorphic gods altogether."[1097] He ridiculed the traditional polytheism of Homer and Hesiod, noting that they "have ascribed to the gods all things that are a shame and a disgrace among mortals, stealings and adulteries and deceivings of one another."[1098] Xenophanes claimed there is "one god, the greatest among gods and men, neither in form like unto mortals nor in thought."[1099]

A specific argument offered by Aristotle for the existence of God was that all motion is started by other motion. "Nothing is moved at random, but there must always be something present to move it."[1100] If the chain of causation is traced back indefinitely, logic compels us to admit "there is something which moves without being moved, being eternal, substance, and actuality."[1101] This first mover, Aristotle identified as God. "And life also belongs to God; for the actuality of thought is life, and God is that actuality; and God's self-dependent actuality is life most good and eternal. We say therefore that God is a living being, eternal, most good, so that life and duration continuous and eternal belong to God; for this is God."[1102] Aristotle's God was an impersonal, ambiguous, spiritual principle, not an anthropomorphic deity interested in the human condition.

## Logic and Ethics

Aristotle was the inventor of *logic*, the study of the principles of reasoning. He wrote six books (that survive) on the subject of logic: *Categories*, *On Interpretation*, *Prior Analytics*, *Posterior Analytics*, *Topics*, and *Sophistical Refutations*. Together, these six works are known as the *Organon*, meaning instrument. The *Organon* is the instrument to be employed for correct reasoning, it provides us with the "universal key to philosophic and scientific discussion."[1103]

Aristotle did not consider logic to be a subdiscipline of philosophy. Rather, he regarded it as a discipline incorporating formal rules that must be obeyed. The ancients did not study rhetoric as an end in itself; they sought to learn the art of oral persuasion so that they could apply their rhetorical skills in activities such as political debate. Similarly, Aristotle viewed logic as a tool to be employed in the sciences.

Chief among the tools to be found in the *Organon* is the syllogism. Aristotle defined the syllogism as a "discourse in which, certain things being stated, something other than what is stated follows of necessity from their being so."[1104] A syllogism is a form of deductive reasoning that consists of three parts: a major premise, a minor premise, and a conclusion. The major and minor premises are assumed to be true, and the conclusion proven to be true by deduction. The classic example of a syllogism is the major premise "all men are mortal," the minor premise "Socrates is a man," and the conclusion "Socrates therefore must be mortal."

Although Aristotle's logic emphasized deductive reasoning, he recognized that induction is necessary to reach the premises upon which the syllogism is based. "It is clear we must get to know the primary premises by induction."[1105] And induction is impossible without observation of phenomena made through and by the senses. "Induction is impossible for those who have not sense-perception. For it is sense-perception alone which is adequate for grasping the particulars: they cannot be objects of scientific knowledge, because neither can universals give us knowledge of them without induction, nor can we get it through induction without sense-perception."[1106] "It is induction that leads the learner up to universal principles, while syllogism starts from these. There are principles, then, from which syllogism starts, which are not arrived at by syllogism, and which, therefore, must be arrived at by induction."[1107]

*Ethics* is the branch of philosophy that studies moral values and choices. People apply ethics to derive a code of values that guide choices and actions in life.[1108] Although Socrates probably has the strongest claim to be the founder of ethics, Aristotle made substantial contributions.

Ethics is primarily concerned with the question, "what is the best way to live?"[1109] In Aristotle's view, to answer this question we have to have self-knowledge. Every living thing has a purpose, a function, a role in life. Fish swim, birds fly, and cows graze. What is the unique function of a human being? Aristotle argued that the distinctive characteristic of human beings is that they thought and reasoned better than any other living thing. The power of thought is our glory; therefore the best life must involve active use of the intellect. "He who exercises his reason and cultivates it seems to be both in the best state of mind and most dear to the gods."[1110]

To be happy, we must use our intellect to appreciate and choose virtue. What is virtuous? Aristotle's answer was typically Greek. He said that any virtue was the result of making the correct choice between two extremes. Courage is a virtue; if we have too little, we exhibit the vice of cowardice. But too much courage is also a vice, the vice of rashness. Thus we must seek the golden mean. "Excess and defect are characteristic of vice, and the mean of virtue."[1111]

The virtuous mean is not necessarily an arithmetic average, because the degree to which we should exhibit a virtue depends on the circumstances. There are no hard and fast rules, because every situation is different. In some cases, it may be appropriate to exhibit extreme courage. In others, the virtuous course may be to turn and walk away. The virtuous mean is "determined by a rational principle, and by that principle by which the man of practical wisdom would determine it."[1112]

Furthermore, virtue and good behavior cannot be taught theoretically so much as they have to be habituated by practice. One cannot learn to play baseball by attending lectures. Thus the virtuous human being is the person who uses reason to make correct moral choices. In making these choices, moderation is the guiding principle.

For Aristotle, happiness was not so much a state to be realized, as a state of harmonious activity. "Happiness is an activity of soul in accordance with perfect virtue."[1113] But Aristotle was also a realist. He realized that certain external factors are a necessary adjunct to happiness. Among these are wealth and beauty; it is better to be born beautiful and rich than ugly and poor. "The man who is very ugly in appearance or ill-born or solitary and childless is not very likely to be happy."[1114] Aristotle considered friendship to be the noblest aid to happiness. However, he warned that he who has too many friends has none at all. "Those who have many friends and mix intimately with them all are thought to be no one's friend."[1115]

## *Poetry and Art*

Aristotle had views on both poetry and art. In *Poetics*, he expressed the view that poetry is more revealing of universal truths than history. Historical narratives deal with facts, but poetry is concerned with fundamental truths regarding human nature. "The true difference is that one relates what has happened, the other what may happen. Poetry, therefore, is a more philosophical and a higher thing than history: for poetry tends to express the universal, history the particular."[1116]

Aristotle defined art by noting that "art imitates nature."[1117] However by "imitate," Aristotle did not mean "copy." The artist looks into nature and sees a truth or universal form. It is his job to express this truth through his art in a manner that is more direct and comprehendible than the original that is present in nature. To Aristotle, art was the distillation of eternal truths. In *Physics* he explained that art completes nature. "Art partly completes what nature cannot bring to a finish, and partly imitates her."[1118]

## *Politics*

Aristotle's political conceptions were set forth in his book, *Politics*. George Sarton expressed the view that Aristotle's *Politics* was thousands of years ahead of its time. There is no work in all of antiquity or the Middle Ages that is comparable to *Politics*. Only in modern works do we find a similar level of understanding and sophistication.[1119]

Aristotle's political ideas were quite different from Plato's. Plato was a dogmatic utopian who planned the model state down to the last detail. In *The Republic*, Plato set out to perfect human nature; through education and training he proposed to produce the ideal ruler, the philosopher-king. Aristotle's approach was more pragmatic. He stated that politics must deal with men as they are, not engage in utopian fantasies. "Political science does not make men, but takes them from nature and uses them."[1120]

Starting from first principles, Aristotle posed the question of how does a political entity, a *polis*, arise? He said that it is an inevitable product of human nature. The basic human unit is the family. Families unite in tribes, and tribes come together to form states. The most important characteristic that separates man from other animals is the use of language. Use of language makes man a social animal; he is destined by the virtue of his nature to live with other men, and it is to his benefit to do so. "Hence it is evident that the state is a creation of nature, and that man is by nature a political animal ... he who is unable to live in society, or who has no need because he is sufficient for himself, must be either a beast or a god: he is no part of a state. A social instinct is implanted in all men by nature."[1121]

The state is not only necessary, it is also a beneficial institution because it promotes justice and virtue:

> He who first founded the state was the greatest of benefactors. For man, when perfected, is the best of animals, but, when separated from law and justice, he is the worst of all; since armed injustice is the more dangerous, and he is equipped at birth with the arms of intelligence and with moral qualities which he may use for the worst ends. Wherefore, if he have not virtue, he is the most unholy and the most savage of animals, and the most full of lust and gluttony. But justice is the bond of men in states, and the administration of justice, which is the determination of what is just, is the principle of order in political society.[1122]

Aristotle had a realistic, if not cynical, appraisal of human nature. "The avarice of mankind is insatiable ... men always want more and more without end; for it is the nature of desire not to be satisfied, and most men live only for the gratification of it."[1123]

Having established both the necessity and desirability of states, Aristotle considered the question of what type of government is the best. Before deciding on the ideal form, he weighed what factors would ensure government that was both good and stable. Aristotle concluded that the secret of provident government lay in the establishment and maintenance of a large middle class:

> Thus it is manifest that the best political community is formed by citizens of the middle class, and that those states are likely to be well-administered, in which the middle class is large, and larger if possible than both the other classes, or at any rate than either singly; for the addition of the middle class turns the scale, and prevents either of the extremes from being dominant. Great then is the good fortune of a state in which the citizens have a moderate and sufficient property; for where some possess much, and the others nothing, there may arise an extreme democracy, or a pure oligarchy; or a tyranny may grow out of either extreme — either out of the most rampant democracy, or out of an oligarchy; but it is not so likely to arise out of a middle and nearly equal condition.[1124]

A large middle class promoted stability, because "where the middle class is large, there are least likely to be factions and dissensions."[1125]

Who should compose the middle class, and how large a percentage of the population should be admitted to this class? Aristotle was considerably more restrictive in his qualifications than twenty-first-century liberal democracies that embrace universal suffrage. Women and slaves were most certainly excluded. Citizens must be those individuals who own property (real estate) and have the right to both possess and carry arms. Aristotle suggested that it generally will be wise to see to it that at least half or more of the population have the full privileges of citizenship and participation in the government.

"Those who possess arms are the citizens,"[1126] and "the government should be confined to those who carry arms. As to the property qualification, no absolute rule can be laid down, but we must see what is the highest qualification sufficiently comprehensive to secure that the number of those who have the rights of citizens exceeds the number of those excluded. Even if they have no share in office, the poor, provided only that they are not outraged or deprived of their property, will be quiet enough."[1127]

If the middle class is diminished or destroyed, Aristotle warned that politics will not be conducted for the good of the people, but will degenerate into a struggle for power between the rich and the poor. "The poor and the rich quarrel with one another, and whichever side gets the better, instead of establishing a just or popular government, regards political supremacy as the prize of victory, and the one party sets up a democracy and the other an oligarchy."[1128]

The preservation of the middle class was not enough by itself. Education was also a necessity. Like Plato, Aristotle believed that the state must undertake the responsibility for education, but he was considerably less rigid in describing the particulars. He maintained that the primary role of education was to inculcate virtue in citizens. This process united the community by instilling common values. "The state, as I was saying, is a plurality, which should be united and made into a community by education."[1129]

Although Aristotle did not believe in women being admitted to citizenship, he advocated their education, implicitly recognizing their importance to the perpetuation of the state. "Women and children must be trained by education with an eye to the state, if the virtues of either of them are supposed to make any difference in the virtues of the state. And they must make a difference: for the children grow up to be citizens, and half the free persons in a state are women."[1130]

Aristotle divided governments into three possible types: monarchy, aristocracy, and constitutional republic. Each of these three types was also capable of assuming a perverted and harmful form. Monarchy could degenerate into tyranny, aristocracy into oligarchy, and a republic into a democracy.[1131]

The best form of government was "that which is administered by the best [persons.]"[1132] Aristotle was not disposed to monarchies, because "a multitude is a better judge of many things than any individual,"[1133] and "the many are more incorruptible than the few ... [because] the individual is liable to be overcome by anger or by some other passion, and then his judgment is necessarily perverted."[1134]

In fact, Aristotle regarded all forms of government as flawed. Monarchy, aristocracy, and the constitutional republic "all fall short of the most perfect form of government."[1135] The constitutional republic however is to be preferred, because of all the perverted forms, democracy is the least offensive. "Tyranny, which is the worst of governments, is necessarily the farthest removed from a well-constituted form; oligarchy is little better, for it is a long way from aristocracy, and democracy is the most tolerable of the three."[1136]

Aristotle also preferred democracies because they were more stable and more accommodating of his overarching goal of preserving and maintaining a large middle class. "Democracies are safer and more permanent than oligarchies, because they have a middle class which is more numerous and has a greater share in the government; for when there is no middle class, and the poor greatly exceed in number, troubles arise, and the state soon comes to an end."[1137]

Aristotle explained that the common downfall of all perverted and degenerate forms of government was that self-interest replaced the interest of the common good. "For tyranny is a kind of monarchy which has in view the interest of the monarch only; oligarchy has in view the interest of the wealthy; democracy, of the needy: none of them the common good of all."[1138]

In *Laws*, Plato had advocated the communal ownership of "women and children and property,"[1139] noting "the first and highest form of the state and of the government and of the law is that in which there prevails most widely the ancient saying, that 'friends have all things in common.'"[1140]

Aristotle explicitly rejected Plato's communism. He claimed that the ills which are supposed to arise from private property in fact originated in human nature. "Some one is heard denouncing the evils now existing in states, suits about contracts, convictions for perjury, flatteries of rich men and the like, which are said to arise out of the possession of private property. These evils, however, are due to a very different cause — the wickedness of human nature."[1141]

Aristotle found a multitude of potential problems in communal ownership of property. "For that which is common to the greatest number has the least care bestowed upon it. Every one thinks chiefly of his own, hardly at all of the common interest; and only when he is himself concerned as an individual."[1142] Also, "we see that there is much more quarrelling among those who have all things in common, though there are not many of them when compared with the vast numbers who have private property."[1143]

Aristotle appeared to be hard-hearted when he said that public money distributed to the poor would be wasted. "Where there are revenues the demagogues should not be allowed after their manner to distribute the surplus; the poor are always receiving and always wanting more and more, for such help is like water poured into a leaky cask."[1144]

Yet again he sought moderation, warning that the condition of the poor must not be allowed to deteriorate to such a condition that it weakened the society as a whole. Aristotle counseled that the best help was given in the form of seed money that would help individuals become productive through farming or by conducting a trade. "Yet the true friend of the people should see that they be not too poor, for extreme poverty lowers the character of the democracy; measures therefore should be taken which will give them lasting prosperity; and as this is equally the interest of all classes, the proceeds of the public revenues should be accumulated and distributed among them, if possible, in such quantities as may enable them to purchase a little farm, or, at any rate, make a beginning in trade or husbandry."[1145]

The hallmark of Aristotle's politics was moderation and reasonableness. He was ready to compromise, and he recognized that government and politics must be based upon a sensible and realistic appraisal of human nature. In a sense, his politics was scientific, in that it was apparently based in part on an analytical study of a number of constitutions from Greek city states. According to Diogenes Laërtius, Aristotle authored a lost book "containing an account of the constitutions of a hundred and fifty-eight cities."[1146]

## *Slavery and Women*

Aristotle unabashedly defended slavery. However he drew a distinction between "natural" and "unnatural slavery." Some men, he maintained, are inherently inferior and therefore born slaves. "But is there any one thus intended by nature to be a slave, and for whom such a condition is expedient and right, or rather is not all slavery a violation of nature? There is no difficulty in answering this question, on grounds both of reason and of fact. For that some should rule and others be ruled is a thing, not only necessary, but expedient; from the hour of their birth, some are marked out for subjection, others for rule."[1147]

Furthermore, not only is slavery of this type natural, it is also beneficial for both the slave and his master. The human slave is little better than a domestic animal; he acts on the basis of instinct, not rational thought, and is capable of working only with his body, not his mind.

> The lower sort are by nature slaves, and it is better for them as for all inferiors that they should be under the rule of a master. For he who can be, and therefore is another's, and he who participates in reason enough to apprehend, but not to have, reason, is a slave by nature. Whereas the lower animals cannot even apprehend reason; they obey their instincts. And indeed the use made of slaves and of tame animals is not very different; for both with their bodies minister to the needs of life.[1148]

Aristotle concluded, "it is clear, then, that some men are by nature free, and others slaves, and that for these latter slavery is both expedient and right."[1149]

But if someone becomes a slave not because of their inherent inferiority, but rather as an unfortunate circumstance such as defeat in war, then the slavery is unnatural and the slave and master are at odds. "Where the relation between them [master and slave] is natural they are friends and have a common interest, but where it rests merely on law and force the reverse is true."[1150]

In ancient Greek culture, women were generally considered to be inferior to men and accordingly enjoyed fewer privileges. In Greek mythology, Pandora, the first woman, was responsible for unleashing evil in the world. Pandora had within her "lies and crafty words and a deceitful nature."[1151] Before Pandora arrived, "the tribes of men lived on earth remote and free from ills and hard toil and heavy sicknesses ... but the woman [Pandora] took off the lid of the jar with her hands and scattered all these and her thought caused sorrow and mischief to men."[1152]

Democritus wrote, "to be ruled by a woman is the ultimate outrage for a man."[1153] The Athenian lawmaker Solon (c. 638–558 B.C.) "forbade the sale of daughters or sisters into slavery by fathers or brothers; a prohibition which shows how much females had before been looked upon as articles of property."[1154] Solon made rape punishable by the fine of "a hundred drachmas."[1155] According to Thucydides, the daily wage for a hoplite was "two drachmae a day, one for himself, and one for his servant."[1156] Thus rape was apparently treated as a serious misdemeanor rather than a felony.

In Aeschylus' [c. 525–456 B.C.] play, *Seven Against Thebes*, Eteocles inveighed against the character of women. "Insufferable creatures that ye are!... Neither in evil days nor in gladsome prosperity may I have to house with womankind. Has she the upper hand,—'tis insolence past living with; but, if seized with fear, to home and city she is still greater bane."[1157]

In the *Timaeus*, Plato said that men who "were cowards or led unrighteous lives" were reincarnated as women.[1158] In *The Republic*, Plato commented on the "general inferiority of the female sex,"[1159] noting humorously that women were only superior in "the management of pancakes and preserves."[1160] Plato concluded "all the pursuits of men are the pursuits of women also, but in all of them a woman is inferior to a man."[1161] However Plato made it clear that this conclusion was a general rule to which there were individual exceptions by noting "many women are in many things superior to many men."[1162]

Aristotle also considered women to be markedly inferior to men. "The male is by nature superior, and the female inferior; and the one rules, and the other is ruled; this principle, of necessity, extends to all mankind."[1163] In *Politics*, Aristotle made it clear that it was a woman's duty to meekly do what she was told. "The courage of a man is shown in commanding, of a women in obeying ... all classes must be deemed to have their special attributes; as the poet says of women, 'silence is a woman's glory,' but this is not equally the glory of man."[1164]

## *End of Life*

Aristotle taught at the Lyceum for twelve or thirteen years. In June of 323 B.C., Alexander died. It is likely that the death of Aristotle's nephew, Callisthenes, had alienated Aristotle from Alexander. Nevertheless, Aristotle was still resented in Athens as a representative of the occupying Macedonian force. "The suppressed anti–Macedonian sentiment burst forth in powerful tide,"[1165] and Aristotle's political enemies indicted him for impiety.

Declining Socrates' example, Aristotle chose exile. "Reflecting on the death of Socrates, and his own danger," Aristotle explained that he was leaving Athens "because he would not [let] the Athenians ... sin twice against philosophy."[1166] Accordingly, Aristotle packed his belongings and relocated to the city of Chalcis in Euboea, a province where the Macedonian influence was still in force.[1167]

Soon thereafter, Aristotle fell ill and died. Diogenes Laërtius gave his age at death as sixty-three.[1168] It seems that Aristotle's death was not sudden, but the result of a chronic illness. According to Aulus Gellius, "Aristotle the philosopher, being sixty-two years of age, became sick, and weak in body, and there remained little hope of his life."[1169] Diogenes Laërtius quoted a lengthy excerpt from Aristotle's will that is indicative of the careful and conscientious reflection of a man anticipating his imminent demise.[1170]

In *De Die Natali* (*The Natal Day*), the Roman grammarian Censorinus (fl. 3rd century A.D.), wrote that Aristotle's constitution was inherently weak. "Such was said to be the natural feebleness of his [Aristotle's] temperament and the continuance of the maladies which assailed his debilitated body ... that it is more surprising that his life was prolonged to sixty-three years."[1171]

CHAPTER 2

# Hellenistic Science (c. 300 B.C.–A.D. 200)

## *The Ptolemaic Kings*

### *PTOLEMY SOTER (367–283 B.C.)*

When Alexander the Great died in 323 B.C., his empire was divided among his generals. Egypt fell into the hands of Ptolemy I (367–283 B.C.), or Ptolemy Soter. "Ptolemy ... without any difficulty possessed himself of Egypt, and carried himself with great mildness and winning behavior towards the people."[1] Wisely deferring the immediate exploitation of Egypt for his personal gratification, Ptolemy I concentrated on consolidating his power. In his *History*, the Roman historian Tacitus (c. A.D. 56–117) said, "Ptolemy, the first Macedonian king who consolidated the power of Egypt ... [set up] in the newly-built city of Alexandria fortifications, [and] temples. ... he [Ptolemy I] continued to increase the dignity of his embassies, the number of his ships, and the weight of his gold."[2]

The introduction of Grecian technology and innovation revitalized Egyptian agriculture, industry, and trade. The navy was greatly strengthened, securing Egypt from attack by sea and protecting trade. Ptolemy I started a cultural renaissance in Alexandria that endured for hundreds of years.[3] He made efforts to bring Greek philosophers, writers, and artists to his court.[4] When he was in Greece, Ptolemy invited the philosopher Stilpo to Egypt. "Ptolemy Soter, it is said, received him [Stilpo] with great honor; and when he had made himself master of Megara, he gave him money, and invited him to sail with him to Egypt."[5] Under Ptolemy and his successors, Egypt became prosperous and art and science flourished.[6]

### *MUSEUM AND LIBRARY AT ALEXANDRIA*

The most important and enduring scientific legacy of the Ptolemaic kings was the creation of the museum and library in Alexandria. Although museums today are largely known as buildings that house collections of artifacts, in Hellenistic times a museum was a building or institution "dedicated to the Muses."[7]

The nine Greek Muses were the daughters of the gods Mnemosyne (memory) and Zeus. Each Muse was the patron of a separate art or science. In *Theogony*, Hesiod said that the Muses were begotten by the union of Zeus and Mnemosyne. "For nine nights did wise Zeus lie with her [Mnemosyne], entering her holy bed remote from the immortals. And when a year was passed and the seasons came round as the months waned, and many days were accomplished, she bare nine daughters ... Cleio and Euterpe, Thaleia, Melpomene and

Terpsichore, and Erato and Polyhymnia and Urania and Calliope, who is the chiefest of them all."[8]

The museum created in Alexandria was a research institution for the study and advancement of the arts and sciences; it was the first state institution for the promotion of science, and had a faculty of scholars who received a salary from the state. By the time of Ptolemy II, the total number of scholars in residence at the museum was about one hundred.[9] Although scholars at the museum occasionally gave lectures, the museum was not a university—there were no regular classes and no students enrolled. The physical layout of the museum was described briefly by Strabo. "The museum is a part of the palaces. It has a public walk and a place furnished with seats, and a large hall, in which the men of learning, who belong to the Museum, take their common meal."[10] The Museum existed for seven centuries. In A.D. 415, a mob of Christian monks murdered the last head of the Museum, a pagan woman named Hypatia.[11]

More important than the museum was the great library. Egypt was a providential site for a library because it was the home of papyrus manufacture. The manufacture of writing paper from the papyrus plant was described by Pliny the Elder in *Natural History*. "Before quitting Egypt, however, we must make some mention of the nature of the papyrus, seeing that all the usages of civilized life depend in such a remarkable degree upon the employment of paper."[12]

One of the philosophers invited to Egypt by Ptolemy I was Demetrius Phalerus (c. 350–280 B.C.). Demetrius had governed Athens for ten years, but the political climate changed and he was "impeached" and "condemned."[13] "On that account [Demetrius] fled to Ptolemy Soter ... [and] he remained at his court for a long time."[14] Strabo described Demetrius as "a disciple of Theophrastus the philosopher ... [who] was obliged to fly into Egypt."[15]

Demetrius is likely to have been the person who responsible for convincing Ptolemy I to found the library at Alexandria. Aristotle was a collector of books, and Demetrius was a student of Aristotle's successor at the Lyceum, Theophrastus. "Demetrius of Phalerum, being keeper of the king's library, received large grants of public money with a view to his collecting, if possible, all the books in the world; and by purchases and transcriptions he to the best of his ability carried the king's purpose into execution. Being asked once in our presence, about how many thousands of books were already collected, he replied, 'more than two hundred thousand, sire: but I will ere long make diligent search for the remainder, so that a total of half a million may be reached.'"[16]

The preceding account, found in the *Letter of Aristeas* (c. 2nd century B.C.), is probably confused as to the timeline, as it is unlikely that the library at Alexandria grew to any size until the reign of Ptolemy II. Demetrius died in prison during the reign of Ptolemy II, as he had recommended to Ptolemy I the succession of another son.[17]

Aulus Gellius (c. A.D. 125–180) stated that "a prodigious number of books were in succeeding times collected by the Ptolemies in Egypt, to the amount of near seven-hundred-thousand volumes."[18] John Tzetzes (c. A.D. 1110–1180) described the size of the Alexandrian collection as "400,000 mixed books and 90,000 single, unmixed books."[19] However, it is not clear how many of these "books" were one volume divided into multiple rolls, or how many were duplicates. In any event, all accounts place the size of the Alexandrian library at several-hundred-thousand rolls.

Strabo wrote that "Aristotle was the first person with whom we are acquainted who made a collection of books, and suggested to the kings of Egypt the formation of a library."[20]

As Aristotle died in 322 B.C., and Ptolemy I came into Egypt only after the death of Alexander in 323 B.C., it is unlikely that Aristotle was able to directly advise Ptolemy I. The influence of Aristotle was indirect, through the scholarly lineage of his students, Theophrastus and Demetrius.

Contrary to Strabo's claim that Aristotle was the first person to have a library of books, there were several significant collections that preceded both Aristotle and the Alexandrian library. "Ashurbanipal, King of Assyria in the seventh century B.C.," had a royal library, one of many in Mesopotamia.[21] In describing the zeal of Ptolemy II, Athenaeus said he sought "such a library of ancient Greek books as to exceed in that respect all those who are remarkable for such collections; such as Polycrates of Samos, and Pisistratus who was tyrant of Athens, and Euclides who was himself also an Athenian, and Nicorrates the Samian, and even the kings of Pergamos, and Euripides the poet, and Aristotle the philosopher, and Nelius his librarian."[22]

The great library at Alexandria was not the only collection of books in the ancient world, but it was the largest and most important. According to Strabo, Eumenes II (reigned 197–159 B.C.), ruler of the city of Pergamum in Asia Minor, also started a library. "Eumenes embellished the city, he ornamented the Nicephorium with a grove, enriched it with votive offering and a library, and by his care raised the city of Pergamum to its present magnificence."[23] Vitruvius implied that the institution established at Pergamum was a public library when he wrote that it was constructed "to give pleasure to the public."[24]

Pliny the Elder is the source of the well-known story that the library at Pergamum was such a successful rival to the Alexandrian institution, that one of the latter Ptolemies attempted to stop its expansion by forbidding the exportation of papyrus. "A rivalry having spring up between King Ptolemy and King Eumenes, in reference to their respective libraries, Ptolemy prohibited the export of papyrus; upon which ... parchment was invented for a similar purpose at Pergamus."[25]

Parchment is a writing material made from the skin of a sheep or goat. By the first century B.C., the collection of books at Pergamum had grown to 200,000. Plutarch related that the Roman general Mark Antony (83–30 B.C.) gave the library of Pergamum to Cleopatra as a gift. "He [Antony] had given her [Cleopatra] the library of Pergamus, containing two hundred thousand distinct volumes."[26]

## *Ptolemy Philadelphus (308–246 B.C.)*

In 285 B.C., Ptolemy I abdicated the throne to his youngest son, Ptolemy II, or Ptolemy Philadelphus. Ptolemy I died two years later, "at the age of eighty-four, leaving his mark upon the world, and affording us a striking example of great and permanent success attained by the exercise of moderate abilities, good temper, good sense, and reasonable ambition."[27]

Ptolemy II, also known as Ptolemy Philadelphus, was assessed by Athenaeus as "that most admirable of all monarchs."[28] "Philadelphus surpassed most kings in riches; and he pursued every kind of manufacturing and trading art so zealously, that he also surpassed every one in the number of his ships ... and concerning the numbers of his books, and the way in which he furnished his libraries, and the way in which he collected treasures for his Museum, why need I speak? for every one remembers all these things."[29]

Although a patron of learning, Ptolemy II was also described by Athenaeus as an egoist and hedonist. "Ptolemy the Second, king of Egypt, the most admirable of all princes, and the most learned and accomplished of men, was so beguiled and debased in his mind by his unseasonable luxury, that he actually dreamed that he should live for ever, and said that he alone had found out how to become immortal."[30] "Ptolemy Philadelphus ... had a great many mistresses, — namely, Didyma ... [and] ... Bilisticha ... and, besides them, Agathoclea, and Stratonice ... and Myrtium ... and Myrtium was one of the most notorious and common prostitutes in the city."[31] Strabo described Ptolemy II as a "lover of science," but related that "on account of bodily infirmities [he was] always in search of some new diversion and amusement."[32]

Ptolemy II was fanatical in his pursuit of book collecting. According to Galen, all ships that docked at Alexandria were searched for books. The originals were seized, copies made, and the owners given the copies in surfeit of what had been confiscated. Ptolemy II also allegedly borrowed from Athens official copies of the works of Aeschylus, Sophocles and Euripides. The works were loaned to Alexandria only after a significant deposit of 15 talents had been made to ensure the return of the manuscripts. But Ptolemy II happily kept the originals, returned copies, and forfeited his deposit. In the time of Ptolemy II, the search and demand for rare books reached such a stage of intensity that prices rose and forgeries became common.[33]

Around 250 B.C., Ptolemy II invited some Jewish scholars to come to Alexandria and translate the first five books of the Hebrew bible into Greek. The translation of the books of Moses (the *Pentateuch*), *Genesis*, *Exodus*, *Leviticus*, *Numbers*, and *Deuteronomy*, became known as the *Septuagint*. The title Septuagint derives from the Latin word for the number seventy, although in fact the number of original translators was likely to have been seventy-two. Furthermore, the translation of the entire Hebrew Bible into Greek likely took another two or three centuries to complete.[34]

The primary source that describes the origin of the Septuagint is the *Letter of Aristeas* (c. 2nd century B.C.), a document that is viewed by present day scholars as "a work typical of Jewish apologetics, aiming at self-defense and propaganda."[35] According to the *Letter of Aristeas*, the number of translators was "in all seventy-two persons."[36] The translation was reportedly "accomplished in seventy-two days, as though this coincidence had been intended."[37] The translators collaborated on the translation, "arriving at an agreement on each point by comparing each others' work."[38] However, the legend later became that the translators, working in separate groups, produced versions that were identical. This was interpreted as a divine sign that the translation was accurate and inspired. Describing the work of the Septuagint translators, Philo (20 B.C.–A.D. 50), a Hellenized Jew, claimed "every one of them employed the self-same nouns and verbs, as if some unseen prompter had suggested all their language to them."[39]

In the harbor of Alexandria, on the island of Pharos, Ptolemy I began the construction of a lighthouse that was completed by Ptolemy II. The lighthouse was one of the seven wonders of the ancient world. It had a total height in the range of 120 to 140 meters, but was destroyed by an earthquake in the 13th century A.D.[40] Strabo related that the tower was designed by Sostratus of Cnidus and constructed of white marble. "Pharos is a small oblong island, and lies quite close to the continent, forming towards it a harbor with a double entrance.... [There is] a tower upon it of the same name as the island, admirably constructed of white marble, with several stories. Sostratus of Cnidus, a friend of kings, erected it for the safety of mariners."[41]

The tower of Pharos was also described by Pliny the Elder. "There is another building, too, that is highly celebrated; the tower that was built by a king of Egypt, on the island of Pharos, at the entrance to the harbor of Alexandria. The cost of its erection was eight hundred talents, they say; and, not to omit the magnanimity that was shown by King Ptolemaeus [Ptolemy II] on this occasion, he gave permission to the architect, Sostratus of Cnidos, to inscribe his name upon the edifice itself."[42]

There is some controversy as to whether or not the tower was used as a lighthouse from its inception, or if night fires were added later. The earliest descriptions of the Pharos tower do not mention a lighthouse function.[43] However, writing in the first century A.D., Pliny said "The object of it is, by the light of its fires at night, to give warning to ships, of the neighboring shoals, and to point out to them the entrance to the harbor."[44] The Jewish historian, Josephus (c. A.D. 37–100), claimed that the fires of Pharos could be seen at a distance of "three hundred furlongs" [60 kilometers].[45]

Ptolemaic Alexandria was famous for its wealth and luxury. Diodorus Siculus (fl. 1st century B.C.) wrote "it is reported to be one of the greatest and most noble cities in the world; for beauty, rich revenues, and plentiful provision of all things for the comfortable support of man's life, far excelling all others; and far more populous than any other: for, when I was in Egypt, I was informed by them that kept the rolls of the inhabitants, that there were above three hundred thousand freemen who inhabited there."[46]

In the words of a poet who flourished third century B.C., "In the land of Egypt: wealth, power, fame, repose, gymnasium, gold, young men, philosophers, shows, the shrine of brother and sister, the good king, wine, the museum — every pleasant thing the heart of man can wish for [can be found]."[47]

It is likely that the wealth of Ptolemaic Egypt was enjoyed primarily by the upper class. Society was stratified into two tiers, "the upper stratum constituted by a European ruling race and the lower stratum constituted by the great subject mass of Egyptians."[48] Over time, racial differences were blurred by intermarriage and social stratification "became more a matter of culture and tradition than of physical race."[49]

Polybius (c. 203–120 B.C.) described Alexandrian society as composed of three races:

> A personal visit to Alexandria filled me with disgust at the state of the city. It is inhabited by three distinct races, — native Egyptians, an acute and civilized race; secondly, mercenary soldiers (for the custom of hiring and supporting men-at-arms is an ancient one), who have learnt to rule rather than obey owing to the feeble character of the kings; and a third class, consisting of native Alexandrians, who have never from the same cause become properly accustomed to civil life, but who are yet better than the second class; for though they are now a mongrel race, yet they were originally Greek, and have retained some recollection of Greek principles.[50]

It is likely that the Alexandrian population was more heterogeneous than Polybius recognized. By the time the Christian era began, "the Jews in Egypt had come to number about a million out of a total population of about seven and a half millions."[51]

Under the Ptolemaic kings, the Egyptian economy was centrally planned and run by the state. The king owned all land.[52] The monarch also claimed a monopoly on several industries including textile manufacture, goldsmithing, brewing, manufacture of papyrus, banking, salt, spices and perfumes, timber, vegetable oil, and mining and quarrying.[53] The administration of this government necessitated an "immense bureaucratic machine."[54] There was a poll tax, and taxes on personal property (including slaves), commercial transactions, and agricultural production. The Ptolemaic government derived additional revenue from professional licensing, import and export duties, and criminal fines.[55]

## *Decline of the Ptolemaic Dynasty*

The decline of Ptolemaic Egypt began with the rule of Ptolemy IV (238–205 B.C.), or Ptolemy Philopator. "No member of the [Ptolemaic] dynasty was more criminally worthless, none so careless, none so fatal to the greatness and prosperity of Egypt."[56] Like his grandfather, Ptolemy II, Philopator was a hedonist, but without his ancestor's redeeming interest in promoting intellectual pursuits.[57]

Polybius relates that "immediately after his father's death, Ptolemy Philopator put his brother Magas and his partisans to death, and took possession of the throne of Egypt ... and began conducting his reign as though it were a perpetual festival ... even his regents abroad found him entirely careless and indifferent ... [he was] absorbed in unworthy intrigues, and senseless and continuous drunkenness."[58] Ptolemy IV followed the murder of his brother by the killing of his uncle and mother.[59]

Strabo stated that "all these kings, after the third Ptolemy, were corrupted by luxury and effeminacy, and the affairs of government were very badly administered by them; but worst of all the fourth, [and] the seventh."[60] Diodorus described Ptolemy VII as "hated by all," for "his cruelty, murders, filthy lusts, and deformed body."[61] Ptolemy VII in particular was infamous for his obesity. Athenaeus said "his whole body was eaten up with fat, and with the greatness of his belly, which was so large that no one could put his arms all round it; and he wore over it a tunic which reached down to his feet."[62] Persecution by Ptolemy VIII in 145–144 B.C. encouraged many intellectuals to emigrate, and diffused Alexandrian knowledge through the Mediterranean basin.[63]

The Ptolemaic succession ended in 30 B.C. when Augustus (63 B.C.–A.D. 14), first Roman emperor, entered and occupied Egypt. Augustus had seventeen-year-old Ptolemy XIV killed, and Egypt became a Roman province.[64]

## *Euclid (c. 325–265 B.C.)*

### *Life*

Among the most important Alexandrian scholars was the mathematician Euclid (c. 325–265 B.C.). We know little to nothing of Euclid's life. In his *Commentaries*, Proclus (A.D. 411–485) simply said "he [Euclid] lived in the times of the first Ptolemy."[65] According to a thirteenth century (A.D.) Arabic writer, Euclid was "born at Tyre," but the accuracy of claim is dubious.[66] Euclid was probably educated at the Academy in Athens, at that time the leading mathematical school in the world, and had his own school at Alexandria.[67]

Two apocryphal stories are told about Euclid. Ptolemy I asked Euclid if there was not an easier way to learn mathematics. Euclid answered that the study of mathematics was as difficult for a monarch as anyone else, "there was no other royal path which led to geometry."[68] The second story illustrates the disdain Greek philosophers had for practical applications of mathematics. A student beginning his study of geometry pestered Euclid by inquiring what he could gain by studying such things. Disgusted, Euclid summoned his slave and instructed him to award the student with a token remuneration. "Give him threepence, since he must make gain out of what he learns."[69]

### *Elements of Geometry*

Although he wrote other books, Euclid's monumental contribution was his mathematics textbook, *Elements of Geometry*. The *Elements* is divided into thirteen separate chapters or books. The first six chapters deal with plane (two dimensional) geometry. Chapters seven, eight, and nine cover number theory. The tenth chapter discusses the theory of irrational numbers, and chapters eleven through thirteen are devoted to three-dimensional geometry. The *Elements* thus dealt with more than just geometry; it was essentially a compendium of all Greek work in mathematics.[70]

Most of what appears in any textbook is derivative, but Euclid's book was the first known synthesis and logical ordering of extant mathematical knowledge. The *Elements* also contains original material. In some cases, Euclid gave original proofs; in other cases he improved on earlier work.

The single outstanding characteristic of the *Elements* is the clarity with which the mathematical theorems are presented and proven. English mathematician Thomas Heath characterized it as containing a "rigor and exactitude which have so long excited the admiration of men of science."[71] The *Elements* begins with definitions, "a point is that which has no part," and "a line is breadthless length." After definitions, Euclid next introduces five postulates, essential propositions that cannot be proven and must therefore be assumed. Heath concluded that the postulates were "Euclid's own work."[72]

The most significant of the postulates is the fifth, "if a straight line falling on two straight lines make the interior angles on the same side less than two right angles, the two straight lines, if produced indefinitely, meet on that side on which are the angles less than the two right angles."[73] It follows that if two lines are parallel, the "interior angles" made by an intersecting line that is perpendicular will be "less than the two right angles" on neither side. Thus, the fifth postulate is equivalent to stating that two parallel lines never intersect.

The fifth postulate was regarded by subsequent workers for "more than twenty centuries" as not a postulate, but a hypothesis that could be proved.[74] But all attempts failed. Hence, Heath surmised, "we cannot but admire the genius of the man [Euclid] who concluded that such a hypothesis, which he found necessary to the validity of his whole system of geometry, was really indemonstrable."[75] The fifth postulate is a reflection of the nature of the space within which we live; it is the basis for what has come to be known as Euclidean Geometry.[76]

The *Elements* was widely used as a mathematics textbook until the middle of the nineteenth century. From the time the printing press was invented (c. A.D. 1450), the *Elements* went through approximately a thousand editions; it is second only to the *Bible* in its longevity and influence. Thomas Heath translated the *Elements* and assessed it as the greatest mathematical text of all time. "This wonderful book, with all its imperfections, which are indeed slight enough when account is taken of the date it appeared, is and will doubtless remain the greatest mathematical textbook of all time."[77] The fact that the *Elements* remained a primary textbook in its field for twenty-two hundred years is testimony to the intellectual ascendancy of the Greeks.

## *Strato of Lampsacus (c. 340–270 B.C.)*

### *The Mechanistic Philosopher*

Very little is known of the details of Strato's life; all of his writings have been lost. Strato's importance derives largely from his rejection of supernaturalism and his embracement of the experimental method.

Strato was the second successor of Aristotle at the Lyceum. He became head of the school after the death of Theophrastus (c. 371–287 B.C.). Strato must have also spent part of his life at Alexandria, because he was the "preceptor of Ptolemy Philadelphus [308–246 B.C.]."[78]

Strato "devoted himself to the study of every kind of philosophy, and especially of that branch of it called natural philosophy."[79] Polybius (c. 203–120 B.C.) referred to him as "the physicist."[80] Strato's entire approach was mechanistic. George Sarton characterized Strato's philosophy by noting, "even his psychology is mechanistic."[81] Marcus Tullius Cicero (106–43 B.C.) complained that Strato "abandoned the most essential part of philosophy, that which relates to virtue and morality, and dedicated himself entirely to the study of physics."[82] Cicero also indicated that Strato did not believe in a personal deity that directed the world, and was an atomist.

> You deny that any thing can be done without God: but here Strato of Lampsacus opposes you to the teeth, for he gives your God an immunity from toils.... [Strato] denies that there was any occasion for divine art in framing the world: and teaches that every thing whatsoever that exists, is the product of nature ... that all things are made out of rough and smooth, crooked and hooked corpuscles, acting in a vacuum. He takes all these to be dreams of Democritus, who conceived them.... Whereas he [Strato], upon examining the several constituent parts of the terraqueous system, is of [the] opinion, that every thing which either has happened or can happen, is the effect and result of certain properties of motion and gravitation inherent to nature.[83]

## *Experiments Demonstrating the Existence of the Vacuum*

*Pneumatica* is a text authored during the first century A.D. by Hero of Alexandria. Most of the book is devoted to a practical exposition of "various mechanical contrivances worked by air, water and steam."[84] However, the introduction is largely theoretical in nature, and believed to have been copied by Hero from an earlier work of Strato.[85] It is "a jumbled but otherwise faithful version of an extract from a book by Strato."[86]

Aristotle had argued that a vacuum does not exist and can not be found in nature. But the author of the *Pneumatica*'s Introduction not only argued for the existence of the vacuum, but also maintained that experiment and observation were superior to reasoned argument as an epistemological method.

Strato began by stating that vacuums are not continuous, but are scattered through natural substances, and that he will demonstrate their existence through an experiment. "Some assert that there is absolutely no vacuum; others that, while no continuous vacuum is exhibited in nature, it is to be found distributed in minute portions through air, water, fire, and all other substances: and this latter opinion, which we will presently demonstrate to be true from sensible phenomena, we adopt."[87]

Strato first established that air is matter by invoking the ancient demonstration of Empedocles that water will not flow into an empty container unless the air inside is allowed to escape.

> Vessels which seem to most men empty are not empty, as they suppose, but full of air.... [T]he air ... is composed of particles minute and light, and for the most part invisible. If, then, we pour water into an apparently empty vessel, air will leave the vessel proportioned in quantity to the water which enters it. This may be seen from the following experiment. Let the vessel which seems to be empty be inverted, and, being carefully kept upright, pressed down into the water; the water will not enter it even though it be entirely immersed: so that it is manifest that the air, being matter, and having itself filled all the space in the vessel, does not allow the water to enter. Now, if we bore the bottom of the vessel, the water will enter through the mouth, but the air will escape through the hole ... hence it must be assumed that the air is matter.[88]

Strato argued that the compressibility and elasticity of air was a demonstration that vacua existed between the air particles. If the vacua did not exist, there would be no place for the air particles to go when compressed. Strato also asserted that an experimental proof was sufficient to overcome any theoretical argument to the contrary.

> They, then, who assert that there is absolutely no vacuum may invent many arguments on this subject, and perhaps seem to discourse most plausibly though they offer no tangible proof. If, however, it be shown by an appeal to sensible phenomena that there is such a thing as a continuous vacuum, but artificially produced; that a vacuum exists also naturally, but scattered in minute portions; and that by compression bodies fill up these scattered vacua, those who bring forward such plausible arguments in this matter will no longer be able to make good their ground.[89]

Strato described the construction of an experimental apparatus, an air-tight metallic sphere, that could be used to demonstrate the compressibility of air, and thus the existence of vacua between air particles. By forcibly blowing into the vessel, the air inside it could be compressed. Thus there had to exist empty spaces, vacua, between the air particles.

> Provide a spherical vessel, of the thickness of metal plate so as not to be easily crushed, containing about 8 cotylae [1.9 liters]. When this has been tightly closed on every side, pierce a hole in it, and insert a siphon, or slender tube, of bronze, so as not to touch the part diametrically opposite to the point of perforation, that a passage may be left for water. The other end of the siphon must project about 3 finger's breadth [5 centimeters] above the globe, and the circumference of the aperture through which the siphon is inserted must be closed with tin applied both to the siphon and to the outer surface of the globe, so that when it is desired to breathe through the siphon no air may possibly escape from the vessel. Let us watch the result.... If any one, inserting the siphon in his mouth, shall blow into the globe, he will introduce much wind without any of the previously contained air giving way. And, this being the uniform result, it is clearly shown that a condensation takes place of the particles contained in the globe into the interspersed vacua. The condensation however is effected artificially by the forcible introduction of air. Now if, after blowing into the vessel, we bring the hand close to the mouth, and quickly cover the siphon with the finger, the air remains the whole time pent up in the globe; and on the removal of the finer the introduced air will rush out again with a loud noise, being thrust out, as we stated, by the expansion of the original air which takes places from its elasticity.... By this experiment it is completely proved that an accumulation of vacuum goes on in the globe.[90]

## *Acceleration of Falling Bodies*

Strato was also one of the first to realize that objects did not fall as Aristotle described, with a constant velocity proportional to their weight. He recognized that falling objects accelerate. Strato's original work has been lost, but Simplicius (c. A.D. 490–560), in his *Commentary on Aristotle's Physics*, quoted Strato as stating that the simple observation of a column of falling water was sufficient to prove the acceleration of a falling body. "If one observes water pouring down from a roof and falling from a considerable height, the flow at the top is seen to be continuous, but the water at the bottom falls to the ground in discontinuous parts. This would never happen unless the water traversed each successive space more swiftly."[91]

An additional proof was offered by Strato: identical objects falling from greater heights struck the ground harder, thus they had to be moving faster. "If one drops a stone or any other weight from a height of about an inch [2.54 centimeters], the impact made on the ground will not be perceptible, but if one drops the object from a height of a hundred feet [30.5 meters] or more, the impact on the ground will be a powerful one."[92]

The fragments of Strato's work clearly indicate that the experimental method was known and adopted by at least a minority of Greek philosophers.

# *Aristarchus of Samos (310–230 B.C.)*

## *The Heliocentric Hypothesis*

Aristarchus of Samos (310–230 B.C.) was an Alexandrian astronomer who first proposed the heliocentric system, the theory that the Earth revolves around the sun. Aristarchus "was a pupil of Strato of Lampsacus, a natural philosopher of originality, who succeeded Theophrastus as head of the Peripatetic school in 228 or 287 B.C. and held that position for eighteen years."[93]

The Roman architect and engineer Vitruvius (c. 90–20 B.C.) characterized Aristarchus as one of a handful of brilliant polymaths whose depth of knowledge and understanding led them beyond the practical application of the arts to pure mathematics. "As for men upon whom nature has bestowed so much ingenuity, acuteness, and memory that they are able to have a thorough knowledge of geometry, astronomy, music, and the other arts, they go beyond the functions of architects and become pure mathematicians. Hence they can readily take up positions against those arts because many are the artistic weapons with which they are armed. Such men, however, are rarely found, but there have been such at times; for example, Aristarchus of Samos."[94]

Unfortunately, whatever Aristarchus wrote concerning a Sun-centered universe has been lost. Our primary knowledge of his heliocentric hypothesis is a clear and unmistakable reference to it in Archimedes' (287–212 B.C.) book, *The Sand Reckoner.*

> Aristarchus of Samos brought out a book consisting of some hypotheses, in which the premises lead to the result that the universe is many times greater than that now so called. His hypotheses are that the fixed stars and the sun remain unmoved, that the earth revolves about the sun in the circumference of a circle, the sun lying in the middle of the orbit, and that the sphere of the fixed stars, situated about the same center as the sun, is so great that the circle in which he supposes the earth to revolve bears such a proportion to the distance of the fixed stars as the center of the sphere bears to its surface.[95]

Aristarchus must have also believed that the Earth not only revolved around the sun, but rotated daily about its axis. In a sun-centered system there is no other way to explain the occurrence of night and day.

In originating the heliocentric hypothesis, Aristarchus may have been influenced by a student of Plato's Academy. There is a sentence in the *Timaeus* that suggests the Earth moves. The "earth, our nurse going to and fro on its path round the axis which stretches right through the universe."[96] And Aristotle's successor at the Lyceum, Theophrastus, claimed "Plato in his old age repented of having given the earth the central place in the universe."[97] But Thomas Heath pointed out that a heliocentric system requires an Earth that rotates, and "the idea of the Earth rotating at all on its axis is quite inconsistent with the whole astronomical system described in the Timaeus."[98] Heath concluded that the tradition of associating Plato with heliocentrism is "due to a misunderstanding and is unworthy of credence."[99]

In his essay, *The Face Appearing Within the Orb of the Moon*, Plutarch stated that the stoic philosopher Cleanthes (331–232 B.C.) objected to Aristarchus' theory as being impious. "Cleanthes thought that the Greeks ought to have called Aristarchus the Samian into question and condemned him of blasphemy against the Gods, as shaking the very foundations of the world, because this man, endeavoring to save the appearances, supposed that the heavens remained immovable, and that the earth moved through an oblique circle, at the same time turning about its own axis."[100] There is no record of Aristarchus being prosecuted for his theory.

Aristarchus' heliocentric system was almost universally rejected by ancient astronomers. "The very few and scanty references to the system of Aristarchus by classical authors prove that it can never have been favorably received."[101] A significant reason was that in certain respects the theory was not in accordance with the observations nearly so well as the geocentric hypothesis. The primary problem was that the apparent position of the stars in the sky did not change when the Earth was on opposite sides of the sun every six months. When the position of an observer changes, the apparent position of an observed object should change. The magnitude of the change, or *parallax*, is large for a close object and is diminished as the distance of the observed object increases. Lacking telescopes, astronomers in Aristarchus' day could not detect any parallax of the stars due to the Earth's supposed revolution around the sun. They therefore rejected the heliocentric hypothesis. Aristarchus alone thought it was reasonable to suppose that the distance to the stars was so great that the parallax was small enough to escape detection. Aristarchus was correct, but his colleagues may have been drawing more reasonable conclusions based on the information available to them.

## *Estimates of the Sizes and Distances of the Sun and Moon*

In addition to being the originator of the heliocentric theory, Aristarchus calculated the relative sizes and distances to the Moon and Sun. An exposition of the calculations is given in his book, *On the Sizes and Distances of the Sun and Moon*.[102] Aristarchus was not the first of the Greek astronomers to make these sorts of calculations. In *Meteorologica*, Aristotle noted "if astronomical demonstrations are correct and the size of the Sun is greater than that of the Earth and the distance of the stars from the Earth many times greater than that of the Sun (just as the Sun is further from the Earth than the Moon), then the cone made by the rays of the Sun would terminate at no great distance from the Earth, and the shadow of the Earth (what we call night) would not reach the stars."[103]

Aristarchus reasoned that when the Moon was exactly half-full, the Moon, Sun, and Earth formed the apices of a right triangle. The Moon occupied the corner of the triangle with a right angle, and the Earth and Sun the other two corners. If the angle between the Moon and Sun could be measured, then all three angles of the triangle would be known, as the sum of the angles in any triangle must sum to one-hundred-and-eighty degrees. Aristarchus also estimated relative sizes for the Earth and Moon. The basis for his calculations was the estimation that the size of the Earth's shadow on the Moon during an eclipse was twice the size of the Moon.[104]

By comparing the ratio of the angles, Aristarchus determined the ratio of the lengths of the triangle sides that represented the distances from the Earth to the Moon and Sun. His method was geometrically elegant and correct, but sensitive to small errors. When the Moon was half-full, Aristarchus measured an angle between the Earth and Sun of 87 degrees. The true angle is 89.87 degrees. In *On the Sizes and Distances of the Sun and Moon*, Aristarchus mistakenly estimated that both the Sun and Moon subtended an angular distance of about two degrees in the sky.[105] But in *The Sand Reckoner*, Archimedes cited Aristarchus as the source of the true value of 1/720 of a circle, or one-half degree. "Aristarchus discovered that the sun appeared to be about 1/720th part of the circle of the zodiac."[106]

Aristarchus' results were as follows, with the modern numbers[107] in parentheses.

Aristarchus estimated that the distance from the Earth to the Sun was about 19 (390) times the distance from the Earth to the Moon; the distance from the Earth to the Moon was 52.5 (221) times the radius of the Moon; the radius of the Sun was 6.75 (109) times the radius of the Earth; and the radius of the Earth was 2.8 (3.8) times the radius of the Moon.[108]

There were significant errors in Aristarchus' numbers, but his method was both original and correct. He also correctly concluded that the Sun was considerably larger than both the Moon and the Earth. This result may have led him to the idea that the Earth revolves around the Sun, for it is counterintuitive to suppose that a larger body revolves around a smaller one.

## *Eratosthenes (276–195 B.C.)*

### *Librarian*

Eratosthenes (276–195 B.C.) was an Alexandrian scientist and polymath who made significant contributions in mathematics, geography, and astronomy. He was born and received his early education in the Greek city of Cyrene. Eratosthenes' higher education was completed at the Academy and Lyceum in Athens.[109] According to Strabo, "at Athens he [Eratosthenes] became a disciple of Zeno of Citium [c. 335–263 B.C.] ... [but] makes no mention of his [Zeno's] followers.... he [Eratosthenes] seems to have held a middle course between the man who devotes himself to philosophy, and the man who cannot make up his mind to dedicate himself to it."[110]

Around 246 B.C., Eratosthenes was called to Egypt by Ptolemy III.[111] He was appointed a fellow of the Museum, and undertook the responsibility of tutoring the kings' son, Ptolemy IV.[112] In 235 B.C., the chief librarian at Alexandria died, and Eratosthenes was appointed to fill the vacant position. None of Eratosthenes' writings have survived, except in fragments and quotations by later authors. Strabo, in particular, exhaustively quoted, criticized, and commended Eratosthenes' work in geography.

Among his colleagues, Eratosthenes acquired the nicknames of *beta* and *pentathlos*.[113] *Beta* is the Greek word for "second," and *pentathlos* refers to the five events of the Olympic pentathlon. The derogatory implication was that Eratosthenes was second-best in a number of fields, but primary in none. The wider implication was that even at this early point in history scientific studies were becoming specialized, and a polymath such as Eratosthenes was misunderstood and underrated.[114] But Eratosthenes must have been held in high regard by many of his contemporaries, as Archimedes' book, *The Method*, is dedicated to him.[115]

### *Contributions to Mathematics*

A prime number is a number that is not evenly divisible by any other number. Eratosthenes was the first person to devise a systematic method of finding prime numbers. The method is called "the sieve of Eratosthenes," and involves marking out all numbers that are multiples of other numbers.[116] Any number that is not crossed out is prime.

Eratosthenes' sieve works in the following way. A sequence of integers is written down; the list can be as long as desired. Starting from the number "2," all multiples of "2," such as 4, 6, 8, 10, etc., are marked out until the last number of the sequence is reached. Multiples of the next number, 3, are then struck from the list. The procedure is repeated until the starting point is the end of the sequence. The remaining numbers that have not been marked through are prime. These integers are not multiples of any whole number and therefore are not evenly divisible by any whole number.

Another mathematical problem that Eratosthenes solved was the duplication of the cube. The name of the problem refers to how to lengthen the side of a cube so as to double its volume. Expressed mathematically, the problem reduces to finding a solution to the equation $x^3 = 2[a^3]$, where $x$ is the length of a side of the larger cube, and $a$ is the (known) length of a side of the smaller cube.

There are a number of apocryphal stories relating to the origin of the problem.[117] Theon of Smyrna (c. A.D. 70–135) cited Eratosthenes' version:

> Eratosthenes in his work entitled *Platonicus* relates that when the god proclaimed to the Delians by the oracle that, if they would get rid of a plague, they should construct an altar double of the existing one, their craftsmen fell into great perplexity in their efforts to discover how a solid could be made double of a (similar) solid; they therefore went to ask Plato about it, and he replied that the oracle meant, not that the god wanted an altar of double the size, but that he wished, in setting them the task, to shame the Greeks for their neglect of mathematics and their contempt for geometry.[118]

Eratosthenes' solution was geometrical and mechanical. To commemorate Eratosthenes' work, a stone column was erected. A bronze instrument for achieving the solution was attached to the column, and the monument was inscribed with an epigram written by Eratosthenes that noted "this is the gift of Eratosthenes of Cyrene."[119]

### *Calculation of the Size of the Earth*

Eratosthenes most significant contributions were made in the areas of geography and geodesy. The most memorable of these is his estimate of the circumference of the Earth. Although Eratosthenes' original work has been lost, his measurement of the Earth's circumference was described by the Greek astronomer and stoic philosopher, Cleomedes [c. A.D. 200], in his book *Caelestia* (*The Heavens*.)[120]

Eratosthenes was not the first Greek to estimate the size of the Earth. In *De Caelo* (*On the Heavens*), Aristotle noted that "mathematicians who try to calculate the size of the Earth's circumference arrive at the figure of 400,000 stades."[121] In *The Sand Reckoner*, Archimedes wrote "some have tried ... to prove that the said perimeter [of the Earth] is about 300,000 stadia."[122] A chronic problem in evaluating these estimates is that we do not know precisely how long the Greek unit of length, the stadium, or stade, was.

Eratosthenes' method was geometrical, elegant, simple, and entirely correct. He judged the city of Syene [Aswan] to be located on the latitudinal line known as the Tropic of Cancer. This is the latitude at which the Sun is directly overhead on the summer solstice. At the Tropic of Cancer on the date of the summer solstice a perpendicular stick casts no shadow.

However, on the summer solstice, a perpendicular stick located at the city of Alexandria cast a small but measurable angle. Eratosthenes estimated Alexandria to be 5,000 stadia north of Syene. By assuming (correctly) that the Sun was distant enough from the Earth that its rays were essentially parallel to each other, Eratosthenes was able to deduce geometrically that the angle made by the shadow at Alexandria was identical to the angle subtended at the center of the Earth between Syene and Alexandria. The angle made by the shadow at Alexandria was one-fiftieth of a circle of 360 degrees, or 7.2 degrees. Accordingly, the distance between Syene and Alexandria had to be one-fiftieth of the Earth's circumference. It followed immediately that the circumference of the Earth was 250,000 stadia.

Other authorities claimed that Eratosthenes later corrected his estimate to 252,000 stadia.[123] Strabo stated "the whole circle of the equator according to Eratosthenes contains 252,000 stadia."[124] The numbers used by Eratosthenes (5000 stadia and 1/50 of a circle), are clearly rounded approximations, thus "he felt himself at liberty to add 2,000 stadia to the 250,000 ... in order to have a number that would be readily divisible into sixty parts, or into degrees of 360 to a great circle."[125]

Unfortunately, the Greeks were considerably ambiguous in their definition of units. Scholars have argued for centuries precisely how long Eratosthenes' stade was. If, as a recent authority argues, the stade employed by Eratosthenes was 184.98 meters in length,[126] then his estimate of the circumference of the Earth (252,000 stadia) was 46,615 kilometers, or about 16 percent higher than the modern value of 40,030 kilometers. Employing an alternative value for the stade of 157.5 meters[127] results in an estimate of 39,690 meters (for 252,000 stadia), which is within one percent of the modern value.

However the true angle subtended at the center of the Earth between Syene and Alexandria was not precisely 1/50 of a circle (7.2 degrees) as Eratosthenes estimated, but 7.08 degrees.[128] Furthermore, Syene was not precisely on the Tropic of Cancer at 23.73 degrees (in 3rd century B.C.) latitude north, but at 24.07 degrees north.[129] Nevertheless, Eratosthenes' method was elegant and his result accurate for the age.

### *Geographer*

Eratosthenes' most significant contributions were made in the area of geography. He wrote a book, *Geography*, published in three parts. Eratosthenes' "work was the first scientific attempt to put geographical studies on a sound mathematical basis, and its author may be said to have been the founder of mathematical geography."[130]

Eratosthenes' accuracy was limited by the scarcity of precise measurements of latitude and longitude. The Greeks had no way to determine longitude with any precision, because they lacked a portable and accurate clock. "Eratosthenes placed Carthage and the Sicilian Strait on the same meridian with Rome, though the one lies more than two degrees to the west, the other more than three degrees to the east of the city."[131] The Greeks knew how to accurately determine latitude using a gnomon, but in practice few reliable measurements were made.[132]

Strabo said that Eratosthenes published a map of the known world. "In the Third Book of his *Geography* Eratosthenes furnishes us with a chart of the habitable earth."[133] To "the Greek geographers ... the known or habitable world was conceived as a definite and limited portion of the earth's surface, situated wholly within the northern hemisphere, and comprised within about a third of the extent of that section."[134] Eratosthenes estimated the north-south extent of the habitable world to be 38,000 stadia, and its breadth from east to west to be 77,800 stadia.[135] Thus Eratosthenes calculated the breadth (east-west) of the known world to be 31 percent of the Earth's circumference (252,000 stadia).

Strabo did not concern himself with what he had no way of knowing. "What is beyond our habitable earth it is not however the business of the geographer to consider."[136] But he did report that Eratosthenes speculated that it was possible to reach India by sailing west from Spain. "[Eratosthenes says] if the extent of the Atlantic Ocean were not an obstacle, we might easily pass by sea from Iberia to India, still keeping in the same parallel."[137]

But Strabo was even more prescient in anticipating the presence of the American continent. He pointed out that the voyage to India would likely be impeded by the presence of an unknown land mass. "[Eratosthenes'] reasoning is incorrect ... [because] it is quite possible that in the temperate zone there may be two or even more habitable earths, especially near the circle of latitude which is drawn through Athens and the Atlantic Ocean."[138]

In 1592, Christopher Columbus was encouraged to make the voyage westward because he relied upon a smaller estimate of the Earth's diameter, 180,000 stadia, made by Posidonius (c. 135–51 B.C.).[139] Posidonius noted that Alexandria and the island of Rhodes were on the same meridian of longitude. Like Eratosthenes, Posidonius' method was geometrical. He deduced that the diameter of the Earth could be estimated from the fact that when the star Canopus was 7.5 degrees above the horizon at Alexandria, it was flush on the horizon at Rhodes. "When we reach Alexandria by sailing the 5,000 stadia from Rhodes, this star [Canopus], when precisely at the meridian, is determined as being elevated above the horizon ¼ of a zodiacal sign, that is 1/48 of the meridian [1/48 of 360 degrees] through Rhodes and Alexandria."[140]

Estimating the distance between Rhodes and Alexandria to be 5,000 stadia, Posidonius calculated a terrestrial diameter of 240,000 stadia, a value nearly identical to Eratosthenes' 252,000 stadia. But Posidonius' estimate was derived from faulty data. The true distance between Rhodes and Alexandria was not 5,000 stadia, but 3,750. And when Canopus was on the horizon at Rhodes, its elevation at Alexandria was not 7.5 degrees, but 5.25 degrees. Posidonius became aware that he had overestimated the distance between Rhodes and Alexandria, and subsequently reduced his estimate of the Earth's diameter to 180,000 stadia. But he was not aware of the error in his estimate of Canopus' elevation. The two errors in his original estimate had tended to cancel each other. So when one was corrected, the new estimate was less accurate than the original.[141]

Anticipating the discipline of cultural geography, Eratosthenes preached cultural toleration. "Eratosthenes blames the system of those who would divide all mankind into Greeks and Barbarians.... [H]e suggests, as a better course, to distinguish them according to their virtues and their vices, 'since amongst the Greeks there are many worthless characters, and many highly civilized are to be found amongst the Barbarians; witness the Indians and Ariani, or still better the Romans and Carthaginians, whose political system is so beautifully perfect.'"[142]

## *Archimedes (287–212 B.C.) and the Punic Wars (264–146 B.C.)*

### *Mathematical Works*

Archimedes (287–212 B.C.) was a native of Syracuse in Sicily. Syracuse was founded by Greek settlers from Corinth in 734 B.C.[143] According to Thucydides, "Archias, one of the Heracleidae*, came from Corinth and founded Syracuse."[144] It is probable that Archimedes "spent a considerable time at Alexandria, where it may be inferred that he studied with the successors of Euclid."[145] Archimedes father, Pheidas, was an astronomer, and Archimedes was either an intimate friend, or relative of Hieron II, the ruler of Syracuse.[146]

*legendary descendants of Hercules

Archimedes' most significant accomplishments were in the field of mathematics. He wrote a number of treatises, many of which survive. These include *On the Sphere and Cylinder*, *Measurement of a Circle*, *On Spirals*, *The Sand-Reckoner*, *On Floating Bodies*, *The Method*, and others.[147] In *Measurement of a Circle*, Archimedes showed that the ratio of a circle's circumference to its diameter, $\pi$, had a value "less than 3 1/7 but greater than 3 10/71."[148]

Archimedes also showed that the surface area of a sphere was exactly four times as large as the area of a circle with the same diameter. In Book 1 of *On the Sphere and Cylinder*, he wrote "Certain theorems not hitherto demonstrated have occurred to me, and I have worked out the proofs of them. They are these: first, that the surface of any sphere is four times its greatest circle."[149]

Archimedes sought to calculate the area circumscribed by curves. In his efforts, he came close to discovering the integral calculus invented by Isaac Newton and Gottfried Wilhelm Leibnitz in the late 17th century A.D.[150] "He performed in fact what is equivalent to integration in finding the area of a parabolic segment, and of a spiral, [and] the surface and volume of a sphere."[151]

In his frequent correspondence to mathematicians in Alexandria, Archimedes was in the habit of sending theoretical results without proof. Apparently, some of his colleagues found it convenient to claim that they had found the same result independently. Archimedes laid a trap for them by sending two theorems that were false. In the introduction to his book *On Spirals*, he sprang the jaws of the trap. First, he announced that he would give the proofs for previous results. He then noted that two of the results he had announced earlier were "impossible," thus catching the thieves in the act of stealing his work. "I wish now to put them in review one by one, particularly as it happens that there are two included among them which are impossible of realization. This may serve as a warning how those who claim to discover everything but produce no proofs of the same may be confuted as having actually pretended to discover the impossible."[152]

Archimedes was the first person to conceive of and find a way of expressing large numbers. His book, *The Sand Reckoner*, was addressed to Gelon, eldest son of King Hieron II. Archimedes began by stating that some people thought the number of sand grains in the world was a number so large as to be infinite. He then claimed that he would show that there existed a number that was not only larger than all the sand grains in the world, but larger than all the sand grains necessary to fill up a volume equal to the size of the entire universe.

> There are some, king Gelon, who think that the number of the sand is infinite in multitude; and I mean by the sand not only that which exists about Syracuse and the rest of Sicily but also that which is found in every region whether inhabited or uninhabited. Again there are some who, without regarding it as infinite, yet think that no number has been named which is great enough to exceed its magnitude. And it is clear that they who hold this view, if they imagined a mass made up of sand in other respects as large as the mass of the Earth, including in it all the seas and the hollows of the Earth filled up to a height equal to that of the highest of the mountains, would be many times further still from recognizing that any number could be expressed which exceeded the multitude of the sand so taken. But I will try to show you by means of geometrical proofs, which you will be able to follow, that, of the numbers named by me and given in the work which I sent to Zeuxippus, some exceed not only the number of the mass of sand equal in magnitude to the Earth filled up in the way described, but also that of the mass equal in magnitude to the universe.[153]

Archimedes estimated that "the number of grains of sand which would be contained in a sphere equal to the sphere of the fixed stars" was $10^{63}$, or one followed by sixty-three zeroes.[154]

He then showed that larger numbers were possible. No one would attempt to define a larger number until the term *googol* ($10^{100}$) was coined by the nine-year-old nephew of the American mathematician Edward Kasner (1878–1955) in the year 1940.[155]

Archimedes was also an ingenious mechanical inventor. In his book, *The Method,* he described how he would sometimes first attempt to solve a mathematical problem by mechanical means. Only after thereby obtaining some intuitive insight into the nature of the problem would he attempt to find a mathematical proof. "Certain things first became clear to me by a mechanical method, although they had to be demonstrated by geometry afterwards because their investigation by the said method did not furnish an actual demonstration."[156]

In assessing Archimedes mathematical works, Thomas Heath concluded, "The treatises are, without exception, monuments of mathematical exposition; the gradual revelation of the plan of attack, the masterly ordering of the propositions, the stern elimination of everything not immediately relevant to the purpose, the finish of the whole, are so impressive in their perfection as to create a feeling akin to awe in the mind of the reader."[157]

## *Hydrostatics and the Adulterated Crown*

Archimedes "invented the whole science" of hydrostatics in his book *On Floating Bodies.*[158] "*Archimedes' Principle*" is that a solid body suspended or floating in a fluid has its weight reduced by an amount equal to the weight of the displaced liquid. In Archimedes' words, "the solid will be driven upwards by a force equal to the difference between its weight and the weight of the fluid displaced."[159]

The story by which Archimedes became interested in hydrostatics is well known. The king of Syracuse, Hieron II, came to Archimedes with a problem. He had ordered a crown made of gold. When the crown was delivered, it had the same weight as the gold supplied to the artisan. The king suspected the craftsman had dishonestly replaced some of the gold with less-costly silver, but was unable to prove any wrongdoing. He turned the problem over to Archimedes.

Archimedes mulled the problem over in his mind but was baffled. He knew that an equal volume of silver did not weigh as much as a volume of gold, but lacked a technique for precisely measuring the volume of the crown. The solution was to measure the volume of water displaced by the crown when it was immersed. According to the Roman architect, Marcus Vitruvius Pollio (c. 90–20 B.C.), Archimedes hit upon the solution while relaxing in his bath:

> Hiero, after gaining the royal power in Syracuse, resolved, as a consequence of his successful exploits, to place in a certain temple a golden crown which he had vowed to the immortal gods. He contracted for its making at a fixed price, and weighed out a precise amount of gold to the contractor. At the appointed time the latter delivered to the king's satisfaction an exquisitely finished piece of handiwork, and it appeared that in weight the crown corresponded precisely to what the gold had weighed. But afterwards a charge was made that gold had been abstracted and an equivalent weight of silver had been added in the manufacture of the crown. Hiero, thinking it an outrage that he had been tricked, and yet not knowing how to detect the theft, requested Archimedes to consider the matter. The latter, while the case was still on his mind, happened to go to the bath, and on getting into a tub observed that the more his body sank into it the more water ran out over the tub. As this pointed out the way to explain the case in question, without a moment's delay, and transported with joy, he jumped out of the tub and rushed home naked, crying with a loud voice that he had found what he was seeking; for as he ran he shouted repeatedly in Greek, Eureka! Eureka![160]

What Archimedes had realized was that he could precisely measure the volume of the crown by measuring the volume of water it displaced when submerged. He did so, and also measured the volume of a mass of both gold and silver that had the same weight as the crown. Archimedes discovered that the crown had indeed been diluted with silver, and was able to calculate the precise amount of silver that had been added.

### *Archimedes' Orrery and the Antikythera Mechanism*

In *On the Republic*, the Roman author Marcus Tullius Cicero (106–43 B.C.) stated that he had seen an *orrery*, or mechanical model of the solar system, constructed by Archimedes and brought to Rome by the general Marcellus (268–208 B.C.). According to Cicero, Archimedes' orrery had been "deposited by the same Marcellus in the Temple of Virtue at Rome."[161] Cicero maintained that Archimedes was not the original inventor, but "the first model had been originally made by Thales of Miletus," with improvements by Eudoxus.[162] Archimedes' orrery was capable of reproducing "the motions of the sun and moon, and the five planets," and could replicate the motions of each body through a single motion, such as the turning of a crank.[163] "The invention of Archimedes was admirable, because he had calculated how a single revolution should maintain unequal and diversified progressions in dissimilar motions."[164] Cicero implied that Archimedes' machine was capable of predicting eclipses. "The progress of the sun was marked as in the heavens, and that the moon touched the point where she is obscured by the earth's shadow at the instant the sun appears opposite."[165]

Archimedes' orrery may have been the forerunner of the Antikythera mechanism. The Antikythera mechanism is a mechanical device recovered in fragments from a Roman shipwreck in 1901.[166] The name derives from the fact that the shipwreck was located near the island of Antikythera, northwest of Crete. The date of shipwreck is estimated to be 80–50 B.C.[167] The mechanism was probably constructed between 150–100 B.C.[168] Analysis of the corroded bronze fragments, aided by x-ray imaging, indicates that the device was a sophisticated astronomical calculator. "It calculated and displayed celestial information, particularly cycles such as the phases of the moon and a luni-solar calendar ... the Antikythera mechanism is technically more complex than any known device for at least a millennium afterwards."[169] The specific functions of the calculator remain a subject of controversy, because of the degradation of the recovered pieces. But recent analysis has found that calendar names of months on the mechanism "are unexpectedly of Corinthian origin ... suggesting a heritage going back to Archimedes."[170]

### *The Archimedean Screw*

Archimedes also invented a device that came to be known as the *Archimedean screw*. An Archimedean Screw is a tube that encloses a helix. The upper end of the helix contains a handle for rotation. The lower end of the helix extends a half-turn from the enclosing tube. When the lower end of the Archimedean screw is placed in water and the helix rotated, water is transported up the tube. The Archimedean screw was used extensively by the Romans for dewatering mines.

In his *Historical Library*, Diodorus Siculus (fl. 1st century B.C.) said that Archimedes invented the device during his time in Egypt, presumably at Alexandria. "They pump out those floods of water with those instruments called Egyptian pumps, invented by Archimedes the Syracusan, when he was in Egypt."[171] In *The Deipnosophists*, Athenaeus described an enormous ship constructed by Hieron II, whose hold was dewatered by means of a device constructed by Archimedes, most likely an Archimedean Screw. "The hold, although of a most enormous depth, was pumped out by one man, by means of a pulley, by an engine which was the contrivance of Archimedes."[172]

## *Levers and Pulleys*

Although levers have been used from time immemorial, Archimedes was the first person to write an exposition clearly developing the mathematical principles inherent in their operation. His book, *Levers*, is lost.[173] But Archimedes is said to have stated "give me a place to stand and with a lever I will move the whole world."[174]

King Hieron II was skeptical, and asked for a demonstration. Archimedes obliged. He arranged for a large ship in the Syracuse harbor to be loaded with both cargo and crew. He then attached a rope to the ship and arranged a series of pulleys that gave him great leverage. Slowly, pulling hand-over-hand, Archimedes single-handedly drew the ship to shore. The story was told by Plutarch (c. A.D. 45–125) in his *Life of Marcellus*.

> Archimedes, however, in writing to King Hieron, whose friend and near relation he was, had stated that given the force, any given weight might be moved, and even boasted, we are told, relying on the strength of demonstration, that if there were another Earth, by going into it he could remove this. Hieron being struck with amazement at this, and entreating him to make good this problem by actual experiment, and show some great weight moved by a small engine, he fixed accordingly upon a ship of burden out of the king's arsenal, which could not be drawn out of the dock without great labor and many men; and, loading her with many passengers and a full freight, sitting himself the while far off, with no great endeavor, but only holding the head of the pulley in his hand and drawing the cords by degrees, he drew the ship in a straight line, as smoothly and evenly as if she had been in the sea.[175]

Hieron was impressed by Archimedes' engineering skill. Accordingly, he placed Archimedes in charge of Syracuse's defenses, charging him with constructing mechanical devices to defend the city. "The king, astonished at this, and convinced of the power of the art, prevailed upon Archimedes to make him engines accommodated to all the purposes, offensive and defensive, of a siege. These the king himself never made use of, because he spent almost all his life in a profound quiet and the highest affluence. But the apparatus was, in most opportune time, ready at hand for the Syracusans, and with it also the engineer himself."[176]

## *Hannibal Crosses the Alps*

The world was changing, and Archimedes' defensive installations would soon be called upon. The power and influence of the Greeks was long past its prime. The two powers in the Mediterranean world were Rome and Carthage.

Carthage was founded on the northern shore of Africa by Phoenicians from the city of Tyre in the 9th century B.C.[177] The city "was placed on a fertile rising ground ... covered with groves of olive and orange trees, falling off in a gentle slope towards the plain ... the place proved singularly favorable for agriculture and commerce."[178] "The Phoenicians directed all the resources of courage, acuteness, and enthusiasm to the full development of commerce and its attendant arts of navigation, manufacturing, and colonization."[179] As the centuries passed, the Carthaginians lost their Phoenician identity and began to think of themselves purely as Carthaginians. Carthage expanded its sphere of control and influence, founding its own colonies in Africa, and the Italian islands of Sicily and Sardinia.

The Carthaginian rise to power was described by Edward Creasy:

> Carthage was originally neither the most ancient nor the most powerful of the numerous colonies which the Phoenicians planted on the coast of Northern Africa. But her advantageous position, the excellence of her constitution (of which, though ill-informed as to its details, we know that it commanded the admiration of Aristotle), and the commercial and political energy of her citizens, gave her the ascendancy.... The Carthaginians ... had the monopoly of all the commerce of the world that was carried on beyond the Straits of Gibraltar ... the boldness and skill of the seamen of Carthage, and the enterprise of her merchants, may be paralleled with any achievements that the history of modern navigation and commerce can supply.[180]

For three hundred years, the Carthaginians fought the Greeks for control of Sicily. When Greek power began to wane and Rome began to rise, it was inevitable that Carthage and Rome would come into conflict.[181] From 264–241 B.C., Rome and Carthage fought the first of three Punic Wars, the term *Punic* deriving from the Roman word for Carthage.

The Romans were victorious in the First Punic War, and the Carthaginians had to surrender Sicily to Rome. The proposed terms of surrender were recorded in *The Histories* of Polybius (c. 203–120 B.C.). "The Carthaginians shall evacuate the whole of Sicily: they shall not make war upon Hiero, nor bear arms against the Syracusans or their allies. The Carthaginians shall give up to the Romans all prisoners without ransom. The Carthaginians shall pay to the Romans in twenty years 2,200 Euboic talents of silver."[182] When the proposed terms of surrender reached Rome, the Romans increased their severity. "They reduced the time allowed for the payment of indemnity by one half; they added a thousand talents to the sum demanded; and extended the evacuation of Sicily to all islands lying between Sicily and Italy."[183]

Polybius described the First Punic War as "the longest, most continuous, and most severely contested war known to us in history.... [I]n the course of the war, counting what were destroyed by shipwreck, the Romans lost seven hundred quinqueremes, the Carthaginians five hundred.... [N]ever in the whole history of the world have such enormous forces contended for mastery of the sea."[184]

Roman victory in the First Punic War was ascribed by Polybius to superior virtue. "It was not by mere chance or without knowing what they were doing that the Romans struck their bold stroke for universal supremacy and dominion, and justified their boldness by its success. No: it was the natural result of discipline gained in the stern school of difficulty and danger."[185]

As the Carthaginians were by trade and temperament merchants, they tended to rely upon mercenaries, reasoning "that money would indeed be worth nothing ... if they themselves should be compelled to serve as soldiers."[186] At the end of the First Punic War, Carthage was bankrupt and unable to pay its mercenary troops. There ensued a rebellion and "fearful [civil] war," during which "Carthage was brought to the brink of destruction."[187] After a period lasting three years and four months, "the Carthaginians succeeded in suppressing the insurrection and destroying the rebels."[188]

Having lost Sicily to the Romans, the Carthaginians hit upon the strategy of conquering Spain. The southern coast of Spain, in particular, was rich in natural resources including "very rich silver mines."[189] In 238 B.C., "the Carthaginian government collected an army and dispatched it under the command of Hamilcar [Barca, c. 275–228 B.C.] to Iberia.... [W]ith his son Hannibal (247–183 B.C.), then nine years old, [Hamilcar Barca] set about recovering the Carthaginian possessions in Iberia."[190]

Hannibal later related that before departing for Spain, he swore eternal hostility toward Rome.

> When my father was about to go on his Iberian expedition I was nine years old: and as he was offering the sacrifice to Zeus I stood near the altar. The sacrifice successfully performed, my father poured the libation and went through the usual ritual. He then bade all the other worshippers stand a little back, and calling me to join him asked me affectionately whether I wished to go with him on his expedition. Upon my eagerly assenting, and begging with boyish enthusiasm to be allowed to go, he took me by the right hand and led me up to the altar, and bade me lay my hand upon the victim and swear that I would never be friends with Rome ... [and] there is nothing in my power that I would not do against her.[191]

After the death of his father, Hannibal was elected by the Carthaginian troops in Spain to be their commander. Hannibal would turn out to be one of the greatest military geniuses the world has ever seen. The Roman historian Livy (Titus Livius, 59 B.C.–A.D. 17) described Hannibal as a person "made up of virtues and vices."[192]

> His [Hannibal's] fearlessness in encountering dangers, and his prudence when in the midst of them were extreme. His body could not be exhausted, nor his mind subdued, by any toil. He could alike endure either heat or cold. The quantity of his food and drink was determined by the wants of nature, and not by pleasure.... Many have seen him wrapped in a military cloak, lying on the ground amid the watches and outposts of the soldiers.... He was at once by far the first of the cavalry and infantry; and, foremost to advance to the charge, was last to leave the engagement.[193]

But, Livy claimed, "Excessive vices counterbalanced these high virtues of the hero; inhuman cruelty, more than Punic perfidy, no truth, no reverence for things sacred, no fear of the gods, no respect for oaths, [and] no sense of religion."[194]

The Second Punic War began in 218 B.C., instigated by Hannibal's aggressive tactics in Spain. Livy called it "the most memorable of all wars that were ever waged.... [T]he hatred with which they [Rome and Carthage] fought also was almost greater than their resources; the Romans being indignant that the conquered aggressively took up arms against their victors, the Carthaginians, because they considered that in their subjection it had been lorded over them with haughtiness and avarice."[195]

Hannibal designed a bold plan to attack Rome by heading north and east, first crossing the Pyrenees and then the Alps. The crossing of the Alps was difficult. "Hannibal ... immediately found himself involved in the most serious dangers.... [T]he Carthaginians sustained severe losses, not so much at the hands of the enemy, as from the dangerous nature of the ground, which proved especially fatal to the horses and beasts of burden. For as the ascent was not only narrow and rough, but flanked also with precipices, at every movement which tended to throw the line into disorder, large numbers of the beasts of burden were hurled down the precipices with their loads on their backs."[196]

The natives resisted the Carthaginian advance. "The Gauls ... by rolling down boulders, or throwing stones, reduced the troops to a state of the utmost confusion and danger."[197] The passage took fifteen days to accomplish, and Hannibal lost nearly half of his entire force.[198] Polybius states that "when Hannibal crossed the Rhone [before ascending the Alps], he had thirty-eight thousand infantry, and more than eight thousand cavalry."[199] But the surviving army consisted of "twelve thousand Libyans and eight thousand Iberians, and not more than six thousand cavalry in all."[200]

After successfully crossing the Alps, Hannibal rested his troops for two and a half months. The Romans and Carthaginians first clashed near the river Trebia, in December of 218 B.C.[201] Hannibal was content to prepare his position and wait for the Romans. He decided not to cross the Trebia, because the water in December was icy cold.

> [But] the Romans were imprudent enough, during the night, to wade through the river; which had risen so much that they were up to their chins in water; in addition to this, the wind blew

the snow into their faces, so that they were almost frozen when they arrived on the right bank. Hannibal advanced to meet them, and the Romans, although in reality they were already defeated by the elements, yet fought as brave soldiers; they formed indeed an army of 30,000 men against 20,000 enemies, but the Carthaginian cavalry quickly drove back the Romans, whose infantry was in fact worn out: they did all they could, but they were a militia against an army of veterans, and were opposed by the elements. When all had got through the river, the Carthaginians, who had been lying in ambush, rushed forth and attacked their flanks. The loss of the Romans was very great: many were thrown back into the river and perished.[202]

The next battle was fought the following summer, in June of 217 B.C. The Romans had dispatched an army under the command of Gaius Flaminius. Hannibal assessed Flaminius as "a mere mob-orator and demagogue, with no ability for the actual conduct of military affairs."[203] He therefore reasoned that he could lure Flaminius into a trap by provoking him. Hannibal accordingly "burned and wasted the country with a view of rousing the wrath of the enemy and tempting him to come out."[204] Having incited Flaminius to pursue him, Hannibal laid a trap near Lake Trasimene.

As Flaminius closed on Hannibal's forces, Hannibal attacked, and the Romans found themselves ambushed and besieged on every side. According to Polybius, "Flaminius was taken completely by surprise ... [he] was attacked and killed ... as many as fifteen thousand Romans fell in the valley."[205]

Several thousand Roman troops were trapped in the lake itself, weighed down by the weight of their armor. "When the [Carthaginian] cavalry rode in after them, and certain death stared them in the face, they raised their hands and begged for quarter, offering to surrender, and using every imaginary appeal for mercy; but were finally dispatched by the enemy, or, in some cases, begged the favor of the fatal blow from their friends, or inflicted it on themselves."[206] Livy stated that "fifteen thousand Romans were slain in the battle," with the number of Carthaginian casualties at "one-thousand five-hundred."[207]

From the moment that it became apparent that Hannibal had crossed the Alps and was in Italy, "the Romans were very much given to superstition."[208] Superstitious omens, or prodigies, reported by Livy included "shores [that] had blazed with frequent fires ... shields [that] had sweated blood ... red hot stones [that] had fallen from the heavens ... the sun fighting with the moon ... two moons [that] rose in the daytime ... waters ... [that] had flowed mixed with blood ... [and] that the statue of Mars at Rome, on the Appian way, had sweated at the sight of images of wolves."[209]

At one point, the Romans even revived human sacrifice, putting "alive underground a pair of Greeks, one male, the other female; and likewise two Gauls, one of each sex."[210] According to Plutarch, "So much were all things at Rome made to depend upon religion; they would not allow any contempt of the omens and the ancient rites, even though attended with the highest success, thinking it to be of more importance to the public safety that the magistrates should reverence the gods, than that they should overcome their enemies."[211]

The Carthaginians had won two significant victories, but "Hannibal rejected the idea of approaching Rome for the present."[212] "In those times Rome was an extremely strong fortress, protected by steep rocks, walls, banks, and moats.... Hannibal would ... have required very large [siege] engines, of which he had none ... and he had, besides, several reasons for not undertaking anything of the sort. His army was suffering from diseases ... and required rest for the recovery of its health."[213] Polybius specifically listed scurvy as one of the afflictions of Hannibal's army.[214]

Hannibal was merciless in plundering and ravaging the Italian provinces. According to Polybius, he "traversed the country plundering it without resistance ... the order was given, unusual in the storming of cities, to kill all adults who came in their way: an order which Hannibal was prompted to give now by his deep-seated hatred of Rome."[215]

After two stinging defeats, the Romans were content to wage a war of attrition. They recognized that the Carthaginian troops were more seasoned, and counted upon eventually wearing them down by drawing upon superior resources. Polybius explained the strategy of the Roman commander, Fabius.

> [Fabius recognized that] the forces of his opponents had been trained from their earliest youth without intermission in war; had a general who had grown up with them and from childhood had been instructed in the arts of the camp; had won many battles in Iberia, and twice running had beaten the Romans and their allies: and, what was more than all, had thoroughly made up their minds that their one hope of safety was in victory. In every respect the circumstances of the Roman army were the exact opposite of these; and therefore, their manifest inferiority making it impossible for Fabius to offer the enemy battle, he fell back upon those resources in which the Romans had the advantage of the enemy; clung to them; and conducted the war by their means: and they were — an inexhaustible supply of provisions and of men.[216]

In August of 216 B.C., the Romans and Carthaginians fought a third battle at Cannae. Polybius estimated the size of the respective forces as eighty thousand Roman infantry, and six thousand cavalry. The Carthaginians were outnumbered by the Roman infantry, having only half the Roman number. However their cavalry was ten thousand, almost twice the Roman number.[217]

"The battle began in earnest, and in the true barbaric fashion: for there was none of the usual formal advance and retreat; but when they once got to close quarters, they grappled man to man, and, dismounting from their horses, fought on foot."[218] Hannibal had arranged his forces so that the Romans "could derive no advantage from their superiority in numbers."[219] At first, the Romans appeared to be winning. They pressed back the Carthaginian center. But the Romans over extended themselves. They "advanced so far, that the Libyan heavy-armed troops on either wing got on their flanks ... as long as the Romans could keep an unbroken front, to turn first in one direction and then in another to meet the assaults of the enemy, they held out; but the outer files of the circle continually falling, and the circle becoming more and more contracted, they at last were all killed on the field."[220]

Ten thousand Roman infantry that had been held in reserve were taken captive. Livy estimated the Roman casualties as "forty-thousand foot, two-thousand seven-hundred horse."[221] But Polybius said that the number of Roman dead amounted to seventy thousand, with only three thousand escaping, and ten thousand taken prisoner.[222] Little mercy was shown. Livy related that "the vanquished preferring death in their places to flight; and the conquerors, who were enraged at them for delaying the victory, butchering those whom they could not put to flight."[223]

Polybius attributed the Carthaginian victory to a superiority in cavalry. "[It is] a lesson to posterity that in actual war it is better to have half the number of infantry, and the superiority in cavalry, than to engage your enemy with an equality in both."[224]

Having imposed a crushing defeat on the Romans, Hannibal's cavalry commander advised an immediate attack on Rome. Hannibal demurred, causing his general to observe "You know how to conquer, Hannibal, but you do not know how to make use of your victory."[225]

Hannibal sought ransom for the ten thousand Roman infantry he had captured, but the Roman Senate refused to pay. They "established the rule for their own men, that they must either conquer or die on the field, as there was no other hope of safety for them if they were beaten ... so that Hannibal was not so much rejoiced at his victory in the battle, as struck with astonishment at the unshaken firmness and lofty spirit displayed in the resolutions of these senators."[226] Hannibal sold the captive Romans into slavery, and they were scattered throughout the Mediterranean.[227]

## *Siege of Syracuse*

About the time of the Battle of Cannae (216 B.C.), King Hieron II of Syracuse died at the age of ninety. Polybius praised the long and beneficent rule of Archimedes' friend:

> Hieron gained the sovereignty of Syracuse and her allies by his own unaided abilities without inheriting wealth, or reputation, or any other advantage of fortune. And ... was established king of Syracuse without putting to death, banishing, or harassing any one of the citizens,—which is the most astonishing circumstance of all. And what is quite as surprising as the innocence of his acquisition of power is the fact that it did not change his character. For during a reign of fifty-four years he preserved peace for the country, maintained his own power free from all hostile plots, and entirely escaped the envy which generally follows greatness; for though he tried on several occasions to lay down his power, he was prevented by the common remonstrances of his citizens. And having shown himself most beneficent to the Greeks, and most anxious to earn their good opinion, he left behind him not merely a great personal reputation but also a universal feeling of goodwill towards the Syracusans. Again, though he passed his life in the midst of the greatest wealth, luxury, and abundance, he survived for more than ninety years, in full possession of his senses and will all parts of his body unimpaired; which, to my mind, is a decisive proof of a well-spent life.[228]

Hieron II's legacy was shortly squandered by his successor, Hieronymous, a fifteen-year-old grandson.[229] Livy described the boy as a tyrant.

> Those who for so many years had seen Hieron II and his son Gelon differing from the rest of the citizens neither in the fashion of their dress nor any other mark of distinction, now beheld the purple, the diadem, and armed guards, and their king sometimes proceeding from his palace in a chariot drawn by four white horses, according to the custom of the tyrant Dionysius. This costliness in equipage and appearance was accompanied by corresponding contempt of everybody, capricious airs, insulting expressions, difficulty of access, not to strangers only, but even to his guardians also, unheard of lusts, [and] inhuman cruelty.[230]

Polybius, ever the careful, objective, and analytical student of history, assessed other historians' depictions of Hieronymous as a cruel tyrant to be exaggerations. "One would think from their description ... [no] other tyrant was every fiercer than he. Yet he was a mere boy when he succeeded to power, and only lived thirteen months after. In this space of time it is possible that one or two men may have been put to the rack, or certain of his friends, or other Syracusan citizens, put to death; but it is improbable that his tyranny could have been extravagantly wicked, or his impiety outrageous."[231]

Nevertheless, Hieronymous' short rule was marked by instability and imprudent politics. Polybius admitted that the boy was "reckless and unscrupulous in disposition."[232] Shortly after his accession to the throne of Syracuse, Hieronymous fell under the influence of advisors who favored Carthage over Rome. He opened negotiations with Carthage, and insulted the Roman ambassadors. Hannibal sent ambassadors to Syracuse. These included "Hippocrates and Epicydes, the grandsons of an exiled Syracusan who had settled at Carthage."[233] Under their baneful influence, the alliance between Syracuse and Rome that Hieron II had so carefully nurtured for decades was scuttled. Polybius said "Hippocrates got the young Hieronymous entirely into his hands."[234]

After little more than a year of rule, Hieronymous was assassinated.[235] Syracuse was in "tumult and confusion."[236] The Carthaginian faction prevailed, and "the government was placed in the hands of [the Carthaginian ambassadors] Hippocrates and Epicydes, and all the members of Hieron's family were butchered."[237] Commenting on the revolution at Syracuse, Livy concluded, "Such is the nature of the populace; they are either cringing slaves or haughty tyrants. They know not how with moderation to spurn or to enjoy that liberty which holds the middle place."[238]

Syracuse was now in open revolt, and began to make war upon neighboring cities that were still allied with Rome. "Hippocrates began to ravage the adjoining parts of the Roman province, at first by stealthy excursions, but afterwards, when [the Roman general] Appius had sent a body of troops to protect the lands of the [Roman] allies, he made an attack with all his forces upon the guard posted over against him, and slew many."[239]

The killing of their troops was more than the Romans could bear. They dispatched the general Marcus Claudius Marcellus (c. 268–208 B.C.) to Syracuse to put down the rebellion. Combining his forces with those of Appius, Marcellus attacked the Sicilian city of Leontini and "took the town by storm on the first assault."[240] Anxious to punish the rebellion, Marcellus scourged and decapitated "two thousand deserters."[241] This action had the unforeseen consequence of convincing the Syracusans to resist the Romans with all possible tenacity, as they feared their surrender would be met by wholesale execution.

Marcellus marched to Syracuse where "he proceeded to attack the city both by land and sea."[242] But the entire Roman army and navy was defeated by one old man. At the request of King Hieron II, now deceased, Archimedes had long ago prepared defenses that were now called into use. In Polybius' succinct assessment, the Romans "did not take into account the abilities of Archimedes; nor calculate on the truth that, in certain circumstances, the genius of one man is more effective than any numbers whatever. However they now learnt it by experience."[243]

Plutarch described the battle scene:

> When, therefore, the Romans assaulted the walls in two places at once, fear and consternation stupefied the Syracusans, believing that nothing was able to resist that violence and those forces. But when Archimedes began to ply his engines, he at once shot against the land forces all sorts of missile weapons, and immense masses of stone that came down with incredible noise and violence; against which no man could stand; for they knocked down those upon whom they fell in heaps, breaking all their ranks and files. In the meantime huge poles thrust out from the walls over the ships sunk some by the great weights which they let down from on high upon them; others they lifted up into the air by an iron hand or beak like a crane's beak and, when they had drawn them up by the prow, and set them on end upon the poop, they plunged them to the bottom of the sea; or else the ships, drawn by engines within, and whirled about, were dashed against steep rocks that stood jutting out under the walls, with great destruction of the soldiers that were aboard them. A ship was frequently lifted up to a great height in the air (a dreadful thing to behold), and was rolled to and fro, and kept swinging, until the mariners were all thrown out, when at length it was dashed against the rocks, or let fall.... [The Romans] then took a resolution of coming up under the walls, if it were possible, in the night; thinking that as Archimedes used ropes stretched at length in playing his engines, the soldiers would now be under the shot, and the darts would, for want of sufficient distance to throw them, fly over their heads without effect. But he [Archimedes], it appeared, had long before framed for such occasions engines accommodated to any distance, and shorter weapons; and had made numerous small openings in the walls, through which, with engines of a shorter range, unexpected blows were inflicted on the assailants. [When the Romans] came close up to the walls, instantly a shower of darts and other missile weapons was again cast upon them. And when stones came tumbling down perpendicularly upon their heads, and, as it were, the whole wall shot out arrows at them, they retired ... [but] as they were going off, arrows and darts of a longer range inflicted a great slaughter among them, and their ships were driven one against another; while they themselves were not able to retaliate in any way. For Archimedes had provided and fixed most of his engines immediately under the wall; whence the Romans, seeing that indefinite mischief overwhelmed them from no visible means, began to think they were fighting with the gods."[244]

The attack on Syracuse and its repulsion by the defenses of Archimedes is attested to by Livy, Polybius, and Plutarch. The historical accuracy is unquestioned. But one aspect of Archimedes' defenses is more uncertain. Some later authors claimed that Archimedes used mirrors to focus sunlight and set the Roman ships on fire.

Thomas Heath stated that "the story he [Archimedes] set the Roman ships on fire by an arrangement of burning-glasses or concave mirrors is not found in any authority earlier than Lucian."[245] Lucian was a 2nd century A.D. rhetorician who wrote "Archimedes ... burned the ships of the enemy by means of his science," but did not mention mirrors or the focusing of sunlight.[246]

The best-known description of Archimedes' mirrors is perhaps that by John Tzetzes (c. A.D. 1110–1180), a Byzantine poet. According to Tzetzes, "[Archimedes] constructed a kind of hexagonal mirror, and at an interval proportionate to the size of the mirror he set similar small mirrors with four edges, moved by links and by a form of hinge, and made it the center of the sun's beams ... when the beams were reflected in the mirror, a fearful kindling of fire was raised in the ships, and at the distance of a bow-shot he turned them into ashes."[247]

Historians are skeptical of Archimedes' burning mirrors, the story "is generally considered to be a work of pure fiction."[248] One writer concluded, "the historical evidence for Archimedes' burning mirrors is feeble, contradictory in itself, and the principal and very late authorities for the story are unreliable, while the standard and contemporary authority was silent.... [T]here are ample historical, scientific, and military grounds for concluding that Archimedes did not use a burning mirror as a weapon of war."[249]

The primary problem in authenticating the story is that there is no mention of burning mirrors in any of the accounts by Polybius, Livy, or Plutarch. These are the primary historical sources for the siege of Syracuse.

On the other hand, no one has doubted Archimedes' theoretical ability to construct a "burning mirror." It has also been demonstrated that the concept can work. In 1747, Georges-Louis Leclerc Comte de Buffon (1707–1788) conducted experiments with mirrors and concluded "it is certain that Archimedes was able to do with metal mirrors what I have done with glass ones."[250] In 1973, the *New York Times* reported that a Greek engineer, Ioannis Sakkas, used about "200 flat bronze-coated mirrors, measuring 5 feet by 3 [1.52 by 0.91 meters]," to set afire "a small rowboat afloat 165 feet [50.3 meters] from the pier." Sakkas claimed that Archimedes likely used "a large number of flat mirrors," to "create the effect of a vast concave mirror."[251]

Burning mirrors or not, Marcellus was frustrated by Archimedes' formidable defenses. "Deriding his own artificers and engineers," he asked "must we give up fighting with this geometrical Briareus, who plays pitch-and-toss with our ships, and, with the multitude of darts which he showers at a single moment upon us, really outdoes the hundred-handed giants of mythology?"[252]

The reference to Briareus derived from the *Iliad*, where Homer described a monster with a hundred hands, so frightful that he even scared the gods.

> The monster whom [the] gods [call] Briareus,
> Shakes the solid Earth so strong:
> With giant-pride at Jove's high throne he stands,
> And Brandishes round him all his hundred hands;
> The frightened Gods confessed their awful Lord,
> They dropped the fetters, trembled and adored.[253]

Marcellus resolved to stop the assault and take Syracuse by long-term siege. Indeed, he had no other choice. Polybius concluded, "one man and one intellect, properly qualified for the particular undertaking, is a host of itself and of extraordinary efficacy ... the Romans [were] confident that their forces by land and sea would enable them to become masters of the town, if only one old man could be got rid of ... [but] as long as he remained there, they did not venture even to think of making the attempt."[254]

Unfortunately, Archimedes considered engineering to be vulgar in comparison to the pure science of mathematics. Therefore he never wrote about his mechanical inventions or described them in any way. Plutarch related that Archimedes considered engineering to be "sordid and ignoble" in comparison to the pursuit of pure mathematics.

> Archimedes possessed so high a spirit, so profound a soul, and such treasures of scientific knowledge, that though these inventions had now obtained him the renown of more than human sagacity, he yet would not deign to leave behind him any commentary or writing on such subjects; but, repudiating as sordid and ignoble the whole trade of engineering, and every sort of art that lends itself to mere use and profit, he placed his whole affection and ambition in those purer speculations where there can be no reference to the vulgar needs of life; studies, the superiority of which to all others is unquestioned, and in which the only doubt can be whether the beauty and grandeur of the subjects examined, of the precision and cogency of the methods and means of proof, most deserve our admiration. It is not possible to find in all geometry more difficult and intricate questions, or more simple and lucid explanations.[255]

It was Archimedes' intellectual preoccupation that eventually led to his demise. Plutarch said that Archimedes was in the habit of entering into mental concentration so intense that he would forget to eat and bathe. "And thus it ceases to be incredible that (as is commonly told of him) the charm of his familiar and domestic Siren made him forget his food and neglect his person, to that degree that when he was occasionally carried by absolute violence to bathe or have his body anointed, he used to trace geometrical figures in the ashes of the fire, and diagrams in the oil on his body, being in a state of entire preoccupation, and, in the truest sense, divine possession with his love and delight in science."[256]

Marcellus laid siege to Syracuse for eight months.[257] Waiting patiently, the Roman general noted that there was "a tower into which a body of men might be secretly introduced, as the wall near to it was not difficult to surmount, and it was itself carelessly guarded."[258] Eventually, the Syracusans became careless. During a festival dedicated to the goddess Diana, the Syracusans gave themselves up "entirely to wine and sport."[259] As the Romans scaled "the wall they found no one to oppose them, for the guards of the several towers, owing to it being a time of public sacrifice, were either still drinking or were gone to sleep again in a state of drunkenness."[260]

By the time the populace had been alarmed, the Romans were in complete control of the city. According to Livy, Marcellus had mixed feelings concerning his victory. "When Marcellus, having entered the walls, beheld this city as it lay subjected to his view from the high ground on which he stood, a city the most beautiful, perhaps, of any at that time, he is said to have shed tears over it; partly from the inward satisfaction he felt at having accomplished so important an enterprise, and partly in consideration of its ancient renown.... [I]n an instant every thing before him would be in flames, and reduced to ashes."[261]

The Roman soldiers were allowed to plunder the city. "Marcellus ... ordered his soldiers ... not to offer violence to any free person, but told them that every thing else might be their booty ... [and] they did not desist from plunder till they had gutted the houses of all the property which had been accumulated during a long period of prosperity."[262]

In the tumult that followed, Archimedes was killed. Marcellus was aggrieved by Archimedes' death.

> Nothing afflicted Marcellus so much as the death of Archimedes, who was then, as fate would have it, intent upon working out some problem by a diagram, and having fixed his mind alike and his eyes upon the subject of his speculation, he never noticed the incursion of the Romans, nor that the city was taken. In this transport of study and contemplation, a soldier, unexpectedly coming up to him, commanded him to follow to Marcellus; which he declining to do before he had worked out his problem to a demonstration, the soldier, enraged, drew his sword and ran him through.[263]

Plutarch related that Archimedes "is said to have requested his friends and relations that, when he was dead, they would place over his tomb a sphere containing a cylinder, inscribing it with the ratio which the containing solid bears to the contained."[264] Cicero (106–43 B.C.) visited Syracuse and found that Archimedes was forgotten and his tomb overgrown:

> Archimedes ... tomb [was] over grown with shrubs and briars.... [T]hey Syracusans knew nothing of it, and even denied that there was any such thing remaining.... I remembered some verses ... [that] set forth [that] on the top of it; there was placed a sphere with a cylinder. When I had carefully examined all the monuments ... I observed a small column standing out a little above the briars, with the figure of a sphere and a cylinder upon it ... I found the inscription, though the latter half part of all the verses were effaced almost half way.[265]

There is some indication that not only did Marcellus allow Syracuse to be plundered, but he also dealt harshly with the citizens. "The town was not set on fire, but thoroughly plundered, and those of its inhabitants who were not sold as slaves were driven out of the town into the open fields, where they tore up the grass from the earth to satisfy their hunger, or died from starvation, so that the free Syracusans had to envy the lot of slaves, and many of them pretended to be slaves, merely to obtain the means of satisfying their hunger."[266]

## *Hasdrubal Defeated*

In Italy, the war continued, albeit without major battles. Polybius explained that drawing out the conflict favored the Romans. The Carthaginians "have their hopes of freedom ever resting on the courage of mercenary troops: the Romans on the valor of their own citizens and the aid of their allies. The result is that even if the Romans have suffered a defeat at first, they renew the war with undiminished forces, which the Carthaginians cannot do. For, as the Romans are fighting for country and children, it is impossible for them to relax the fury of their struggle; but they persist with obstinate resolution until they have overcome their enemies."[267] The Carthaginians and Romans were at a stalemate. "Though Hannibal could not be driven out of Italy ... it was clear that the unaided resources of his army were unequal to the task of her [Rome's] destruction."[268]

Although sorely pressed in Italy, from the second year of the Second Punic War, the Romans had sent troops into Spain to harass the Carthaginians. After some years, the Romans started to gain the initiative in Spain, and Hannibal's brother, Hasdrubal, resolved to lead his troops over the Alps into Italy. In Italy, he would join forces with Hannibal. "Whether Spain in his [Hasdrubal's] absence would remain faithful to the Carthaginians or not, was to him a matter of indifference, for he thought that it would be easy to re-conquer it after having gained possession of Italy."[269]

Hasdrubal followed the footsteps of his brother through the Alps and arrived in Italy in the spring of 207 B.C. In Rome, the Senate assembled and chose two consuls to lead armies against the Carthaginians: Caius Claudius Nero and Marcus Livius. Marcus Livius had been condemned in a trial years earlier, "a disgrace which he took so much to heart, that he retired into the country, and for many years absented himself from the city, and avoided all public assemblies."[270] Upon being told of his election to consul, Marcus Livius asked, "if they esteemed him a good man, why had they thus passed a sentence of condemnation upon him as a wicked and guilty one?"[271] The Senate called upon the consuls' patriotism, and "succeeded in uniting Marcus Livius in the consulate with Caius Claudius."[272]

The two consuls marched out of Rome in opposite directions. Marcus Livius was to oppose Hasdrubal in the north, Caius Claudius Nero to face Hannibal in the south. It was now apparent that the Second Punic War was approaching a crisis point. "The whole weight of the danger and the entire burden pressed upon one point. Whichever of these generals should be first victorious, he would in a few days unite his camp with the other."[273]

At this point, fate intervened. Hasdrubal dispatched messengers to his brother with a plan of uniting their armies and attacking Rome. The messengers never reached Hannibal; they were intercepted and the letters delivered to Caius Claudius Nero. Nero realized that the fate of Rome hung in precarious balance. If the two Carthaginian armies managed to join, the combined force might be unstoppable.[274]

Caius Claudius Nero hit upon a bold and risky plan. He decided to send reinforcements north to Livius and attack Hasdrubal before Hannibal became aware of the situation. Trusting no one, Nero "selected the choicest troops out of the whole army of the Romans and allies, to the amount of six-thousand infantry and one-thousand horse."[275] The troops marched hurriedly out of camp at night. Knowing that Hannibal probably had spies in his camp, Nero marched south, but as soon as his army was out of sight, he turned and marched north.

Once the force was well on its way, Caius Claudius Nero gathered his troops and announced the true nature of their mission. He told them that the Roman troops in the north were now deadlocked with the Carthaginians, and that their added force would be sufficient to defeat the enemy. Nero convinced his troops that they would receive all the credit and glory for the inevitable victory. He told them "there never was a measure adopted by any general which was in appearance more daring than this, but in reality more safe ... [and] he was leading them to certain victory."[276]

Nero was surreptitious in his approach to his Roman allies. His troops moved into the Roman camp "secretly by night" and hid themselves in existing tents. No new tents were erected, and pains were taken so that Hasdrubal would have no way of knowing that the Roman forces had swelled by 7,000 men in the middle of the night.[277]

When word of Caius Claudius Nero's plan reached Rome, the Senate was alarmed, "nor could people make up their minds whether they should commend, or censure, this so bold march of the Consul."[278] Nero had left his army weakened, with no general in command. "What would be the consequence if that should be discovered, and Hannibal should think proper to either to pursue Nero with his whole army, who had gone off with only six thousand armed men, or to assault the camp, which was left as a prey for him, without strength, without command, without auspices?"[279]

The morning following the arrival of Nero's reinforcements, the Romans prepared for an immediate attack. Hasdrubal was ready to fight; however as he moved forward his keen powers of observation convinced him that all was not as it should be. "The Roman camp showed no change of size, [but] it had not escaped the quick ear of the Carthaginian general that the trumpet which gave the signal to the Roman legions sounded that morning once oftener than usual, as if directing the troops of some additional superior officer. Hasdrubal, from his Spanish campaigns, was well acquainted with all the sounds and signals of Roman war, and from all that he heard and saw, he felt convinced that both the Roman consuls were before him."[280]

Hasdrubal could only conclude that Nero had defeated Hannibal and marched north to join Livius with his entire force. Not wanting to face two Roman armies at once, Hasdrubal decided to retreat to Gaul until he could assess his brother's fate. He waited until dark and then led his troops north toward the Metaurus river.

Entering Italy, the Carthaginians had easily forded the Metaurus. Now they unexpectedly found the stream swollen with spring rains, impossible to cross. It was the worst of all possible positions. After searching all night in vain for a way across the river the Carthaginians "lost all discipline and subordination."[281] The next morning, the Roman cavalry began to approach and Hasdrubal did the best he could to arrange his troops into a defensible position.

The battle was joined. "A furious contest arose, and the slaughter on both sides was dreadful."[282] Nero then suddenly deserted his position. "Drawing several cohorts from the right wing ... [he] led them round the rear of the line, and, to the surprise not only of the enemy but his own party, charged their [the Carthaginians'] right flank.... [T]hus on all sides, in front, flank, and rear, the Spaniards and Ligurians were cut to pieces."[283]

"Up to this point the victory had been doubtful; for both sides fought with desperation ... but as soon as Claudius fell upon the rear of the enemy the battle ceased to be equal: for the Iberians found themselves attacked on front and rear at once, which resulted in the greater part of them being cut down on the ground."[284] Polybius estimated the casualties as "taking Carthaginians and Celts together, not less than ten thousand were killed, and about two thousand Romans."[285] When Hasdrubal saw that defeat was inevitable, he charged into the Roman columns on horseback, preferring to die fighting rather than face capture.[286]

After defeating the Carthaginians, Caius Claudius Nero returned south as quickly as he had come. "Nero set out on the night following the battle, and marching at a more rapid rate than when he came, arrived at his camp before the enemy on the sixth day.... [O]n his return to camp, [he] ordered the head of Hasdrubal, which he had carefully kept and brought with him, to be thrown before the advanced guards of the enemy, and the African prisoners to be shown to them bound just as they were. Two of these also he unbound, and bid them go to Hannibal and tell him what had occurred."[287]

> Ten years had passed since Hannibal had last gazed on those [Hasdrubal's] features. The sons of Hamilcar had then planned their system of warfare against Rome which they had so nearly brought to successful accomplishment. Year after year had Hannibal been struggling in Italy, in the hope of one day hailing the arrival of him whom he had left in Spain, and of seeing his brother's eye flash with affection and pride at the junction of their irresistible hosts. He now saw that eye glazed in death, and in the agony of his heart the great Carthaginian groaned aloud that he recognized his country's destiny.[288]

## *Carthage Defeated*

Carthaginian power was broken, and in 204 B.C., Roman troops under the command of Publius Scipio (236–183 B.C.) went on the offensive in Africa. Confronting the Carthaginian forces, Scipio hit upon the plan of burning their camp. "The winter huts of the Carthaginians, which were constructed from materials hastily collected out of the fields, were almost entirely of wood. The Numidians, particularly, lay for the most part in huts formed of interwoven reeds, and covered with mats.... [T]hese circumstances having been reported to Scipio, gave him hopes that he might have an opportunity of burning the enemy's camp."[289]

Accordingly, in a surprise attack, Scipio positioned part of his troops to set fire to the huts of the Carthaginians. The rest of the Romans lay in ambush, ready to slaughter Carthaginians fleeing from the conflagration. "The fire caught all the first row of huts fiercely, and soon got beyond all control, from the closeness of the huts to each other, and the amount of combustible material which they contained."[290]

The Carthaginians "got trampled to death by their own friends at the exits from the camp; many were caught by the flames and burnt to death; while all those who escaped the flame fell into the hands of the enemy, and were killed without knowing what was happening to them or what they were doing."[291] Livy estimated the Carthaginian casualties in the tens of thousands. "Forty-thousand men were either slain or destroyed by the flames, and above five thousand captured."[292] In Polybius' assessment, it was "impossible for the imagination to exaggerate the dreadful scene, so completely did it surpass in horror everything hitherto recorded. Of all the brilliant achievements of Scipio this appears to me to have been the most brilliant and the most daring."[293]

The Carthaginians subsequently approached the Romans with peace negotiations. "They sent thirty of their principal elders as deputies to solicit peace."[294] But in Livy's opinion, "their only object was to gain time for Hannibal to cross over into Africa."[295] Hannibal was unhappy at being recalled to Africa:

> When Hannibal heard the message of the [Carthaginian] ambassadors he gnashed his teeth, groaned, and scarcely refrained from shedding tears.... [A]fter they [the ambassadors] had delivered the commands with which they were charged, he [Hannibal] said: "those who have for a long time been endeavoring to drag me home, by forbidding the sending of supplies and money to me, now recall me, not indirectly but openly. Hannibal, therefore, hath been conquered, not by the Roman people, who have been so often slain and routed, but by the Carthaginian senate, through envy and detraction."[296]

The final battle of the Second Punic War occurred at Zama in 202 B.C. "To the Carthaginians it was a struggle for their own lives and the sovereignty of Libya; to the Romans for universal dominion and supremacy."[297] The battle ended in a decisive victory for Rome, the Carthaginian defeat being attributable to a lack of unity in their ranks. "The Roman soldiers charged the enemy, shouting as usual their war-cry, and clashing their swords against their shields."[298] The Romans pressed forward, but at first the advantage was to the Carthaginian mercenaries. "The dexterity and daring of the mercenaries ... enabled them to wound a considerable number of the Romans ... [however] the Carthaginians did not keep up with their mercenaries nor support them, but showed a thoroughly cowardly spirit."[299]

"Believing that they had been shamelessly abandoned by their own side," the Carthaginians mercenaries "fell upon the men on their rear as they were retreating, and began killing them."[300] The Carthaginians thus had to "fight with their own men and the Romans at the same time."[301] The conflict was so intense, and the casualties so high, that at one point Scipio had to call the battle to a temporary halt so that the battlefield could be cleared of corpses that might impede the final assault. Polybius described the ground as "slippery with gore, the corpses lying piled up in bloody heaps, and with the corpses' arms flung about in every direction."[302]

When the battle resumed, "the two lines charged each other with the greatest fire and fury. Being nearly equal in numbers, spirit, courage, and arms, the battle was for a long time undecided."[303] At last, some Roman divisions managed to flank the Carthaginians. "Upon their [the Romans] charging Hannibal's rear, the greater part of his men were cut down in their ranks ... on the Roman side there fell over fifteen hundred, on the Carthaginian over twenty thousand, while the prisoners taken were almost as numerous."[304]

The Carthaginians pleaded for peace, and though the Romans "all were impelled by just resentment to demolish Carthage, yet, when they reflected upon the magnitude of the undertaking, and the length of time which would be consumed in the siege of so well fortified and strong a city ... the minds of all were inclined to peace."[305]

Under the terms of the peace treaty, Carthage was allowed to remain independent in north Africa and retain territories it had occupied before the war. However Carthage had to surrender a hundred hostages, was allowed to retain only ten triremes in its navy, and had to pay Rome 200 talents of silver a year for fifty years. An additional term was that Carthage was forbidden from waging war "in or out of Africa" without Rome's permission.[306]

Following the establishment of peace, Hannibal was elected to the office of praetor and discovered that the Carthaginian government was corrupt. "He [Hannibal] made heavy charges ... on the whole order of judges; in consequence of whose arrogance and power, neither the magistracy nor the laws availed anything."[307]

Hannibal attempted to break up the oligarchy by instituting term limits. "He proposed a law, and procured it to be enacted, that the judges should be elected annually; and that no person should hold the office two years successively."[308] Furthermore, Hannibal discovered that a substantive portion of the public tax revenues were being embezzled. He put a stop to this, and "those persons who, for several years past, had maintained themselves by plundering the public, were greatly enraged; as if this were ravishing from them their own property, and not as dragging out of their hands their ill-gotten spoil. Accordingly, they instigated the Romans against Hannibal, who were seeking a pretext for indulging their hatred against him."[309]

The Romans, "though they had granted peace to the Carthaginians, their war against him [Hannibal], individually, remained irreconcilable."[310] Hannibal surreptitiously fled Carthage, sailing to Tyre, and then to Antioch, "lamenting the misfortunes of his country oftener than his own."[311] Pursued relentlessly by the Romans, in 183 B.C. Hannibal found himself surrounded and hopelessly trapped. Accordingly, he decided to kill himself. "He [Hannibal] called for poison, which he had long kept in readiness to meet such an event, and said 'Let us release the Romans from their long anxiety, since they think it too long to wait for the death of an old man'.... This was the end of the life of Hannibal."[312]

## *Cato the Censor (234–149 B.C.)*

Carthage never recovered from her defeat in the Second Punic War. The treaty with Rome was taken advantage of by the Numidian general, Masinissa. "Being incensed against the Carthaginians and relying upon the friendship of the Romans, [Masinissa] seized a considerable part of the [Carthaginian] territory ... on the ground that it had once belonged to himself."[313] Carthage was unable to defend itself, because it was forbidden by the terms of its treaty with Rome from waging war. So "the Carthaginians appealed to the Romans to bring Masinissa to terms. The Romans accordingly sent arbitrators, but told them to favor Masinissa as much as they could."[314]

With the Romans backing him up, Masinissa continued harassing the Carthaginians. The Romans were called upon repeatedly to mediate. But upon one trip to Carthage, the Roman delegation was astonished at how strong the Carthaginian nation had become since its defeat in the Second Punic War.

> They carefully observed the country; they saw how diligently it was cultivated, and what valuable resources it possessed. They entered the city too and saw how greatly it had increased in power and population since its overthrow by Scipio not long before; and when they returned to Rome they declared that Carthage was to them an object of apprehension rather than of jealousy, a great and hostile city, near at hand, and growing thus easily. Cato especially said that even the liberty of Rome would never be secure until Carthage was destroyed. When the Senate learned these things it resolved upon war, but still needed a pretext.[315]

Among those who had visited Carthage was Marcus Porcius Cato (234–149 B.C.), commonly known as Cato the Censor. Cato became Carthage's most bitter enemy. In the Roman Senate, he called repeatedly for the total destruction of Carthage. Cato was not of noble birth, but had risen to power on the basis of natural talents. According to Livy, Cato was a man of formidable abilities and incorruptible integrity:

> So great powers of mind and energy of intellect were in this man, that, no matter how lowly the position in which he was born, he appeared capable of attaining to the highest rank.... [T]his man's genius was so versatile, and so well adapted to all things, that in whatever way engaged, it might be said, that nature formed him for that alone.... His temper, no doubt, was austere, his language bitter and unboundedly free, but his mind was never conquered by his passions, his integrity was inflexible, and he looked with contempt on popularity and riches. In spare diet, in enduring toil and danger, his body and mind were like iron; so that even old age, which brings all things to dissolution, did not break his vigor.[316]

Cato was a skilled orator. Plutarch called him "the Roman Demosthenes."[317] Cato lived plainly and simply, eschewing wealth and luxury. In his government, he "showed most inflexible severity and strictness in what related to public justice, and was righteous and precise in what concerned the ordinances of the commonwealth; so that the Roman government never seemed more terrible, nor yet more mild than under his administration."[318]

As Censor, Cato had the power to conduct "an inquisition into every one's life and manners."[319] He was severe and inflexible in renewing the ancient Roman morality. Cato expelled Manilius from the Roman Senate, a man who had been expected to be selected as the next consul. Cato ejected Manilius from the Senate "because, in the presence of his [Manilius'] daughter, and in open day, he [Manilius] had kissed his wife." Cato the Censor also heavily taxed items of luxury. He cut illicit water lines to private residences that siphoned water from the public utility. Cato "demolished within thirty days all buildings or sheds, in possession of private persons, that projected into public ground."[320]

Cato's relentless drive to purge corruption and immorality from Rome earned him many enemies. But he triumphed over all of them. Plutarch states that "he [Cato] is reported to have escaped at least fifty indictments."[321]

When Cato visited Carthage, he was likely disgusted by the wealth and decadence of the Carthaginian aristocrats. It is also possible that a man of Cato's temperament may have loathed the Carthaginian's practice of sacrificing their own children. The Carthaginians worshipped the deities Baal and Moloch, pagan gods to whom they would sacrifice small children.

"The Carthaginians sacrificed their offspring to Moloch. The children were laid on the hands of a calf-headed image of bronze, from which they slid into a fiery oven, while the people danced to the music of flutes and timbrels [tambourines] to drown the shrieks of the burning victims."[322] The Carthaginians "knowingly and wittingly themselves devoted their own children; and they that had none of their own bought of some poor people, and then sacrificed them like lambs or pigeons, the poor mother standing by the while without either a sigh or tear; and if by chance she fetched a sigh or let fall a tear, she lost the price of her child, but it was nevertheless sacrificed. All the places round the image were in the mean time filled with the noise of hautboys [oboes] and tabors [drums], to drown out the infants' crying."[323]

After the Carthaginians were defeated by the tyrant of Syracuse, Agathocles (361–289 B.C.), in 310 B.C., they resorted to wholesale slaughter of their children.

> [The Carthaginians concluded] that this miserable misfortune was brought upon them by the gods.... [I]n former times they used to sacrifice ... the sons of the most eminent persons; but of later times they secretively bought up and bred children for that purpose.... [W]eighing these things in their minds, and now seeing that the enemy lay before their walls, they were seized with such a pang of superstition ... [that] they offered as a public sacrifice two hundred of the sons of the nobility; and no fewer than three hundred more voluntarily offered themselves up: for among the Carthaginians there was a brazen statue of Saturn putting forth the palms of his hands bending in such a manner towards the earth, as that the boy was laid upon them, in order to be sacrificed, should slip off, and so fall headlong into a deep fiery furnace.[324]

In the ninth century B.C., the Hebrew prophet Elijah had put an end to the worship of Baal in Israel by taking three hundred priests of Baal and putting them all to death. "And Elijah said unto them, take the prophets of Baal; let not one of them escape. And they took them: and Elijah brought them down to the brook Kishon, and slew them there."[325]

It is likely that Cato's motivations were more practical than moral. He was alarmed by how quickly Carthage had recovered from her defeat in the Second Punic War. "He came to the conclusion that Rome could not be secure until Carthage had disappeared from the face of the earth."[326] Cato became obsessed with Carthage, and acquired the habit of ending every speech in the Senate with the same phrase, "Carthage, methinks, ought utterly to be destroyed."[327]

## *Destruction of Carthage*

The Romans gave the Carthaginians no relief from the attacks of the Numidians. In Carthage, a patriotic faction came to power and declared war on the Numidians. The Carthaginian forces were led by Hasdrubal,* a man far removed from the military genius of Hannibal. "Hasdrubal ... was one of the usual army-destroyers whom the Carthaginians were in the habit of employing as generals; strutting about in his general's purple like a theatrical king, and pampering his portly person even in the camp ... [he was] a vain and unwieldy man."[328] In Polybius' opinion, Hasdrubal was "a vain ostentatious person, very far from possessing real strategic ability."[329]

Carthage's decision to engage the Numidians militarily was a miscalculation. The Carthaginian commander, Hasdrubal, was badly defeated by Masinissa, losing perhaps as many as 50,000 men.[330] The war between the Numidians and the Carthaginians also gave Rome the pretext it needed for the destruction of Carthage. The Romans promptly declared war on Carthage.[331]

Carthage was in no condition to fight a Third Punic War with Rome. The Roman army and navy were now immeasurably larger and more powerful than any force possessed by

*not to be confused with the Hasdrubal who was Hannibal's brother

the Carthaginians. When the Carthaginians learned that Rome had declared war on them, they became alarmed. "They had neither allies, nor mercenaries, nor supplies for enduring a siege, nor anything else in readiness for this sudden and unheralded war."[332] Carthage dispatched a team of thirty ambassadors to Rome in an attempt "to dispel the storm by complete submission."[333]

The Romans dealt with the Carthaginians in a manner that was both brutal and deceitful. The Carthaginians in turn were completely submissive, "because they had not the courage fully to realize the import of surrendering themselves beforehand to the arbitrary will of a mortal foe."[334]

The Roman Senate at first appeared to grant the Carthaginian entreaty for peace. They told the Carthaginian ambassadors that if they would give up "300 children of their noblest families as hostages, and would obey their orders in other respects, the freedom and autonomy of Carthage would be preserved and they should retain their lands in Africa ... but they sent word privately to the consuls that they should carry out their secret instructions."[335]

When the Roman army arrived in Africa, the hostages were collected. "The mothers ... clung to their little ones with frantic cries and seized hold of the ships and of the officers who were taking them away, even holding the anchors and tearing the ropes, and throwing their arms around the sailors in order to prevent the ships from moving; some of them even swam out far into the sea beside the ships, shedding tears and gazing at their children. Others on the shore tore out their hair and smote their breasts as though they were mourning the dead."[336]

Once the hostages had been taken, the Roman consuls in Carthage demanded that the city turn over all of its armaments. The Carthaginians obsequiously complied. The Romans "received complete armor for 200,000 men, besides innumerable javelins and darts, and about 2,000 catapults for throwing pointed missiles and stones.... [I]t was a remarkable and unparalleled spectacle to behold the vast number of loaded wagons."[337]

Having secured the Carthaginian hostages and arms, the Roman consuls then presented their final demand. "Yield Carthage to us, and betake yourselves where you like within your own territory at a distance of at least ten miles [16.1 kilometers] from the sea, for we are resolved to raze your city to the ground."[338]

When they heard this, the Carthaginian representatives realized that they were surrendering to Rome, without a fight, everything it could have expected from waging and winning another war. There would be no peace, and they had been artfully deceived. The Carthaginian ambassadors "cursed the Romans ... flung themselves to the ground and beat it with their hands and heads. Some of them even tore their clothes and lacerated their flesh as though they were absolutely bereft of their senses."[339]

After this display, the Carthaginians appealed to the Romans for mercy and pity. They received none. The consul Censorinus stated "the Senate ... has issued its decrees and they must be carried out."[340] His only advice to the Carthaginians was to note that "the healing drug for all evils is oblivion."[341]

When the Carthaginians received the news that their city was to be destroyed, they went mad with anger and grief.

> [There] followed a scene of blind, raving madness ... some fell upon the senators who had advised giving the hostages and tore them in pieces ... others treated in a similar way those who had favored giving up arms. Some stoned the ambassadors for bringing the bad news, and others dragged them through the city ... the city was full of wailing and wrath, of fear and threatenings

> ... [the Carthaginians] upbraided their gods for not even being able to defend themselves. Some went into the arsenals and wept when they found them empty ... most of all was their anger kindled by the mothers of the hostages who, like Furies in a tragedy, accosted those whom they met with shrieks, and reproached them with giving away their children.[342]

After venting their anger and frustration, the Carthaginians resolved to fight. "A wonderful change and determination came over them, to endure everything rather than abandon their city.... All the sacred places, the temples, and every other wide and open space, were turned into workshops, where men and women worked together day and night.... Each day they made 100 shields, 300 swords, 1000 missiles for catapults, 500 darts and spears, and as many catapults as they could. For strings to bend them the women cut off their hair for want of other fibers."[343]

The Carthaginian resistance was aided by immensely strong fortifications, constructed over hundreds of years. "On the sea side, where the city faced a precipice, it [Carthage] was protected by a single wall ... [but] towards the south and the mainland ... there was a triple wall. The height of each wall was forty-five feet [13.7 meters]."[344] Furthermore, the Carthaginian general, Hasdrubal, was at large in the countryside with an army of 20,000 men.[345]

There followed a long siege of three years. The Carthaginians were resupplied by sea. "Ships with provisions still ran into the harbor, partly bold merchantmen allured by the great gain."[346] To cut off the Carthaginians' supplies, the Romans had to construct large-scale engineering works, and this took time.

At one point, the Romans managed to briefly occupy Megara, described by Appian as "a very large suburb adjacent to the city wall."[347] But a number of Roman soldiers were captured, and taken behind Carthage's walls. The next day, "Hasdrubal, enraged at the attack upon Megara, took the Roman prisoners whom he held, brought them upon the walls, in full light of their comrades, and tore out their eyes, tongues, or private parts with iron hooks; of some he lacerated the soles of the feet, of others he cut off the fingers, and some he flayed alive, hurling them all, still living, from the top of the walls. He intended to make reconciliation between the Carthaginians and Romans impossible."[348] In this, Hasdrubal was successful.

Eventually, the Romans were able to cut off the Carthaginians' supply lines, and the city was weakened by "famine and pestilence."[349] Having weakened the Carthaginians by starvation, the Romans finally managed to breach the city walls. The Carthaginians were capable of "only a feeble resistance because they were weak from hunger and downcast in spirit."[350] There followed six days of fighting, street-by-street.[351]

> All places were filled with groans, shrieks, shouts, and every kind of agony. Some were stabbed, others were hurled alive from the roofs to the pavement, some of them falling on the heads of spears or other pointed weapons, or swords.... Then came new scenes of horror. The fire spread and carried everything down, and the soldiers did not destroy the buildings little by little, but pulled them down together ... [so that] old men, women, and young children who had hidden in the inmost nooks of the houses, some of them wounded, some more or less burned, and uttering horrible cries [were killed] ... others ... were torn asunder into all kinds of horrible shapes, crushed and mangled.[352]

Hasdrubal surrendered to the Romans in cowardly fashion. As his wife watched from afar, she reportedly denounced him, screaming "wretch, traitor, most effeminate of men, this fire will entomb me and my children." The spouse of the Carthaginian general then immolated herself and her children. "Having reproached him [Hasdrubal] thus, she slew her children, flung them into the fire, and plunged in after them."[353]

What remained of Carthage was given up to plunder by the common soldiers. The survivors, 30,000 men and 25,000 women, represented less than ten percent of the original population of Carthage. They were mostly disposed of by being sold into slavery.[354] Despite the ferocity of the final battle, "the larger portion of the city remained standing."[355] The Roman Senate ordered that the remaining portions of the city be totally destroyed "and thereafter to pass the plough over the site of Carthage so as to put an end in legal form to the existence of the city, and to curse the soil and site for ever, that neither house nor cornfield might ever re-appear on the spot. The command was punctually obeyed. The ruins burned for seventeen days.... [W]hen the remains of the Carthaginian city wall were excavated [in modern times], they were found to be covered with a layer of ashes from four to five feet [1.2–1.5 meters] deep, filled with half-charred pieces of wood, fragments of iron, and projectiles."[356]

Thus Carthaginian civilization was destroyed.

## *Hipparchus (c. 190–120 B.C.)*

### *Epicycles and Eccentrics*

Hipparchus was born in Nicaea in northwest Asia Minor, but spent most of his career at Rhodes, "a rival of Alexandria as a center of literary and intellectual life."[357] Hipparchus was perhaps the most important of the Greek astronomers, but his only surviving work is *Commentary on Aratus*. Nearly all of what we know concerning him must be pieced together from fragmentary references by other writers. We are "singularly ill-informed" on the details of Hipparchus' work.[358]

Hipparchus constructed a table of chord functions, and is credited by G. J. Toomer as being "the founder of trigonometry ... [and] the man who transformed Greek astronomy from a purely theoretical into a practical, predictive science."[359] In ancient Greek geometry, a *chord* was the linear distance between two points on a circle separated by a specific angle. The chord function was thus similar to the modern sine function. Chords were applied in the elaborate geometrical calculations employed in Greek astronomy.

By the time of Hipparchus, it was well established that the astronomical system of Eudoxus and Aristotle, based on a system of concentric rotating spheres, could not account for the observations. Two prominent problems were the retrograde motions of the planets, and changes in the size and brightness of Mars and Venus. In his commentary on Aristotle's *De Caelo* (*On the Heavens*), Simplicius (c. A.D. 490–560) quoted the astronomer Sosigenes[360] (fl. 1st century B.C.) as stating that the problems were known even in Aristotle's time.

> The theories of Eudoxus and his followers fail to save the phenomena, and not only those which were first noticed at a later date, but even those which were before known and actually accepted by the authors themselves ... one fact which is actually evident to the eye ... [is] that the planets appear at times to be near to us and at times to have receded. This is indeed obvious to our eyes in the case of some of them; for the star called after Aphrodite [Venus] and also the star of Ares [Mars] seem, in the middle of their retrogradations, to be many times as large, so much so that the star of Aphrodite actually makes bodies cast shadows on moonless nights ... we must conclude that the apparent difference in the sizes of the two bodies observed under the same atmospheric conditions is due to the inequality of their distances ... Aristotle too, shows that he is conscious of it when, in the *Physical Problems*, he discusses objections to the hypotheses of astronomers arising from the fact that even the sizes of the planets do not appear to be the same always. In this respect Aristotle was not altogether satisfied with the revolving spheres.[361]

In order to "save the phenomena," or account for the observations, Greek astronomers began to introduce more complicated variations of circular motion. The first variation proposed was the introduction of *eccentrics*, or circles of revolution in which the Earth was offset from the center. Although it is conceivable that the concept may have originated with a member of the Pythagorean School, it is usually associated with Heraclides of Pontus (c. 388–310 B.C.) and Apollonius of Perga (c. 262–190 B.C.). In addition to proposing eccentric circles, Heraclides suggested that the Earth rotated, and that Venus and Mercury revolved around the Sun.[362] "Heraclides of Pontus, by supposing that the earth is in the center, and rotates, while the heaven is at rest, thought in this way to save the phenomena."[363]

Eccentrics by themselves were insufficient, so epicycles were introduced. An *epicycle* is a small circle of revolution superimposed upon a larger circle. In the twelfth book of the *Almagest*, Claudius Ptolemy (c. A.D. 100–170) introduced "the epicyclic theory," in conjunction with Apollonius.[364] But it was Hipparchus who fully applied eccentric and epicyclic motion to develop a theory that described the motions of the Sun and the Moon.[365] Hipparchus' theory of lunar and solar motions allowed him to predict eclipses. He was assisted in his work by having "a complete list of lunar eclipses observed at Babylon ... since 747 B.C."[366] Writing in the first century A.D., Pliny the Elder said "it was discovered two hundred years ago, by the sagacity of Hipparchus, that the moon is sometimes eclipsed after an interval of five months, and the sun after an interval of seven ... Hipparchus calculated the course of both these stars [the Sun and Moon] for the term of 600 years."[367]

Hipparchus also calculated the sizes of the Moon and Sun, and their distances from Earth. He estimated that the average distances of the Moon and the Sun from the Earth were, respectively, 67.3 and 490 Earth radii.[368] The corresponding modern values are 60.3 and 23,481 Earth radii. Hipparchus estimated that Moon's volume was 1/27 that of Earth, and that the Sun's volume was 12.3 times larger than Earth's.[369] The modern values for the sizes of the Moon and Sun in terms of Earth volume are, respectively, 1/50 and 1.3 million.

Writing in the ninth book of the *Almagest*, Claudius Ptolemy stated that Hipparchus was unable to devise a successful theory to describe the motions of the five planets (Mercury, Venus, Mars, Jupiter, and Saturn), although he tried combinations of eccentric and epicyclic motion. "Hipparchus ... did not attempt to give the principle of the hypotheses of the five planets ... but only arranged the observations in a more useful way and showed the appearances to be inconsistent with the hypotheses of the mathematicians of that time ... he also thought that these movements could not be effected either by eccentric circles, or by circles concentric with the ecliptic but bearing epicycles, or even by both together."[370]

## Precession of the Equinoxes

Hipparchus' most important discovery was the *precession of the equinoxes*. An equinox occurs twice a year (fall and spring) when the Earth is on opposite sides of the Sun, and the length of the day and night are equal everywhere on the planet. The winter solstice occurs when the Sun is at its lowest point on the horizon; the summer solstice when the Sun is at its maximum altitude. The precession of equinoxes refers to the fact that the date of the equinoxes (and thus the solstices) moves. In other words, the position of the Earth relative to the background of fixed stars is not the same on each equinox or solstice, but changes systematically. It is understood today that this is due to the Earth wobbling on its axis of rotation. It takes about 26,000 years for one complete wobble. Precession causes the stars to change their longitude by about one degree in a century.[371]

The cause of the precession of the equinoxes was discovered by Isaac Newton in 1687. The mass of the Earth is not distributed in perfect symmetry, but is somewhat concentrated in a band around the equator. The gravitational pull of the Sun and Moon exert a torque on this extra equatorial mass that attempts to pull the Earth's equator into alignment with the plane of the solar system. The situation is somewhat analogous to pushing on a spinning top. If one attempts to push over a top, the top will not fall but will begin to wobble.

Hipparchus estimated the length of the solar or tropical year, to be 365.25 days minus 1/300 of a day, by comparing "the solstice observed by himself in 135 B.C. with those observed by Aristarchus in 280 B.C. and Meton in 432 B.C."[372] The solar year is the time spanned from solstice to solstice, or equinox to equinox. Hipparchus noted that the solar year was not of the same length as the sidereal year, the time taken for the Sun to return to the same position relative the fixed background of stars. Thus, he estimated that the precession or movement of the equinoxes was "at least 1/100th" of a degree per year.[373]

Hipparchus' discovery was noted by Ptolemy in the seventh book of the *Almagest*:

> That the sphere of the fixed stars has a movement of its own in a sense opposite to that of the revolution of the whole universe ... is made clear to us especially by the fact that the same stars have not kept the same distances from the solstitial and equinoctial points in earlier times ... for Hipparchus, in his work *On the Displacement of the Solstitial and Equinoctial Points*, comparing the eclipses of the moon, on the basis both of accurate observations made in his time, and of those made still earlier by Timocharis, concludes that the distance of Spica from the autumnal equinoctial point ... was in his own time 6 degrees, but in Timocharis' time 8 degrees.[374]

It is evident that Hipparchus was meticulous and accurate in his observations. In *Commentary on Aratus*, he corrected Eudoxus. "About the north pole Eudoxus is in error, for he says: 'there is a certain star which remains always in the same spot; this star is the pole of the universe,' the fact being that at the pole there is no star at all, but there is an empty space, with, however, three stars close to it."[375]

### *Star Catalog*

In *Natural History*, Pliny the Elder wrote that Hipparchus observed a nova, or new star, and was therefore motivated to create a catalog of stars so that new stars could be distinguished from existing ones. "Hipparchus ... discovered a new star that was produced in his own age ... [so he] attempted what might seem presumptuous even in a deity, viz. to number the stars for posterity and to express their relations by appropriate names; having previously devised instruments, by which he might mark the places and the magnitudes of each individual star."[376]

Hipparchus' star catalog has been lost. In books 7 and 8 of the *Almagest*, Ptolemy gave the positions of 1,022 stars.[377] Ptolemy's catalog is believed to be derived from Hipparchus, but it is not clear if Hipparchus gave a location for every star in his listing, or what coordinate system he employed. In *Commentary on Aratus*, Hipparchus indicated star positions "in various ways."[378]

## *Claudius Ptolemy (c. A.D. 100–170)*

### *The* Syntaxis *or* Almagest

Claudius Ptolemy (c. A.D. 100–170) was the most important astronomer of late antiquity, and is primarily remembered for writing a compendium that synthesized the astronomical knowledge of his predecessors. Little is known of Ptolemy's life. His surname indicates that he was of Greek ancestry, but the personal name of "Claudius" suggests he was a Roman citizen. So far as we know, Ptolemy spent his entire life in Alexandria.[379]

Ptolemy's great work on astronomy was the *Syntaxis*. The *Syntaxis* later became better known as the *Almagest*, the latter name deriving from a corruption of an Arabic phrase meaning "the greatest." The *Syntaxis* (or *Almagest*) became the standard textbook in astronomy almost as soon as it was published, and it endured for fourteen hundred years as the definitive summation of all astronomical knowledge.[380]

The lasting value of the *Syntaxis* derived from several facets. It was clearly written, logically organized, and contained a summary of previous work by other astronomers. The reader needed no prerequisites other than a generalized knowledge of geometry and basic astronomical terminology. The book was also of considerable practical value, because Ptolemy included tables that allowed the computation of the positions of the Sun, Moon, and planets. A modern translator of the *Syntaxis*, G. J. Toomer, assessed it as a superior work. "As a didactic work the *Almagest* is a masterpiece of clarity and method, superior to any ancient scientific textbook and with few peers from any period. But it is much more than that. Far from being a mere 'systematization' of earlier Greek astronomy, as it is sometimes described, it is in many respects an original work."[381]

In the preface to the *Syntaxis*, Ptolemy began by noting that Aristotle had divided theoretical philosophy into three main branches: theology, physics, and mathematics. Theology dealt with that which cannot be observed but only imagined. Physics was also difficult to study quantitatively, because it treated objects of the sublunary, corruptible world that were in a perpetual state of change.

Ptolemy concluded that most theology and physics ought to be properly termed "guesswork" instead of knowledge. It followed that mathematics was best suited to be applied to a study of the celestial realm, because the planets were eternal and unchanging. Echoing Pythagoras, Ptolemy claimed that the study of astronomy was the path to spiritual purification. "This science, above all things, could make men see clearly; from the constancy, order, symmetry and calm which are associated with the divine, it makes its followers lovers of this divine beauty, accustoming them and reforming their natures, as it were, to a similar spiritual state."[382]

In the beginning of the *Syntaxis*, Ptolemy stated that the "first order of business" was to establish the place of the Earth in the universe.[383] He started by establishing that both the heavens and Earth are spherical in shape. Ptolemy explained that if the stars were not on a sphere with the Earth at its center they would appear to increase and decrease in apparent size during the rotation of the heavens. He conceded that heavenly bodies appear larger when close to the horizon, but attributed this (falsely) to a distortion caused by their light passing through moisture-laden air near the Earth's surface ("just as objects placed in water appear bigger than they are").[384] We now know that the apparent enlargement of bodies such as the Moon when they are near the horizon is a pure optical illusion — measurements show the size of the images is the same regardless of their elevation.

Ptolemy gave some arguments as to why the Earth must be spherical in shape. He noted that observers of lunar eclipses find that the same eclipse occurs at different local times, and that "the differences in the hour are proportional to the distances between the places of observation."[385] As travelers move north, the southern stars disappear below the horizon, and vice versa for travelers who move southward, "hence it is clear that here too the curvature of the Earth cuts off the heavenly bodies in a regular fashion in a north-south direction and proves the sphericity of the Earth in all directions."[386]

After presenting the evidence for the Earth's spherical shape, Ptolemy argued that the Earth must be at the center of the universe and motionless. "The Earth has been proved to be spherical and situated in the middle of the universe."[387] Here, his Aristotelean physics led him to the wrong conclusion. He first conceded that a rotating Earth would explain the astronomical observations. But he then said "to judge by the conditions affecting ourselves and those in the air about us, such a hypothesis must be seen to be quite ridiculous."[388] Having no understanding of the concepts of inertia, gravity, or centripetal acceleration, Ptolemy concluded that if the Earth were rotating from west to east, clouds and other objects not directly connected to the ground surface would be left behind. "Clouds and any of the things that fly or can be thrown could never be seen traveling towards the east, because the earth would always be anticipating them all and forestalling their motion towards the west."[389]

## *The Ptolemaic System*

Having established the basic precepts of astronomy as known to him, Ptolemy constructed an Earth-centered (geocentric) model of the universe that became known as the *Ptolemaic System*. Most of the elements of the *Ptolemaic system* were not original with Claudius Ptolemy; he merely presented the best and last synthesis of a geocentric system in a well-written form that was widely distributed. A systematic and mathematical geocentric cosmology dated back at least to Eudoxus (c. 400–347 B.C.). Later refinements were made by Heraclides of Pontus (c. 388–310 B.C.), Apollonius of Perga (c. 262–190 B.C.), and Hipparchus (c. 190–120 B.C.).

Aristarchus' (310–230 B.C.) heliocentric system remained an eccentric obscurity. Greek astronomy reached its culmination of originality and discovery with Hipparchus in the second century B.C. "In the two hundred and sixty years between Hipparchus and Ptolemy astronomy does not seem to have made any progress."[390]

The central tenet of the geocentric system was derived from Aristotelean physics (and Pythagorean philosophy) which maintained that while earthly motion was linear, heavenly motion was circular. But it is impossible to explain the apparent motion of the planets with a simple model of circular motion around the Earth. One difficulty is that the planets appear at times to stop their nightly progression across the sky, reverse their motion, then stop again, and resume their original path of motion. This phenomenon of retrograde motion is caused by the Earth periodically overtaking other planets in its revolution around the Sun.

To explain the motion of the planets in a geocentric system, Hipparchus employed both *epicycles* and *eccentrics*. An epicycle is a secondary circular motion superimposed on a primary circular motion with a larger diameter. The introduction of epicycles helped to explain the apparent retrograde motion of the planets, but it was not enough by itself—a further complication was necessary. An eccentric was the displacement of the center of a planet's rotation from the center of the universe, the Earth.

Ptolemy found that even these two innovations were not sufficient and introduced a third geometrical complexity to the system, the *equant*.[391] If all this sounds confusing, it is and was. Alphonso X (A.D. 1221–1284), King of Castile, was introduced to the Ptolemaic System and reportedly noted that, had he been asked, he would have recommended a simpler system to God.* "It is certain that at that time astronomers explained the motion of the heavens by such intricate and confused hypotheses, that they did no honor to God, nor answered in any way the idea of an able Worker. So that it is likely that it was in considering that multitude of spheres of which Ptolemy's System is composed, so many eccentric circles, so many epicycles, so many librations, etc., that he [Alphonso] happened to say, *That if God had asked his Advice when he made the World, he would have given him good Counsel* [to design it otherwise]."[392]

Ptolemy apparently never considered that the motion of the planets (including the Moon and Sun) could be anything but both uniform and circular. In the *Syntaxis* he stated explicitly that his purpose was to demonstrate that the motion of the planets could be explained by circular motion, not test the possibility of circular motion as a hypothesis.

Ptolemy explained, "We think that the mathematician's task and goal ought to be to show all the heavenly phenomena being reproduced by uniform circular motions."[393] Later, he added "Now it is our purpose to demonstrate for the five planets, just as we did for the Sun and Moon, that all their apparent anomalies can be represented by uniform circular motions, since these are proper to the nature of divine beings, while disorder and non-uniformity are alien [to such beings]. Then it is right that we should think success in such a purpose a great thing, and truly the proper end of the mathematical part of theoretical philosophy."[394]

Ptolemy apparently considered "uniform circular motion" for heavenly bodies to be as axiomatic as Euclid's assumption that parallel lines do not intersect. When he spoke of considering "hypotheses," he meant only hypotheses involving various arrangements of circular motions. When Ptolemy advocated parsimony, it was a qualified endorsement. "We consider it a good principle to explain the phenomena by the simplest hypotheses possible, in so far as there is nothing in the observations to provide a significant objection to such a procedure."[395]

What Ptolemy meant was that he would consider not the simplest type of motion, but the simplest combination of *circular* motions. Ironically, the inflexible desire on Ptolemy's part to explain the motion of the planets by combinations of simple circular motions led to compound motions that were neither simple or circular. Arthur Koestler (1905–1983) characterized the complexity as offensive. "There is something profoundly distasteful about Ptolemy's universe; it is the work of a pedant with much patience and little originality, doggedly piling 'orb in orb.'"[396]

Ptolemy himself was aware of the complexity of the system, and evidently felt the need to defend it, because he wrote, "now let no one, considering the complicated nature of our devices, judge such hypotheses to be over-elaborated."[397] Ptolemy was apparently led into error by Aristotelean physics which maintained that the heavens and the Earth were governed by separate and distinct sets of physical laws. Ptolemy explained that valid analogies could not be drawn between the terrestrial realm in which we lived and the heavens. "For it is not appropriate to compare human [constructions] with divine, nor to form one's beliefs about such great things on the basis of very dissimilar analogies."[398]

In a masterpiece of rationalization, Ptolemy finally concluded that all heavenly motions

*The quote is apocryphal and appears to originate in its present form with Pierre Bayle's *Historical and Critical Dictionary*, first published in 1697. The quote from Alphonso is from an English translation of the *Dictionary* published in London in 1710. See: Goldstein, B. R., 1991, "The Blasphemy of Alfonso X: History or Myth?," in, Barker, P. and Ariew, R., Editors, *Revolution and Continuity*, vol. 24 of *Studies in Philosophy and the History of Philosophy*: Catholic University of America Press, Washington, DC, p. 143–153.

are simple, forgetting that the challenge was not to the heavens, but to his own model. "[We should judge 'simplicity'] from the unchangingness of the nature of things in the heavens and their motions. In this way all [motions] will appear simple, and more so than what is thought 'simple' on Earth, since one can conceive of no labor or difficulty attached to their revolutions."[399]

Although complex, one reason the Ptolemaic system remained ascendant for fourteen-hundred years was that it produced the right answers within the limits of the existing precision. The epicycles, eccentrics, and equants used in the system had no physical existence, but the inclusion of greater degrees of mathematical freedom allowed Ptolemy to make more precise predictions of the future movements of the planets.

It has been argued that Ptolemy's "epicyclic theory was merely a means of calculating the apparent places of the planets without pretending to represent the true system of the world," and that Ptolemy himself did not regard the complicated geometry of his system as a true representation of physical reality.[400] If that is the case, then it is difficult to explain why Ptolemy thought it necessary to defend the system on the basis of its perceived complexity. He also made physical arguments for the location of the Earth at the center of the universe and for its stationarity.

## *Astrology*

Ptolemy wrote a four-volume work on astrology, the *Tetrabiblos*, that he considered to be a natural companion to the astronomical *Syntaxis*. The *Syntaxis* allowed one to calculate the positions of the heavenly bodies while the *Tetrabiblos* instructed as to the influences of the same bodies upon terrestrial affairs.

Ptolemy did not consider astrological influences to be in any sense of the word supernatural, but purely physical.[401] His view was influenced by Aristotelean cosmology. Ptolemy explained, "that a certain power, derived from the ethereal nature, is diffused over and pervades the whole atmosphere of the earth, is clearly evident to all men. Fire and air, the first of the sublunary elements, are encompassed and altered by the motions of the ether. These elements in their turn encompass all inferior matter, and vary it as they themselves are varied; acting on earth and water, in plants and animals."[402]

To substantiate the reality of astrological influences, Ptolemy described the undeniable influences of the Sun on earthly events. "The Sun ... contributes to the regulation of all earthly things: not only by the revolution of the seasons does he bring to perfection the embryo of animals, the buds of plants, the springs of waters, and the alteration of bodies, but by his daily progress also he operates other changes in light, heat, moisture, dryness and cold; dependent upon his situation with regard to the zenith."[403]

The association of the Moon with phenomena such as tides also was evidence of the occult influence of heavenly bodies on earthly events. "The Moon, being of all the heavenly bodies the nearest to the Earth, also dispenses much influence; and things animate and inanimate sympathize and vary with her. By the changes of her illumination, rivers swell and are reduced; the tides of the sea are ruled by her risings and settings; and plants and animals are expanded or collapsed, if not entirely at least partially, as she waxes and wanes."[404]

It wasn't hard to believe in occult, or hidden forces in nature. The lodestone, or magnet, had been known since the most ancient times, and it produced action at a distance through a means that was completely mysterious and without any apparent physical agency.

If a small, insignificant stone could produce observable effects, a heavenly body naturally could be expected to produce a proportionately larger influence.

Belief in astrology was universal throughout the Roman Empire in Ptolemy's time. The *Satyricon*, a first century A.D. work of fiction, described a banquet with an astrological theme. Dinner guests were presented with a dish configured with the twelve signs of the zodiac. "The oddity of the thing drew the eye of all. An immense circular tray bore the twelve signs of the zodiac displayed round the circumference."[405] The banquet host lectured his guests on the influences of the zodiac over men's characters and talents:

> I was born under the Crab myself. Wherefore I stand on many feet, and have many possessions both by sea and land; for the Crab is equally adapted to either element.... Under the Lion are born great eaters and wasters, all who love to domineer; under the Virgin, women and runaways and jailbirds; under the Scales, butchers and perfumers and all retail traders; under the Scorpion, poisoners and cut-throats ... under Pisces, or the Fishes, fine cooks and fine talkers. Thus the world goes round like a mill, and is for ever at some mischief, whether making men or marring them.[406]

In *Natural History*, Pliny the Elder (A.D. 23–79) affirmed the influence of the heavens over the Earth. "The nature of the celestial bodies is eternal, being interwoven, as it were, with the world, and, by this union, rendering it solid; but they exert their most powerful influence on the earth. This, notwithstanding its subtlety, may be known by the clearness and the magnitude of the effect."[407]

Seneca (c. 4 B.C.–A.D. 65), writing in *Natural Questions*, reflected the Stoic belief in an organic universe in which all events, objects, and living things were intertwined. This was the justification for both divination and astrology.

> The roll of fate ... sends ahead in all directions intimations of what is to follow, which are in part familiar, in part unknown to us. Everything that happens is a sign of something that is going to happen: mere chance occurrences uncontrolled by any rational principle do not admit of the application of divination.... [T]here is no living creature whose movement or meeting with us does not foretell something.... [E]ver those stars that are either stationary or, from their velocity being the same as that of the world as a whole, seem to be so, are not without sway and dominion over us.[408]

In *Decline and Fall of the Roman Empire*, Edward Gibbon (1737–1794) commented on the superstitious practices of the Romans. The Romans were possessed of "a puerile superstition that disgraces their understanding. They listen with confidence to the predictions of haruspices [soothsayers], who pretend to read in the entrails of victims the signs of future greatness and prosperity; and there are many who do not presume either to bathe, or to dine, or to appear in public, till they have diligently consulted, according to the rules of astrology, the situation of Mercury and the aspect of the moon."[409]

The practice of astrology is still popular today. In 1988, the *New York Times* reported that the President of the United States had employed astrology in making important decisions.[410]

## *Geography*

Ptolemy was also a geographer. He wrote a work titled *Geography* that contained a map of the world and a table listing the latitude and longitude of various places. Ptolemy adopted "the scientific conception of Hipparchus that a map of the world could only be laid down correctly by determining the latitude and longitude of all the principal points on its surface — a method of which his own knowledge of astronomy led him fully to appreciate the value."[411]

But the necessary data did not exist in Ptolemy's time, nor did the Roman government seem to recognize the desirability of funding efforts to construct reliable and authoritative maps. Although it would have been simple to collect accurate estimates of latitude, the only data Ptolemy had were those few collected by Hipparchus, who lived and worked about three hundred years before Ptolemy. "Hipparchus alone, and that in the case of a few cities only, in comparison with the vast multitude of those that must find a place in a general map — has transmitted to us the elevations of the north pole, (i.e., observations of latitude)."[412]

Longitude was much more difficult to determine. Hipparchus had noted that accurate estimates could be made if observers at different locations were to note the time at which the same lunar eclipse occurred. Nevertheless, Ptolemy admitted that such observations did not exist, with only one exception. "The distances from one place to another have for the most part been reckoned only in a rough and general way, especially those from east to west, not so much from the carelessness of those who reported them, as from their want of mathematical skill, and the small number of simultaneous observations of lunar eclipses at different places that had been duly recorded — like that which was seen at Arbela at the fifth hour and at Carthage at the second."[413]

By his own admission, Ptolemy lacked accurate data. But he nevertheless proceeded to construct lengthy tables listing the latitude and longitude of many locations, to the precision of half a degree. Three-fourths of *Geography* consists of such tables. "The very definite and positive form in which Ptolemy thus presented his conclusions to the reader, was in itself calculated to disguise the true nature of these statements, and conceal the fact that they were in reality nothing more than the approximate result arrived at by a comparison of authorities, of distances given in itineraries, of the reports of voyagers, and other such material corrected frequently in a very summary and arbitrary manner to suit his own preconceived notions."[414] Thus Ptolemy's *Geography* has been characterized as a "specious edifice ... [that] served by its external symmetry to conceal the imperfect character of its foundations and the rottenness of its materials."[415]

In addition to the imprecision of his coordinate data, Ptolemy underestimated the circumference of the Earth by about thirty percent, and thus introduced a systematic error into all of his calculations:

> [Ptolemy] fell into an error vitiating all his conclusions. Eratosthenes was the first who had attempted scientifically to determine the earth's circumference, and his result of 250,000 (or 252,000) stadia ... was generally adopted by subsequent geographers, including Strabo. Poseidonius, however (c. 135–50 B.C.), reduced this to 180,000, and the latter computation was inexplicably adopted by Marinus and Ptolemy. This error made every degree of latitude or longitude (measured at the equator) equal to only 500 stadia ... instead of its true equivalent of 600 stadia.[416]

In the fifteenth century, Ptolemy's error was a factor in convincing Christopher Columbus to undertake his voyage of discovery. Columbus concluded he could reach India by voyaging westward across the Atlantic Ocean because "he supposed the Land of Spices of which he was in search to be much less distant than it really was."[417]

## *Harmonics and Optics*

Ptolemy made contributions to other sciences. He wrote a book on music titled *Harmonics*. Like Aristotle, Ptolemy was a teleologist who saw purpose in nature. "For in every subject it is inherent in observation and knowledge to demonstrate that the works of nature have been crafted with some reason and prearranged cause and completed not at all in random."[418]

*Harmonics* provides some insight into Ptolemy's scientific philosophy. He described the purpose of the harmonicist as not the testing of hypotheses, but as the "preservation of the hypotheses," in so far as these ideas are consistent with observation. "The purpose of the harmonicist would be then to preserve in every way the reasoned hypotheses of the canon which do not in any way at all conflict with the perceptions as most people interpret them, just as the purpose of the astrologer is to preserve the hypotheses of the heavenly movements concordant with observable paths."[419] Although Ptolemy claimed to want to "preserve the phenomena," he apparently had a superior desire to "preserve the hypothesis." This aspiration is strongly Platonic in its implications, in that a *phenomenon* is considered secondary to a *noumenon*.

Where Ptolemy failed, his successor, Johannes Kepler (1571–1630), succeeded. Kepler was also obsessed with the pursuit of mathematical beauty. But he ultimately abandoned his *noumenon* to the demands of reality, by recognizing that the data demanded abandoning circular motion. An underlying beauty and simplicity in the motion of the planets was finally found by Isaac Newton (1642–1727) in the form of the law of universal gravitation.

Next to his books on astronomy and geography, Ptolemy's most significant work was on the subject of optics in the form of a manuscript with the title *Optics*. Ptolemy's *Optics* dealt with both the reflection and refraction of light. Concave and convex mirrors were discussed, as well as types of optical illusions. Like most of the philosophers of his age, Ptolemy incorrectly believed that vision was due at least partially to the emission of some type of "ray" from the human eyeball.[420]

Ptolemy's proficiencies in both astronomy and optics raise the interesting question of why Ptolemy and his contemporaries failed to invent the telescope. The Egyptians were manufacturing glass as early as 3500 B.C., and "glass-work attained an extraordinarily high degree of excellence among the Romans."[421] Lenses have been found in ancient ruins, but apparently they were used as ornaments, "not as magnifying glasses."[422] The ancient Greeks were aware that "glass spheres filled with water" had the power to magnify images.[423]

Ptolemy understood that light could be bent by glass, and "recognized that refraction was governed by a definite law."[424] But the practical application of this knowledge did not come to fruition until Europeans invented eyeglasses about A.D. 1286. It took more than three hundred additional years before people grasped the principle of the refracting telescope. According to one apocryphal account, the refracting telescope was invented in 1608 when a Dutch spectacle-maker serendipitously held a convex lens in line with a concave lens and noted that the image was magnified. But there are other indications that the invention may have occurred in Holland in 1604, or even as early as 1590 in Italy.[425]

It is apparent that technological and scientific progress is not generally the product of instantaneous insight or unique genius. Innovation appears to be the inevitable result of the slow refinement of knowledge and mechanical techniques. In the case of the telescope, it was first necessary that the principles of optics be studied and applied in the manufacture of eyeglasses for hundreds of years.

## *Ptolemy's Fraud*

No discussion of Claudius Ptolemy can be complete without mentioning the controversy that has swirled about his work for hundreds of years. The first person to raise questions concerning the accuracy and integrity of Ptolemy's work was the Danish astronomer Tycho Brahe (A.D. 1546–1601). Tycho voiced a suspicion that Ptolemy's star catalogue was not the product of Ptolemy's own observations as Ptolemy had claimed, but had been derived by copying the work of Hipparchus.

The French astronomer, Jean Baptiste Joseph Delambre (1749–1822), noted that the stellar positions given by Ptolemy were all wrong. He concluded that Ptolemy had taken Hipparchus' earlier work and updated it by making a correction for precession. However the precession correction used by Ptolemy was wrong. The fraud was revealed by the fact that the values Ptolemy supposedly derived from observation were impossible for his age. J. L. E. Dreyer (1852–1926) reiterated, "[Ptolemy's catalog] is nothing but the catalogue of Hipparchus brought down to his own time with an erroneous value of the constant of precession."[426]

In 1977, the controversy greatly intensified with the publication of a book by Robert R. Newton (1918–1991), *The Crime of Claudius Ptolemy*, that historians found to be stunning in the frankness of its tone and conclusions.[427] In the preface, Newton plainly called Ptolemy a criminal. "This is the story of a scientific crime.... I mean a crime committed by a scientist against his fellow scientists and scholars, a betrayal of the ethics and integrity of his profession that has forever deprived mankind of fundamental information about an important area of astronomy and history."[428]

The gist of Newton's accusation was that so far as he could determine, all of the observations that Ptolemy reported in the *Syntaxis* were fraudulent. Because of the labor involved in the hand-copying of books, Newton concluded that Ptolemy was responsible for much of the genuine work in ancient astronomy being irretrievably lost. Transcribers did not want to go to the trouble of copying earlier authors when all of the relevant information seemed to be masterfully summarized and synthesized in the *Syntaxis*.

Newton pointed out that people should have realized the fraud more than a century earlier. In 1819, Delambre had found irrefutable proof of fraud in the form of "data" that could only have come from fabrication. "Such agreement can never be the result of observations, no matter how bad they may be. It can only come from fabrication.... Delambre's argument is unanswerable."[429]

Newton summarized his feelings on believed Ptolemy's work being fraudulent, and the impact:

> All of his own observations that Ptolemy uses in the *Syntaxis* are fraudulent, so far as we can test them. Many of the observations that he attributes to other astronomers are also frauds that he has committed. His work is riddled with theoretical errors and with failures of comprehension.... His writing of the *Syntaxis* caused us to lose most of the genuine work in Greek astronomy.... The *Syntaxis* has done more damage to astronomy than any other work ever written, and astronomy would be better off it had never existed. Thus Ptolemy is not the greatest astronomer of antiquity, but he is something still more unusual: he is the most successful fraud in the history of science.[430]

Newton's accusation of fraud was coldly received by historians. Writing in the journal *Science* in 1979, Bernard R. Goldstein characterized Newton's rhetoric as excessive. "Newton's arguments in support of [his] charges are marred by all manner of distortions, misunderstandings, and excesses of rhetoric due to an intensely polemical style."[431]

In 1980, Owen Gingerich (b. 1930) came to Ptolemy's defense with a paper titled "Was Ptolemy a Fraud?" published in the *Quarterly Journal of the Royal Astronomical Society*.[432] Gingerich began with the contention that Ptolemy was a person motivated by high ethics. However, his defense quickly became compromised when Gingerich himself raised the question of fabrication. "Did Ptolemy fabricate his purported equinox observation? Perhaps."[433]

Gingerich admitted that examination of the *Almagest* appeared to support Newton's claims. "I shall concede at once that the *Almagest* poses some curious problems to the historians of science, and that Ptolemy's statements regarding observed quantities cannot always be taken at face value."[434]

Gingerich conceded that "[Ptolemy's] observations are surprisingly bad while his final parameters are amazingly good. Finally, we have the disconcerting fact that most of Ptolemy's reported observations, faulty as they are, agree almost perfectly with his theory.... This strongly suggests that the recorded 'observations' depend on the theory, and not vice versa."[435]

After conceding that Ptolemy's "data" had been fabricated, Gingerich then concluded that Ptolemy's fabrications were not "fabrications" but "corrections" made to bring the "observations" into accordance with a theory that Ptolemy believed to be true. "Ptolemy, like many of the brilliant theoreticians who followed him, was perfectly willing to believe that his theory represented Nature better than the error-marred individual observations of the day. As one of America's Nobel laureates remarked to me, any good physicist would do the same today."[436]

Gingerich quoted remarks made by Isaac Newton and Albert Einstein wherein they purportedly proclaimed the superiority of theory over observation. He closed by reasserting the greatness of Ptolemy. "When [Isaac] Newton and Einstein are generally considered frauds, I shall have to include Ptolemy also. Meanwhile, I prefer to think of him as the greatest astronomer of antiquity."[437]

Gingerich's comparison with Isaac Newton and Einstein was inappropriate and misleading. Neither of these men ever fabricated data. In offhand remarks, each noted that if observations tended to disprove their theories, then the observations were most likely wrong. These statements amounted to a recognition that observational data can be of less than ideal quality, and that it is dangerous to place excessive confidence in anecdotal data. This is hardly the same as fraudulently fabricating an entire body of data.

In the past few decades, opinion has started to swing in Robert Newton's favor regarding the fraudulent nature of Ptolemy's work. A recent writer noted "there are many ... examples of reported measurements that disagree with reality but are precisely what Ptolemy needed in order to deduce results that he wanted.... [W]e should stop using data from the *Syntaxis* ... [and] we should reevaluate Ptolemy."[438]

## Chapter 3

# Roman Engineering

## *Lack of Originality in Science and Philosophy*

The Roman Republic and Empire dominated the Mediterranean world for several hundred years, yet produced little to nothing of originality in the way of science or philosophy. Almost all significant Roman manuscripts were derivatives of Greek science. The two most popular philosophies in Rome, Epicureanism and Stoicism, were both developed by the Greeks.

The relevant literature is replete with statements of the following sort: "Rome had little independent science, however great were her engineering achievements,"[1] "the Romans showed no particular originality in that line [science],"[2] "the Latins produced no great creative men of science,"[3] "[there was an] abyss ... between Hellenic science and Roman science, the latter being at best a very imperfect offspring of the former,"[4] "the Romans ... made no scientific discoveries,"[5] and "the Romans themselves ... made hardly any original contributions to science."[6] What works the Romans produced that could be characterized as "science" were "of a distinctly low order" and "essentially derivative in character."[7]

The Romans had little interest in abstractions; they tended to concern themselves only with the practical. "At heart the Roman had little time for theory. He was interested in law and order, an efficient administration, a sanitary city, and his family."[8] The contrast between the Greek and the Roman temperament was stark. Greek scientists and philosophers had a disdain for engineering, even to the point of considering it vulgar. Archimedes' engineering feats were spectacular, yet he considered the accomplishments so plebian that he could not bring himself to describe his methods in writing. He repudiated "as sordid and ignoble the whole trade of engineering."[9]

Greek disparagement for the mechanical arts is illustrated by a passage from Xenophon's *Oeconomicus* (c. 400 B.C.). Xenophon quoted Socrates as stating "those arts which are called handicrafts are objectionable, and are indeed justly held in little repute in communities; for they weaken the bodies of those who work at them or attend to them, by compelling them to sit and to live indoors; some of them, too, to pass whole days by the fire; and when the body becomes effeminate, the mind loses its strength. Such mechanical occupations also, as they are termed, leave those who practice them no leisure to attend to the interests of their friends or the commonwealth."[10]

The Parthenon is a building of legendary beauty and aesthetics; the expense of its construction nearly bankrupted Athens. Yet the Athenians neglected to build a common bridge along a major road leading to the city; travelers were forced to ford a river to reach Athens.[11]

The American engineer David B. Steinman (1886–1960) explained that the theoretical and practical approaches are complementary to each other:

> The Greek civilization produced the first man to think in terms of theoretical abstractions, a thinker who started from a theorem or principle and ended with its application. For this reason the Greek culture was a triumph of the abstract, the ideal, the theoretical. But the Roman reasoned in precisely the reverse manner. He started with the practical application, the pragmatic viewpoint, and through a series of experiments deduced therefrom the theoretical principle. This is the logic of the engineer. The world needs both types of thinkers, for the work of one supplements that of the other and mutually acts as a check.[12]

The practical Roman attitude is illustrated by Pliny the Elder's comparison of Egyptian pyramids to Roman aqueducts. Pliny dismissed the pyramids as foolish vanities that possessed no real value. "The Pyramids of Egypt [are] so many idle and frivolous pieces of ostentation of their resources, on the part of the monarchs of that country ... there was great vanity displayed by these men in constructions of this description."[13]

In contrast to the Egyptian monuments, Pliny described Roman engineering works as having genuine value because of their utility. "The public sewers [of Rome are] ... a work more stupendous than any."[14] Pliny argued that the Roman aqueducts were especially worthy of admiration. "If we only take into consideration the abundant supply of water to the public, for baths, ponds, canals, household purposes, gardens, places in the suburbs, and country-houses; and then reflect upon the distances that are traversed, the arches that have been constructed, the mountains that have been pierced, the valleys that have been leveled, we must of necessity admit that there is nothing to be found more worthy of our admiration throughout the whole universe."[15]

## *Concrete and the Arch*

Although the Romans imported much of their theoretical knowledge from Greece, their practical accomplishments in the field of engineering surpassed the Greeks. The Romans were not only great builders, they were also innovators who invented concrete and perfected the arch.

Concrete is a remarkable and versatile building material: a liquid that can be formed and cast, but one that soon turns into stone, even when under water. Cement was discovered when someone mixed a volcanic ash called *pozzuolana* with burnt limestone (lime) and water and found that the mixture hardened into stone. Vitruvius described pozzuolana as "a kind of powder which from natural causes produces astonishing results.... [T]his substance, when mixed with lime and rubble, not only lends strength to buildings or other kinds, but even when piers of it are constructed in the sea, they set hard under water."[16] Addition of sand, gravel, or other aggregates to the cement made concrete. "From the first century B.C. onwards ... the commonest building material in Rome was concrete."[17]

The Greeks had supported their buildings almost entirely with columns and lintels, a *lintel* being a horizontal beam resting on columns. Although stone is strong under compression, it is relatively weak under tension. The longest distance that can be spanned with a stone lintel is about 20 feet (6.1 meters); thus the interior spaces of Greek buildings were continually interrupted by supporting columns.[18]

"The Romans derived their knowledge of arched construction from their Etruscan forbears, and developed the methods of vaulting and dome-building, in brick, concrete, and masonry, to a very high level under the Empire."[19] In an arch, the stresses are distributed in such a way that the structure is mostly under compression and self-supporting. The great strength of the arch allowed the Romans to span distances as great as 140 feet (42.7 meters), creating buildings with large open spaces.

The singular example of the Romans' use of both concrete and the arch is the *Pantheon*, a temple dedicated to "all the gods." The Pantheon was first built in Rome in 27 B.C.; the original building was a rectangular structure in the Greek style with masonry walls and a wooden roof. The first Pantheon was damaged by fire in A.D. 80 and then again in A.D. 110.

In A.D. 118, the Emperor Hadrian resolved to rebuild the Pantheon; the process of reconstruction took about ten years. Edward Gibbon (1737–1794) described Hadrian as both monarch and artist. "The works of Trajan bear the stamp of his genius. The public monuments with which Hadrian adorned every province of the empire, were executed not only by his orders, but under his immediate inspection. He was himself an artist; and he loved the arts, as they conduced to the glory of the monarch."[20]

Instead of simply reconstituting the Pantheon in its original form, Hadrian envisaged a structure that would be one of the most unique and challenging buildings ever erected. The traditional rectangular front was retained for the portico, but the main building imitated a perfect half-sphere.

The lower section was in the form of a cylinder with concrete walls about twenty feet [6.1 meters] thick. These walls supported a massive concrete dome that spanned the entire 142-foot (43.3 meter) diameter of the building. The apex of the dome was 142 feet [43.3 meters] high; the same as the diameter of the base. Near the bottom, the dome was twenty feet (6.1 meters) thick, but narrowed to 7.5 feet (2.3 meters) at the top.

Concrete in the lower part of the dome was mixed with heavy aggregate, such as pieces of basalt, but the material toward the top was lightened by the use of pumice, a porous volcanic stone. The center of the dome was occupied by an unglazed, circular window or *oculus* 19 feet (5.8 meters) in diameter; this opening is the only source of natural light.

The floor was made of marble, and had a convex shape so that rainwater entering through the oculus drained away. The front portico was supported by sixteen granite columns imported from Egypt. Each of the granite columns was 39 feet (11.9 meters) tall and weighed sixty tons. The entrance to the main building was by way of double bronze doors twenty-one feet (6.4 meters) high. The inside was "lined with a great variety of rich oriental marbles ... [but] has been much injured by alterations."[21]

When Michelangelo (1475–1564) visited the Pantheon, he was so impressed that he exuberantly proclaimed that at least part of the building had to be of angelic, not human, design.[22] The Pantheon has endured despite unfavorable circumstances. It was built on clay, the most unstable and unsuitable substrate to support a building. The Roman architects compensated by enlarging the base of the structure so as to lighten the load on the earth. The concrete dome does not have steel or iron reinforcement anywhere within it; this design is inconceivable to a modern architect. However the Pantheon stands today in a remarkable state of preservation. It "is the most perfect among existing classical buildings in Rome."[23]

## *Roads*

Among the great Roman accomplishments in the field of engineering was the construction of a network of roads throughout the entire Empire. At the height of the Roman

Empire, 372 roads extended over 50,000 miles (80,467 kilometers).[24] These roads were interconnected, thus the saying "all roads lead to Rome."

The Greek idea of a highway between towns consisted of parallel wagon ruts in dirt. Streets in Greek cities were "narrow, muddy, unpaved, and undrained."[25] Most Greek roads were nothing more than footpaths, and could not be successfully navigated by a pack animal, let alone a cart or carriage.[26] But the Romans built roads of such enduring quality that their remains are found today in Britain, throughout Europe, and in Syria and North Africa.[27] "In solidity of construction, they have never been excelled."[28] Construction of the first of the great Roman roads, the *Via Appia* (Appian Way), began in 312 B.C.[29] The *Via Appia* extended southeast of Rome for a distance of 162 miles (261 kilometers).

The first step in building a Roman road was to dig two parallel ditches to "mark the breadth of the road."[30] The roadway was then excavated to a depth of about three feet [0.9 meters], and backfilled with several layers of various materials designed to produce a permanent and substantial substratum. When completed, the final road was higher in elevation than the surrounding terrain. In Britain, segments of surviving Roman roads have embankments as high as eight feet (2.4 meters), although five feet (1.5 meters) is more typical.[31]

The width of Roman roads could be as little as ten feet (3.0 meters) for roads of secondary importance (or important roads passing through difficult terrain), to forty feet (12.2 meters) for main highways. The primary purpose of the embankment was to raise the roadway for proper drainage, for only materials that are properly drained of groundwater can provide a firm foundation that will not shift. Another advantage to an elevated roadway was that it made it harder for an enemy to ambush an advancing legion.

Roman roads were usually constructed of three or four layers. The materials used in each layer varied from place to place, because the Romans made use of locally-available stones, gravels, sands, and earth. The first layer consisted of stones or rubble often obtained by digging ditches at the sides of the embankment. The use of solid stone in the basal layer made a firm foundation for the overlying road.

The second layer consisted of fill in the form of whatever was at hand, usually dirt mixed with gravel or sand to increase permeability and facilitate drainage. The composition of the surface layer varied, depending on the road's location and available materials. In cities, the roads could be paved with flagstone, but in the countryside the use of gravel, sand, or aggregate was typical.

The surface was sloped steeply from the middle toward the sides; the middle of a road fifteen feet (4.6 meters) wide could be as much as a full foot (0.3 meters) higher than the sides. The final step in road construction was to lay parallel lines of stones along the sides to serve as curbs. The total thickness of the completed road was usually about three to five feet (1.0 to 1.5 meters). Roman roads were enormously resistant to wear; a Roman road has been compared to "a wall lying on its side." The operative theory was that a well-built road would require less maintenance.[32]

The Roman poet Publius Papinus Statius (c. A.D. 45–96) described the building of a Roman road in his *Silvae*. "The first task was to prepare the furrow, to open a track and with deep digging hollow out the earth; the next in other wise to re-fill the caverned trench, and prepare a lap on which the convex surface of the road might be erected, lest the ground should sink or the spiteful earth yield an unstable bed for the deep-set blocks: then, with close-knit revetments [retaining walls] on this side and on that, and with many a brace, to gird the road."[33]

Roman roads were typically constructed in short, straight sections. "Roman roads are remarkable for preserving a straight course from point to point regardless of obstacles which might have been easily avoided."[34] The philosophy of the Roman engineers was to solve construction problems in the most efficient way possible. If feasible, they would tunnel through intervening hills or mountains. But if it was easier they would direct the roadway around an obstacle. In swamps and marshes, the Romans supported their roads with timber pilings.

## *Aqueducts*

The Romans knew how to dig wells. Vitruvius described how to "find water,"[35] and discussed the "digging of wells."[36] But the Romans probably discovered that in densely populated urban settings wells became contaminated from the infiltration of sewage. So they constructed aqueducts to bring water from springs in the surrounding countryside into Rome.

The first aqueduct, the *Aqua Appia*, was built c. 300 B.C.; it was eleven miles (17.7 kilometers) long and brought water to Rome entirely through underground pipes.[37] As time went on, both the city and the need for water grew. Convenient sources of water were exhausted and the Romans had to resort to more difficult means to import water.

The primary difficulty in constructing an aqueduct was to build a channel with a uniform downhill slope over a topography that may have been undulating. To compensate for changes in ground elevation in some places, the Romans utilized elegant arched bridges that we commonly associate today with their aqueducts.

The first overhead Roman aqueduct, the *Aqua Marcia*, was built in 144 B.C. Throughout the Empire, the Romans built some 200 aqueducts.[38] To supply Rome and its suburbs, eleven aqueducts were constructed between 312 B.C. and A.D. 226.[39] The combined length of the waterways was about 252 miles (405 kilometers); by the year A.D. 97 they were bringing a total of 58 million U.S. gallons (220 million liters) of water into Rome each day to supply a population of around one million.[40]

Upon arriving in Rome, the water first ran into settling and holding tanks, from there it was distributed to public fountains and baths as well as industries.[41] In theory, there was no provision for private water use or lines. However the emperors and the powerful managed to get private distribution networks. The wealthy arranged for private deliveries by bribing water officials. Sextus Julius Frontinus (c. A.D. 35–103) estimated that the water coming into Rome was distributed as follows: 17 percent to the Emperor, 19 percent to private persons, and 44 percent for public use.[42]

Frontinus, appointed water commissioner in A.D. 97, complained about substantial losses from theft and fraud. "Another variance consists in this: that there is one measure at the intake; another, and by no means smaller one, at the settling reservoir; and the smallest at the distribution. The cause of this is the fraud of the water-men, whom we have detected diverting water from the pubic conduits for private use; but a large number of proprietors of land also, whose fields border on the aqueducts, tap the conduits; whence it comes that the public water-courses are brought to a standstill by private citizens, yea, for the watering of their gardens."[43]

## *Baths*

The supply of water furnished by the aqueducts made the famous Roman baths possible. The baths were elaborate structures and bathing was both a ritual and a social occasion. Although baths varied in size, all had a plumbing system that furnished cold, tepid, and hot water. We are not sure how the Romans heated the water for their baths, but they may have employed boilers made of bronze.[44] Rooms were centrally heated by the *hypocaust*. A hypocaust was constructed by elevating the floor of a room about three feet [0.9 meters], and kindling a fire in the hollow space underneath a thick stone floor.[45]

A typical procedure for a bather was to first shed his clothes and be anointed with oil in "the undressing-room or *apodyterium*."[46] No soap was ever used. After exercising in the nude, the bather would visit a steam room (*caladarium* or *sudatorium*) where his accumulated grime was scraped off with curved pieces of metal. The bath was finished with plunges in the tepid (*tepidarium*) and then the cold pools (*frigidarium*).[47] Presumably, the public baths charged a fee, and services were provided by slaves. Large baths contained not only exercise facilities, but could have had game rooms, gardens, and libraries.[48] "At the time of greatest luxury these baths or *thermae* formed a whole group of buildings containing a great number of rooms whose exact uses can in many cases hardly be ascertained nowadays."[49]

## Sewers

With so much water coming into Rome each day, an efficient drainage system was needed to remove waste water and also to drain the streets of rain water during storms. The Romans were thus great builders of sewers. The original impetus for sewer construction was to drain swampy land. As the city grew in size, the sewers became indispensable for draining effluent from public latrines and dirty water from baths. A more efficient system of water utilities would not be built until the nineteenth century.

The most significant sewer in Rome was the *cloaca maxima*, which "was probably open originally and only later closed."[50] The *cloaca maxima* drained into the Tiber River, and was originally constructed as early as the 6th or 7th century B.C.[51] Writing c. 1828, Barthold Georg Niebuhr (1776–1831) expressed his admiration for the structure:

> Even at the present day there stands unchanged the great sewer, the *cloaca maxima*, the object of which, it may be observed, was not merely to carry away the refuse of the city, but chiefly to drain the large lake which was formed by the Tiber between the Capitoline, Aventine and Palatine.... [T]his work consisting of three semicircles of immense square blocks, which, though without mortar, have not to this day moved a knife's breadth from one another, drew the water from the surface, conducted it into the Tiber, and thus changed the lake into solid ground; but as the Tiber itself had a marshy bank, a large wall was built as an embankment, the greater part of which still exists. This structure equalling the pyramids in extent and massiveness, far surpasses them in the difficulty of its execution. It is so gigantic, that the more one examines it, the more inconceivable it becomes how even a large and powerful state could have executed it.[52]

## Bridges

The Romans were accomplished builders of bridges. Some of the more elegant bridges were those constructed to accommodate aqueducts. One of the most beautiful and impressive of all the aqueduct bridges is not found in Rome, but in the southeast of France. Around 19 B.C., the Romans built an aqueduct 31 miles (50 kilometers) long to bring water from a spring to the city of Nimes. Thirteen miles (21 kilometers) outside of Nimes, a massive, three-story bridge was built to carry the waterway over the River Gard; the structure is known today as the Pont du Gard (bridge on the river Gard). "As a bridge, the *Pont du Gard* has no rival for lightness and boldness of design among the existing remains of works of this class carried out in Roman times."[53]

The superstructure of the Pont du Gard consists of three layers of arches. The arches on the lower two layers have spans of between 50 and 80 feet (15 to 24 meters), the arches on the upper section are about 12 feet (3.7 meters) in length. The total length of the aqueduct/bridge is 900 feet (274 meters); its total height is 160 feet (49 meters). The width at the bottom is nearly 30 feet (9 meters), but narrows to ten feet (3.0 meters) at the top. The Pont du Gard is built entirely of locally quarried stone; the stones comprising the first two tiers of the bridge are held together solely by friction — no mortar or iron clamps were used in the construction.[54]

In *Confessions*, the French philosopher and author Jean-Jacques Rousseau (1712–1778) described his reaction on first visiting the Pont du Gard:

> I had been told to go and see the *Pont-du-Gard*; nor did I fail to do so. After a dinner of excellent figs I took a guide and set out to visit it. 'Twas the first piece of Roman architecture I had ever seen. I had expected to behold a monument worthy of the hands that reared it. For once, the reality passed my expectation, and this was the only time in my life it ever did so. None but Romans could produce this effect. The aspect of this simple and noble work struck me all the more forcibly from being in the midst of a desert where the silence and solitude that reign around render the object more striking and one's wonder more profound, for this pretended bridge is simply an aqueduct. You cannot help asking, what force was able to transport those enormous stones so far from any quarry and unite the arms of so many thousands of men in a place no one inhabits. I went through the three stories of this superb edifice whereon my respect would hardly allow me to tread with my feet. The echo of my steps beneath those immense vaults made me think I heard the lusty voice of the builders thereof. I lost myself like an insect in this immensity. And yet in the comparison of my own littleness I felt my heart dilate, and I said to myself with a sign, "Why was not I born a Roman!" I remained here for several hours in ravishing contemplation. I returned dreamy and pensive, and this reverie was not favorable to Madam de Larnage. She had taken good care to forewarn me against the girls of Montpellier, but not against the *Pont du Gard*. It is impossible to provide for every contingency.[55]

The Pont du Gard displays qualities that are present in much of Roman construction: strength and permanence. The Romans overbuilt their creations for at least two reasons. First, they lacked the mathematics necessary to calculate stresses so they had to error on the side of safety. Secondly, they wanted structures that were strong and enduring for both pragmatic and emotional reasons. The Roman desire for immortality was captured by the poet Horace (65–8 B.C.).

> And now 'tis done: more durable than brass
> My monument shall be, and raise its head
> O'er royal pyramids: it shall not dread
> Corroding rain or angry Boreas,
> Nor the long lapse of immemorial time.
> I shall not wholly die: large residue
> Shall escape the queen of funerals. Ever new
> My after fame shall grow....[56]

On the Roman bridge in Spain named Puente Alcantara, the builder inscribed, "I have left a bridge that shall remain for eternity."[57]

Roman bridges have been called "dams with holes," reflecting the monumental nature of their size and strength. Like modern bridge construction, Roman bridge building usually began with the emplacement in a river of a coffer dam or caisson consisting of timbers arranged in four walls to keep out the river water.

After removing the water inside the caisson, the Romans would excavate until either they reached bedrock or could dig no deeper. They would then make a pier by piling up stones. The piers were connected by arches fabricated of stone, concrete, or brick. The length of the arches was calculated with an eye to the aesthetic. The number of arches was always odd, with the longest arch in the middle so as to draw the eye's attention to the center of the bridge. Arches of shorter span stood symmetrically on each side of the central segment.

Unfortunately, because the Roman bridge piers were so large, they substantially reduced the cross-sectional areas of the rivers they spanned. The piers typically occupied one-third of the extent spanned by a bridge.[58] Forced to flow through a small opening, river water speeds up and tends to scour out a bridge's foundations. Still, several Roman bridges have stood for more than two thousand years, albeit with periodic repairs. Of the eight stone bridges the Romans constructed across the Tiber River in Rome, six are still standing. One of these, the Pons Mulvius (Ponte Molle) carried the heavy tanks of the Italian, German, and Allied forces during World War II.[59]

The Romans also made pontoon bridges. In his *Commentaries*, Julius Caesar described the construction in 55 B.C. of a temporary bridge across the Rhine that was about 1,450 feet (442 meters) long.[60]

> Caesar had decided to cross the Rhine; but he deemed it scarcely safe, and ruled it unworthy of his own and the Romans' dignity, to cross in boats. And so, although he was confronted with the greatest difficulty in making a bridge, by reason of the breadth, the rapidity, and the depth of the river, he still thought that he must make that effort, or else not take his army across. He proceeded to construct a bridge on the following plan. He caused pairs of balks [timber beams] eighteen inches [0.457 meters] thick, sharpened a little way from the base and measured to suit the depth of the river, to be coupled together at an interval of two feet. These he lowered into the river by means of rafts, and set fast, and drove home by hammers; not, like piles, straight up and down, but leaning forward at a uniform slope, so that they inclined in the direction of the stream. Opposite to these, again, were planted two balks coupled in the same fashion, at a distance of forty feet [12.2 meters] from base to base of each pair, slanted against the force and onrush of the stream. These pairs of balks had two-foot [0.61 meter] transoms let into them atop, filling the interval at which they were coupled, and were kept apart by a pair of braces on the outer side at each end. So, as they were held apart and contrariwise clamped together, the stability of the structure was so great and its character such that, the greater the force and thrust of the water, the tighter were the balks held in lock. These trestles were interconnected by timber laid over at right angles, and floored with long poles and wattlework [interlaced branches]. And further, piles were driven in aslant on the side facing down stream, thrust out below like a buttress and close joined with the whole structure, so as to take the force of the stream; and others likewise at a little distance above the bridge, so that if trunks of trees, or vessels, were launched by the natives to break down the structure, these fenders might lessen the force of such shocks, and prevent them from damaging the bridge. The whole work was completed in ten days ... and the army was taken across.[61]

## CHAPTER 4

# The Roman Writers

## *Vitruvius Pollio:* On Architecture *(c. 30 B.C.)*

Vitruvius Pollio (wrote c. 30 B.C.) was the author of a treatise on architecture titled *De Architectura* (*On Architecture*). We know next to nothing about Vitruvius himself. The manuscript was finished around 30 B.C., and appears to have been written toward the end of Vitruvius' life. As was the custom for the time, the book is dedicated and addressed to Gaius Octavius (63 B.C.–A.D. 14), the adopted heir of Julius Caesar who later took the name of Augustus as the first Emperor of Rome. It is evident that Vitruvius worked for both Caesar and Augustus.[1]

*On Architecture* is divided into ten books. Although today architecture is considered to be the science of constructing buildings, the ancients had a considerably looser and more encompassing definition that included what today would be classified as civil and mechanical engineering. Vitruvius himself said that there were three "departments of architecture:" "the art of building, the making of time-pieces, and the construction of machinery."[2] *On Architecture* contains discussions of how to construct aqueducts, wells, and cisterns, as well as catapults, siege machines, and water wheels.

The first topic that Vitruvius discussed in *On Architecture* was the education of the architect. He expressed the opinion that an architect must have a liberal education that includes subjects such as history and philosophy, not just a narrow instruction devoted to technical details and facts. "[The architect] ought, therefore, to be both naturally gifted and amenable to instruction. Neither natural ability without instruction nor instruction without natural ability can make the perfect artist. Let him be educated, skilful with the pencil, instructed in geometry, know much history, have followed the philosophers with attention, understand music, have some knowledge of medicine, know the opinions of the jurists, and be acquainted with astronomy and the theory of the heavens."[3]

Vitruvius argued that an architect should be knowledgeable in areas such as history and music, and also benefited from an acquaintance with philosophy. "As for philosophy, it makes an architect high-minded and not self-assuming, but rather renders him courteous, just, and honest without avariciousness. This is very important, for no work can be rightly done without honesty and incorruptibility."[4]

But Vitruvius largely failed to explain how a liberal education was of practical benefit in the application of his art. "The theme of architecture as one of the liberal arts is ostentatiously picked up and dropped at intervals throughout the work [*On Architecture*], but at very few points can it be said seriously to illuminate the main subject matter."[5]

The "fundamental principles of architecture" are listed as: order, arrangement, eurythmy, symmetry, propriety, and economy.[6] There is some overlap between these different principles; it is clear that Vitruvius considered architecture to be art of a high order. His definition of "eurythmy" mentioned symmetry. "Eurythmy is beauty and fitness in the adjustments of the members. This is found when the members of a work are of a height suited to their breadth, of a breadth suited to their length, and, in a word, when they all correspond symmetrically."[7]

In the beginning of Book 2, Vitruvius discussed the origin of the first human houses. He began by speculating that the discovery and utilization of fire by man fostered the invention of language:

> The men of old were born like the wild beasts, in woods, caves, and groves, and lived on savage fare. As time went on, the thickly crowded trees in a certain place, tossed by storms and winds, and rubbing their branches against one another, caught fire, and so the inhabitants of the place were put to flight, being terrified by the furious flame. After it subsided, they drew near, and observing that they were very comfortable standing before the warm fire, they put on logs and, while thus keeping it alive, brought up other people to it, showing them by signs how much comfort they got from it. In that gathering of men, at a time when utterance of sound was purely individual, from daily habits they fixed upon articulate words just as these had happened to come; then, from indicating by name things in common use, the result was that in this chance way they began to talk, and thus originated conversation with one another.[8]

*On Architecture* included a description of *pozzuolana*, the volcanic ash that the Romans used to make concrete by mixing it with lime and aggregate. "There is also a kind of powder which from natural causes produces astonishing results. It is found in the neighborhood of Baiae and in the country belonging to the towns round about Mt. Vesuvius. This substance, when mixed with lime and rubble, not only lends strength to buildings of other kinds, but even when piers of it are constructed in the sea, they set hard under water."[9]

Vitruvius described how the Romans would kill a tree to season the timber, only felling the tree when the seasoning process was complete. "In felling a tree we should cut into the trunk of it to the very heart, and then leave it standing so that the sap may drain out drop by drop throughout the whole of it. In this way the useless liquid which is within will run out through the sapwood instead of having to die in a mass of decay, thus spoiling the quality of the timber. Then and not till then, the tree being drained dry and the sap no longer dripping, let it be felled and it will be in the highest state of usefulness."[10]

The content of *On Architecture* is for the most part technical and straightforward with few digressions and moralizing. However, in the introduction to Book 6 Vitruvius honored his parents by thanking them for the excellence of his education. "All the gifts which fortune bestows she can easily take away; but education, when combined with intelligence, never fails, but abides steadily on to the very end of life. Hence, I am very much obliged and infinitely grateful to my parents ... for having taken care that I should be taught an art, and that of a sort which cannot be brought to perfection without learning and a liberal education in all branches of instruction."[11]

In Book 6, Vitruvius also discussed climate, and his conviction that human temperaments change systematically from north to south. He showed that he was not free of the universal prejudice: that his own tribe was superior to all others.

> But although southern nations have the keenest wits, and are infinitely clever in forming schemes, yet the moment it comes to displaying valor, they succumb because all manliness of spirit is sucked out of them by the sun. On the other hand, men born in cold countries are indeed readier to meet the shock of arms with great courage and without timidity, but their wits are so slow

> that they will rush to the charge inconsiderately and inexpertly, thus defeating their own devices. Such being nature's arrangement of the universe, and all these nations being allotted temperaments which are lacking in due moderation, the truly perfect territory, situated under the middle of the heaven, and having on each side the entire extent of the world and its countries, is that which is occupied by the Roman people.[12]

Vitruvius was conservative; this is clear from his critique of Roman art. In Chapter 5 of Book 7, titled "The Decadence of Fresco Painting," he decried the modern trend in art away from realistic portrayals of nature. Vitruvius began by claiming that art must imitate nature. "A picture is, in fact, a representation of a thing which really exists or which can exist: for example, a man, a house, a ship, or anything else from whose definite and actual structure copies resembling it can be taken."[13]

Vitruvius then acclaimed the excellent art of the ancients, and compared it with modern art that contained "decadent monstrosities" such as plant stalks with human heads.

> [The paintings of the ancients contained] figures of the gods or detailed mythological episodes, or the battles at Troy, or the wanderings of Ulysses, with landscape backgrounds, and other subjects reproduced on similar principles from real life. But those subjects which were copied from actual realities are scorned in these days of bad taste. We now have fresco paintings of monstrosities, rather than truthful representations of definite things.... Such things do not exist and never have existed. Hence, it is the new taste that has caused bad judges of poor art to prevail over true artistic excellence.... Yet when people see these frauds, they find no fault with them but on the contrary are delighted, and do not care whether any of them can exist or not. Their understanding is darkened by decadent critical principles, so that it is not capable of giving its approval authoritatively and on the principle of propriety to that which really can exist.[14]

The Romans' open aqueducts, characterized by continuous, gentle slopes from source to delivery, could have been replaced by siphons if the Romans had better pipe technology. A pipe carrying water does not have to continuously slope downhill. If the end point is at a lower elevation than the source, water will flow through the pipe, successfully negotiating intermediate rises which do not exceed the elevation of the starting point.

The Romans "occasionally used [siphons] to cross steep valleys, but they only used siphon construction when absolutely necessary, because "siphons were expensive to build and repair."[15] Roman pipes had difficulty withstanding the high water pressures they were subjected to in siphons. In Chapter 6 of Book 8 of *On Architecture*, Vitruvius explained what happens when a lead pipe is subjected to significant water pressure. "The water will break out, and burst the joints of the pipes."[16]

Further comments by Vitruvius revealed that some Romans at least understood the nature of lead poisoning. "Water from clay pipes is much more wholesome than that which is conducted through lead pipes, because lead is found to be harmful for the reason that white lead [lead carbonate] is derived from it, and this is said to be hurtful to the human system."[17]

The scope of *On Architecture* is encyclopedic; not only are there discussions of building methods and techniques, but also topics such as acoustics in a theatre, the construction of catapults, water screws, and siege machines, and astronomy and the construction of sundials. The work was rediscovered in Europe in the fifteenth century and was influential in the revival of classical architecture that occurred during the Italian Renaissance. *On Architecture* remains in print today.[18]

## *Lucius Annaeus Seneca (c. 4 B.C.–A.D. 65) and Nero (A.D. 37–68)*

### *Nero's Tutor*

Lucius Annaeus Seneca (c. 4 B.C.–A.D. 65) was an author, statesman, orator, and poet. In the area of natural philosophy, he contributed an original work in seven books, *Naturales Quaestiones* (*Natural Questions*). Seneca's life was fatefully intertwined with one of the most infamous Roman Emperors, Nero (A.D. 37–68).

Seneca was born in Spain to a distinguished Italian family; his father was a well-known writer and rhetorician.[19] "In his youth [Seneca] was a vegetarian and a water-drinker, but his father checked his indulgence in asceticism."[20] The young Seneca was taken to Rome where he received a first-class education.[21] He had considerable talents for writing, speaking, and politics, and rose rapidly in Rome.[22] But as a result of a court intrigue, in A.D. 41 Seneca was sent into exile on the island of Corsica for eight years.[23]

The exile was a bitter yet productive time for Seneca. He wrote continuously, including a series of essays on morals and some tragic plays. In A.D. 49, Seneca's exile ended when he was recalled to Rome to become the tutor of the Emperor's new son-in-law, Nero, then a youth of about twelve years of age.

Nero was the only son of a woman named Agrippina, a widow who also happened to be the niece of the Emperor Claudius (10 B.C.–A.D. 54). Claudius was a mediocre ruler, due to his inherent weakness of character. "He was deficient in judgment and reflection. He often said and did things which were really stupid."[24]

Claudius' life was marked by a series of unsuccessful marriages. He was forced to divorce his first wife while she was still a virgin, because her parents fell out of favor with the Emperor Augustus. The second wife died on their wedding day of natural causes. Claudius divorced his third and fourth wives.[25]

Claudius' fifth marriage was to Valeria Messalina, a woman described by Roman historians as a viperous nymphomaniac. "Not confining her licentiousness within the limits of the palace, where she committed the most shameful excesses, she prostituted her person in the common stews, and even in the public streets of capital."[26]

> Hear what Claudius endured. As soon as his wife [Messalina] perceived that her husband was asleep, this august harlot was shameless enough to prefer a common mat to the imperial couch. Assuming a night-cowl, and attended by a single maid, she issued forth; then, having concealed her raven locks under a light-colored peruque [wig], she took her place in a brothel reeking with long-used coverlets. Entering an empty cell reserved for herself, she there took her stand, under the feigned name of Lycisca, her nipples bare and gilded.... [H]ere she graciously received all comers, asking from each his fee.[27]

According to Suetonius (c. A.D. 69–130), Messalina went so far as to seduce a young noble and publicly marry him while still married to the Emperor. "As if her conduct was already not sufficiently scandalous, she obliged C. Silius, a man of consular rank, to divorce his wife, that she might procure his company entirely to herself. Not contented with this indulgence to her criminal passion, she next persuaded him to marry her; and during an excursion which the emperor made to Ostia, the ceremony of marriage was actually performed between them."[28]

The story is verified by Tacitus (c. A.D. 56–120), who apologetically introduced the narrative with the disclaimer that he understood the event to be so fantastic that readers might find it unbelievable.

> I am well aware that it will seem a fable that any persons in the world could have been so obtuse in a city which knows everything and hides nothing, much more, that these persons should have been a consul-elect and the emperor's wife; that, on an appointed day, before witnesses duly summoned, they should have come together as if for the purpose of legitimate marriage; that she should have listened to the words of the bridegroom's friends, should have sacrificed to the gods, have taken her place among a company of guests, have lavished her kisses and caresses, and passed the night in the freedom which marriage permits. But this is no story to excite wonder; I do but relate what I have heard and what our fathers have recorded.[29]

Messalina may not have been that atypical of the upper-class women of Rome. Although a traditional morality still prevailed in the rural districts, historian Barthold Georg Niebuhr (1776–1831) noted that "an honest man generally found a more faithful friend and companion in his female slave than in a Roman lady of rank."[30]

Even a person as dull-witted as Claudius could not escape noticing Messalina's flagrant infidelity. "Messalina was ordered into the emperor's presence, to answer for her conduct ... [but] retired into the gardens.... [I]n the extremity of her distress, she attempted to lay violent hands upon herself, but her courage was not equal to the emergency ... [then] a tribune burst into the gardens, and plunging his sword into her body, she instantly expired."[31]

Disconsolate at five failed marriages, the last most disastrous of all, Claudius summoned the Praetorian guard and told them to kill him if he ever married again. "As I have been so unhappy in my unions, I am resolved to continue in future unmarried; and if I should not, I give you leave to stab me."[32]

## *Nero's Rise to Power*

Claudius' resolve to forsake marriage was short lived. He was seduced by his own niece, Agrippina, a woman with a young son named Nero. Claudius was "ensnared by the arts of Agrippina, the daughter of his brother Germanicus, who took advantage of the kisses and endearments which their relationship admitted, to inflame his desires."[33] Anxious to emulate Philip and Alexander, Agrippina sought to obtain a latter-day Aristotle as tutor for the eleven-year-old Nero. Seneca was the only man who could fill the role, and Agrippina had him recalled from exile.

Agrippina used all of her skill to manipulate Claudius and advance the status of her son, Nero. When Nero was sixteen years of age, he married Claudius' daughter, Octavia. The weak-willed Claudius was even convinced to adopt Nero, making him his successor, over his natural son, Britannicus.[34] In time, Claudius came to regret his decision. But before it could be reversed, he was poisoned by Agrippina, "who had long decided on the crime."[35] Nero immediately ascended to the throne. Absolute power over the Roman Empire was now held by a spoiled seventeen-year-old boy under the baneful influence of a domineering mother.

Nero's fourteen-year reign started auspiciously enough, but ended in an affirmation of Lord Acton's (John Emerich Edward Dalberg-Acton, 1834–1902) aphorism, "power tends to corrupt and absolute power corrupts absolutely."[36] Upon assuming the throne, Nero announced that "he designed to govern according to the model of Augustus and omitted no opportunity of showing his generosity, clemency, and complaisance."[37] Burdensome taxes were reduced or repealed. The noblest of the Senators were granted generous allowances. When he was obligated to sign the death warrant for a condemned criminal, the empathetic Nero said that he regretted that he had ever learned how to read and write.[38]

The first years of Nero's rule were beneficent. "He [Nero] was not in general fond of affairs [of state] and was glad to live at leisure."[39] "Anything to do with politics was of no interest to him [Nero].... [T]he sole activity that interested him was his obsession with excelling as a actor and singer."[40] The management of the government was divided between Agrippina, Seneca, and Burrus, the prefect of the Praetorian guard.[41]

Seneca prospered, both politically and financially. The philosopher was charged by the historian Dio Cassius (c. A.D. 150–235) with exhibiting the most flagrant hypocrisy. "He was convicted of doing precisely the opposite of what he taught in his philosophical doctrines. He brought accusations against tyranny, yet he made himself a teacher of tyrants ... while finding fault with the rich, he himself possessed a property of seven-thousand five-hundred myriads; and though he censured the extravagances of others, he kept five-hundred three-legged tables of cedar wood, every one of them with identical ivory feet, and he gave banquets on them."[42]

As time passed, the influence of Agrippina over her son "was gradually weakened."[43] Agrippina became inflamed, and threatened to replace Nero with Britannicus, pointing out that "Britannicus was now of full age ... [and] was the true and worthy heir of his father's sovereignty."[44] Nero responded by poisoning his fourteen-year-old brother, demonstrating that he was an adept student of his mother's lessons.[45] Explaining Britannicus' death to the Roman Senate, Nero portrayed himself as a victim worthy of empathy, because "he had now lost a brother's help."[46]

Perceiving his mother as a threat to his power, Nero banished her from his court, and attempted to poison her as she had poisoned her husband, Claudius.

> [Nero] deprived her of all honor and power, took from her the guard of Rome, and German soldiers, banished her from the palace and from his society, and persecuted her in every way he could contrive; employing persons to harass her when at Rome with law-suits, and to disturb her in her retirement from town with the most scurrilous and abusive language, following her about by land and sea.... [B]eing terrified with her menaces and violent spirit, he resolved upon her destruction, and thrice attempted it by poison.... [H]owever she had previously secured herself by antidotes.[47]

Finally, a convoluted and ingenious murder scheme was suggested by Anicetus, "a freedman, commander of the fleet at Misenum, who had been tutor to Nero in boyhood and had a hatred of Agrippina which she reciprocated."[48] Agrippina was to be lured out to sea on a ship designed to fall apart. The ship foundered as planned, but Nero's mother escaped, swimming until she encountered some boats that safely transported her to shore and her house.[49]

Upon hearing of his mother's escape, Nero called an emergency conference with Seneca and Burrus. It was resolved that they needed to finish the assassination immediately by direct means. Burrus pointed out that, due to their loyalties to all members of the Emperor's household, the Praetorian guard could not be relied upon. Anicetus volunteered for the job and was instructed by Nero to take as many men as he needed to accomplish the task.

Upon arriving at Agrippina's house, Anicetus forced his way in. "The assassins closed in round her couch, and the captain of the trireme first struck her head violently with a club. Then, as the centurion bared his sword for the fatal deed, presenting her person, she exclaimed 'Smite my womb,' and with many wounds she was slain."[50]

After the death of his mother, Nero began to throw off any inhibitions that might have restrained him earlier. Having little interest in government, the young emperor devoted himself to entertainment and the arts. "Among the other liberal arts which he [Nero] was taught in his youth, he was instructed in music."[51] Fancying himself an accomplished musician, Nero began to perform publicly. The young artist was encouraged by the enthusiastic reception he invariably received, for who would dare to criticize him? Suetonius informs us that "during the time of his [Nero's] performance, nobody was allowed to stir out of the theatre upon any account, however necessary; insomuch, that it is said some women with child were delivered there."[52]

A young noble, Marcus Salvius Otho (A.D. 32–69), made the mistake of boasting to Nero of his wife's beauty. Otho's spouse, Poppaea, "had everything but a right mind."[53] Nero promptly appointed Otho governor of a remote province, allowing him to freely purse Poppaea without any interference from her husband.[54]

Poppaea was receptive to the Emperor's advances. "Wherever there was a prospect of advantage, there she transferred her favors."[55] Infatuated with Poppaea, Nero "soon became disgusted with [his wife] Octavia, and ceased from having any intercourse with her.... [H]e made several attempts, but in vain, to strangle her, and then divorced her for barrenness."[56] Twelve days after divorcing Octavia, Nero married Poppaea.[57]

Octavia was sent into exile. "After an interval of a few days, she received an order that she was to die.... [S]he was then tightly bound with cords, and the veins of every limb were opened; but as her blood was congealed by terror and flowed too slowly, she was killed outright by the steam of an intensely hot bath. To this was added the yet more appalling horror of Poppaea beholding the severed head which was conveyed to Rome."[58]

Eventually, Poppaea was also killed by Nero. He "killed her with a kick which he gave her when she was big with child, and in bad health, only because she found fault with him for returning late from driving his chariot."[59]

Nero's crimes grew until "he butchered, without distinction or quarter, all whom his caprice suggested as objects for his cruelty; and upon the most frivolous pretenses."[60] The monster "destroyed all who were allied to him either by blood or marriage."[61] "He came to think that all it was in his power to do was right and gave heed to those whose speech was prompted by fear or flattery as if they told absolute truth."[62]

When the prefect of the Praetorians, Burrus, died, Seneca's influence over Nero began to wane. Nero "began to lean on worse advisers. They assailed Seneca with various charges, representing [to Nero] that he [Seneca] continued to increase a wealth which was already so vast as to be beyond the scale of a subject, and was drawing to himself the attachment of the citizens, while to the picturesqueness of his gardens and the magnificence of his country houses he almost surpassed the emperor."[63]

Aware of the danger, Seneca sought to mollify Nero by voluntarily surrendering his fortune.[64] The emperor's reply was disingenuous:

> "Your gifts to me will, as long as life holds out, be lasting possessions; those which you owe to me, your parks, investments, your country houses, are liable to accidents. Though they seem much, many far inferior to you in merit have obtained more. I am ashamed to quote the names of freedmen who parade a greater wealth. Hence I actually blush to think that, standing as you do first in my affections, you do not as yet surpass all in fortune...." To these words the emperor added embraces and kisses; for he was formed by nature and trained by habit to veil his hatred under delusive flattery. Seneca thanked him ... but he entirely altered the practices of his former greatness ... [he] seldom appeared in Rome, as though weak health or philosophical studies detained him at home.[65]

Nero subsequently tried to poison Seneca, but the attempt was in vain as Seneca maintained himself "on the very simple diet of wild fruits, with water from a running stream when thirst prompted."[66]

## *Natural Questions*

In effective retirement, Seneca wrote *Naturales Quaestiones* (*Natural Questions*) from about A.D. 63 to 65. Seneca identified himself as a Stoic, and most of the text of *Natural Questions* is an exposition of the standard Stoic views on questions in natural philosophy. Seneca was not an original researcher, and *Natural Questions* is largely derivative of Greek science, chiefly the works of Aristotle and Posidonius.[67]

*Natural Questions* is divided into seven sections, or books, on topics related to meteorology, geology, and astronomy. Book 1 opens with a discussion of the merits of natural philosophy. Seneca admitted and justified his love of philosophy. "Life would have been a useless gift, were I not admitted to the study of such themes. What cause for joy would it be to be set merely in the number of those who live? In order to digest food and drink? To repair a diseased, enfeebled body, that would perish unless it were continually refilled, and thus lead the life of a sick man's attendant?"[68]

Seneca discoursed on the nature of God. "But what is God? The universal intelligence. What is God, did I say? All that you see and all that you cannot see. His greatness exceeds the bounds of thought. Render Him true greatness and He is all in all, He is at once within and without His works.... He is wholly reason."[69]

Like Aristotle, Seneca was a teleologist who saw purpose and design in nature. "Mortal eyes are so sealed by error that men believe this frame of things to be but a fortuitous concourse of atoms, the sport of chance. And yet could this universe be fairer, more carefully adjusted, more consistent in plan?"[70]

The discussion of natural phenomena in Book 1 began with Seneca hypothesizing about the nature and causes of fireballs (meteors). He postulated that both meteors and lightning are caused by "violent friction of the atmosphere," just as sparks will fly off steel striking a piece of flint.[71] The same explanation had been advanced by Aristotle in *Meteorologica*.[72]

Seneca concluded that the cause of meteors cannot be the falling of stars because the quantity of stars seen in the sky never diminishes. "If this had been so, they [the stars] would ere this have disappeared."[73]

The text also included a discussion of *coronae* (diffraction rings around the sun or moon) and rainbows. Seneca mistakenly attributed the existence of rainbows to reflected light, instead of refracted (bent) light.[74] Book 1 includes a mention of the *aurora borealis* (northern lights), which Seneca described as "the heavens appeared to be on fire."[75]

Book 1 ends with a discussion of mirrors and morals. Seneca described the erotic use of mirrors by a man who "employed the reflecting power of the mirrors to reveal scenes of revolting and abominable iniquity."[76] The discussion is more extensive and explicit than the tasteful reader will find palatable. Seneca closed by objecting to the growth of luxury and vice in Rome, deploring that men now adorn themselves as richly as women once did. "All that used to be regarded as the decoration of women has become part and parcel of the outfit of man."[77] The discourse is typically Stoic in its rejection of sensuality and luxury.

Book 2 was devoted to a discussion of the nature of air, lightning, and thunder. Seneca had no way of knowing the true causes, but he displayed throughout *Naturales Questiones* a scientific attitude by listing the hypotheses that had been proposed as explanations and then dispassionately commenting on the evidence for each. This was largely the method of Aristotle.

As a Stoic, Seneca was bound to link the study of natural philosophy with morality. The purpose of natural philosophy was to understand the universe. Knowledge of the universe in turn implied a comprehension of man's destiny and moral duty.[78] In Book 2 of *Naturales Questiones*, Seneca was plainspoken in his opinion that the study of nature should be linked with morality. "I quite allow that some moral should be attached to all studies and all discourse."[79]

Book 3 is on the subject of "terrestrial waters." In the beginning, Seneca bemoaned the fact that he had so little time to devote to scientific questions, noting that even an entire lifetime is insufficient. "Old age presses hard on the rear, upbraiding me with the years bestowed on vain pursuits."[80]

The first question that Seneca dealt with was a geological mystery that had perplexed the ancients for thousands of years: the nature of the hydrologic cycle. It is obvious that rivers run continually into the sea, yet the sea never fills up and the rivers never run dry.

Seneca first noted that some people had proposed that rivers are recharged by rainfall. This is the correct explanation, but Seneca rejected it. His first reason was that he did not believe rainfall was sufficient to saturate the ground to great depths. "As a diligent digger among my vines, I can affirm from observation that no rain is ever so heavy as to wet the ground to a depth of more than 10 feet [3.0 meters]. All the moisture is absorbed in the upper layer of earth without getting down to the lower ones. How, then, can rain, which merely damps the surface, store up a supply sufficient for rivers?"[81]

It is unlikely that Seneca ever picked up a spade; it is more likely that he supervised excavations by slaves on his estate. Seneca's second reason was that "some springs well up in the very summit of a mountain. It is plain, therefore, that the water in them is forced up or forms on the spot, since all the rain water runs off."[82] This is inexplicable, in that it is indeed impossible for a spring to arise at the "very summit" of a mountain, but quite plausible for springs to appear anywhere below the apex.

Following Aristotle, Seneca claimed that one source for underground water was condensation. "Underground the forces, whatever they are, that turn air into water, are constant — perpetual darkness, everlasting cold, inert density ... we Stoics are satisfied that the earth is interchangeable in its elements."[83] Having established that air could turn into water, Seneca also speculated that other transubstantiations were possible. "All elements arise from all: air comes from water, water from air; fire from air, air from fire. So why should not earth be formed from water, and conversely water from earth?"[84]

To explain why the sea did not grow from the influx of river water, Seneca invoked a hidden flow of water from the sea to the earth through obscure channels. "The sea, therefore, does not get larger, because it does not assimilate the water that runs into it, but forthwith restores it to the earth. For the sea water returns by a secret path, and is filtered in its passage back. Being dashed about as it passes through the endless, winding channels in the ground, it loses its salinity, and purged of its bitterness in such a variety of ground as it passes through, it eventually changes into pure, fresh water."[85]

The idea that there were numerous unseen channels inside the Earth through which water flowed was an old one. In *Phaedo*, Plato stated "there are diverse regions in the hollows on the face of the globe everywhere.... [A]ll have numerous perforations, and there are passages broad and narrow in the interior of the earth, connecting them with one another; and there flows out of and into them, as into basins, a vast tide of water."[86]

Seneca compared the flow of water and air throughout the Earth to the movement of blood within the human body. "My firm conviction is that the earth is organized by nature much after the plan of our bodies, in which there are both veins and arteries, the former blood-vessels, the latter air-vessels. In the earth likewise there are some routes by which water passes, and some by which air. So exactly alike is the resemblance to our bodies in nature's formation of the earth, that our ancestors have spoken of veins of water."[87]

Seneca's comparison of the earth and the human body is one of the first invocations of the *Doctrine of the Macrocosm and Microcosm*. This is "the belief that there exists between the universe and the individual human being an identity both anatomical and psychical. The macrocosm is the universe as a whole, whose parts are thought of as parts of a human body and mind. The microcosm is an individual human being whose parts are thought of as analogous to the parts of the larger universe."[88]

The Doctrine dates at least to the presocratic natural philosophers. Democritus wrote "man is a universe in little."[89] Another early reference is in Plato's dialogue, *Philebus*, where an analogy is drawn between the cosmos and animal bodies. "The elements which enter into the nature of the bodies of all animals, fire, water, air, and ... earth, reappear in the constitution of the world ... the cosmos ... may be considered to be a body, because made up of the same elements."[90] In Plato's *Timaeus*, the world itself is considered to be "a living creature."[91]

The *Doctrine of the Macrocosm and Microcosm* became a key tenet of Medieval European thought. "It was an attempt to bring a complex universe into some intelligible or simple relation to man himself and thus give a unity to the whole."[92]

In his discussion of why bodies float or sink, Seneca came close to stating *Archimedes' Principle* but instead merely observed that objects heavier than water sink, while those lighter than water float. "Weigh any object and compare it with water while they are equal bulk for bulk. If the water is the heavier, it will bear the object that is lighter than itself, and will raise it above its surface to a height proportionate to its lightness; objects heavier than the water will sink in it."[93]

Seneca noted that there are lakes in which the water has been made more dense by the addition of minerals (salt), and therefore people float in them. "It is well known that there are lakes whose waters bear up those who cannot swim."[94] He also commented on the fact that there exist stones less dense than water. "There are many light pumice stones ... that float."[95]

Seneca closed Book 3 with a moral digression. "Vice quickly creeps in; virtue is difficult to find; she requires ruler and guide. But vice can be acquired even without a tutor."[96]

The first part of Book 4 was devoted to a discussion of the source of the Nile; the second part to the subjects of hail and snow. Seneca's discourse on the source of the Nile is classic. He systematically listed and discussed all the hypotheses that had been proposed for the cause of the Nile's annual flood.

Unfortunately, the discussion ends abruptly before Seneca can tell us what his favored explanation is. The termination is so abrupt that we must conclude that the last part of the discussion has been lost. The subject of Book 5 is wind, and Book 6 discussed earthquakes. In his treatment of temblors, Seneca affirmed naturalism. "It will be useful also to be assured that none of these things is the doing of the gods, and that the moving of heaven and earth is no work of angry deities. Those phenomena have causes of their own."[97]

Before giving the reader what he believed to be the true cause of earthquakes, Seneca justified his work on the problem. "What ... will be the reward of our labor? That reward, I say, which surpasses all others, the knowledge of nature.... [T]he study is pursued, not in hope of gain, but from the wonder it excites."[98]

Seneca noted that the explanations proffered for earthquakes included attributions to fire, water, air, and the Earth itself. He argued that the most logical explanation involved the movement of air within the Earth, an idea apparently derived from Aristotle.[99] Seneca first maintained that the interior of the Earth must contain air. "The Earth does not lack air within; that everybody knows.... Unless the Earth possessed this store of air, how could she infuse it into so many trees and crops, which derive their life from this and no other source?"[100]

Earthquakes, Seneca attested, are due to the sudden and violent movement of air within the Earth. He justified his theory by mentioning the inherent attributes of air, one of the four basic elements. "No one, I suppose, can doubt that there is nothing so restless, so capricious, so fond of disturbance as air."[101]

The last part of *Natural Questions*, Book 7, discussed comets; the stars and planets were also mentioned. The discussion followed the usual pattern. All of the existing theories as to the nature of comets were mentioned and evaluated in terms of their relative strengths and weaknesses. The common view amongst the Stoic School was that comets were an atmospheric phenomenon. "Our Stoic friends, therefore, are satisfied that, like trumpet meteors and beams, and other portents of the sky, comets are formed by dense air."[102]

But Seneca dissented from the standard Stoic interpretation, concluding that comets were a permanent celestial feature, not a transient phenomenon that occurs in the Earth's upper atmosphere. "I do not agree with my school here, for I cannot think a comet is a sudden fire, but I rank it among nature's permanent creations.... [Comets'] return is at long intervals."[103]

Seneca's science was largely confined to a thoughtful discussion of the works he had read, with limited reference to incidental personal observations. He displayed the scientific mindset through his invocation of naturalism, and his employment of multiple working hypotheses. However, Seneca showed no insight into the necessity of a systematic experimental method, nor did he propose any systematic method whereby scientific knowledge might be increased. His scientific thought was philosophical, based primarily on reason and logical analysis.

Seneca closed *Natural Questions* on an optimistic note with some prescient observations concerning the eventual growth of knowledge. "The day will yet come when the progress of research through long ages will reveal to sight the mysteries of nature that are now concealed.... [P]osterity will be amazed that we remained ignorant of things that will to them seem so plain."[104] "Men will some day be able to demonstrate in what regions comets have their paths, why their course is so far removed from the other stars, what is their size and constitution. Let us be satisfied with what we have discovered, and leave a little truth for our descendants to find out."[105] "Many discoveries are reserved for the ages still to be, when our memory shall have perished. The world is a poor affair if it does not contain matter for investigation for the whole world in every age."[106] "Nature does not reveal all her secrets at once. We imagine we are initiated in her mysteries: we are, as yet, but hanging around her outer courts."[107]

At the very end of *Natural Questions*, Seneca indulged in typical Stoic moralizing by bemoaning the moral degradation prevalent in Rome. "There is one object we are bent on, heart and soul — to be as wicked as possible — and we have not yet attained perfection. Vice is still making progress. Luxury is constantly discovering some new outlet for its madness, indecency some new form of insult on itself.... We are still doing out best to extinguish any spark of virtue that is left."[108]

Seneca further noted that his countrymen had little interest in philosophy or science, and that as a result the extant body of knowledge was not being increased, but eroded. "Philosophy gets never a thought. And so it comes to pass that, far from advance being made toward the discovery of what the older generations left insufficiently investigated, many of their discoveries are being lost."[109] Seneca's pessimism was as prescient and realistic as his optimism. The golden age of knowledge that he envisioned would eventually be realized, but not by his age or culture.

## *Nero's Crimes*

While Seneca was writing *Natural Questions* in effective retirement at a country home, back in Rome Nero was scaling new heights of perversion and tyranny:

> Petulancy, lewdness, luxury, avarice, and cruelty, he [Nero] practiced at first with reserve and in private, as if prompted to them only by the folly of youth; but, even then, the world was of [the] opinion that they were the faults of his nature, and not of his age. After it was dark, he used to enter the taverns disguised in a cap or a wig, and ramble about the streets in sport, which was not void of mischief. He used to beat those he met coming home from supper; and, if they made any resistance, would wound them, and throw them into the common sewer. He broke open and robbed shops; establishing an auction at home for selling his booty.[110]
>
> Besides the abuse of free-born lads, and the debauch of married women, he [Nero] committed a rape upon Rubria, a Vestal Virgin. He was upon the point of marrying Acte, his freedwoman, having suborned some men of consular rank to swear that she was of royal descent. He gelded the boy Sporus, and endeavored to transform him into a woman. He even went so far as to marry him, with all the usual formalities of a marriage settlement, the rose-colored nuptial veil, and a numerous company at the wedding. When the ceremony was over, he had him conducted like a bride to his own house, and treated him as his wife. It was jocularly observed by some person, "that it would have been well for mankind, had such a wife fallen to the lot of his [Nero's] father Domitius."[111]
>
> He [Nero] prostituted his own chastity to such a degree, that after he had defiled every part of his person with some unnatural pollution, he at last invented an extraordinary kind of diversion; which was, to be let out of a den in the arena, covered with the skin of a wild beast, and then assail with violence the private parts of both men and women, while they were bound to stakes. After he had vented his furious passion upon them, he finished the play in the embraces of his freedman, Doryphorus, to whom he was married in the same way that Sporus had been married to himself; imitating the cries and shrieks of young virgins, when they are ravished.[112]

"He [Nero] thought there was no other use of riches and money than to squander them away profusely; regarding all those as sordid wretches who kept their expenses within due bounds; and extolling those as truly noble and generous souls, who lavished away and wasted all they possessed."[113]

## *Burning of Rome*

Nero's greatest crime was to burn down the city of Rome. "Pretending to be disgusted with the old buildings, and the narrow and winding streets, he set the city on fire so openly, that many of consular rank caught his own household servants on their property with tow, and torches in their hands, but durst not meddle with them.... [D]uring six days and seven nights this terrible devastation continued."[114] "The whole Palatine hill, the theatre of Taurus, and nearly two-thirds of the remainder of the city were burned and countless human beings perished."[115] While Rome burned, Nero played his lyre and sang. "Nero mounted to the roof of the palace, where nearly the whole conflagration could be taken in by a sweeping glance, and having assumed the lyrist's garb he sang the *Taking of Ilium*, which, to the ordinary vision, however, appeared to be the *Taking of Rome*."[116]

Nero made efforts at public relief by arranging for supplies of food to be brought into the city, setting up temporary housing, and reducing the price of grain.[117] But nevertheless "the populace invoked curses upon Nero without intermission."[118] According to Tacitus, Nero needed a scapegoat. He found one in the Christian community in Rome.

> Nero fastened the guilt and inflicted the most exquisite tortures on a class hated for their abominations, called Christians by the populace.... Accordingly, an arrest was first made of all who pleaded guilty; then, upon their information, an immense multitude was convicted, not so much of the crime of firing the city, as of hatred against mankind. Mockery of every sort was added to their deaths. Covered with the skins of beasts, they were torn by dogs and perished, or were nailed to crosses, or were doomed to the flames and burnt, to serve as a nightly illumination, when daylight had expired. Nero offered his gardens for the spectacle, and was exhibiting a show in the circus, while he mingled with the people in the dress of a charioteer or stood aloft on a car. Hence, even for criminals who deserved extreme and exemplary punishment, there arose a feeling of compassion; for it was not, as it seemed, for the public good, but to glut one man's cruelty, that they were being destroyed.[119]

## *Death of Seneca*

Although Seneca had retired, he was unable to escape from the intrigues of Rome. Nero charged him with participating in a seditious conspiracy. Seneca denied the charges, stating that he no longer had any interest in politics and only wanted peace. Nero was resolute: he demanded that Seneca commit suicide and sent a centurion to enforce the sentence of death. Seneca's death was "a special joy to the emperor, not because he had convicted him of the conspiracy, but [because he was] anxious to accomplish with the sword what poison had failed to do."[120]

Seneca and his wife decided to commit suicide together. "By one and the same stroke they sundered with a dagger the arteries of their arms. Seneca, as his aged frame, attenuated by frugal diet, allowed the blood to escape but slowly, severed also the veins of his legs and knees."[121] But Nero forbade the death of Seneca's spouse, so "at the soldiers' prompting, her slaves and freedmen bound up her arms, and stanched the bleeding."[122]

Seneca's death was not quick, but long and difficult. Having failed to kill himself with knife cuts, Seneca asked for assistance.

> [Seneca] begged Statius Annaeus, whom he had long esteemed for his faithful friendship and medical skill, to produce a poison with which he had some time before provided himself, the same drug which extinguished the life of those who were condemned by a public sentence of the people of Athens. It was brought to him and he drank it in vain, chilled as he was throughout his limbs, and his frame closed against the efficacy of the poison. At last he entered a pool of heated water, from which he sprinkled the nearest of his slaves, adding the exclamation, "I offer this liquid as a libation to Jupiter the Deliverer." He was then carried into a bath, with the steam of which he was suffocated, and he was burnt without any of the usual funeral rites.[123]

## *An Artist Perishes*

In the fourteenth year of Nero's reign, the Gauls revolted. "A Gaul name Gaius Julius Vindex ... delivered a long and detailed speech against Nero, saying that they ought to revolt from the emperor and join him in an attack upon him, 'because he has despoiled the whole Roman world, because he has destroyed all the flower of their Senate, because he debauched and likewise killed his mother, and does not preserve even the semblance of sovereignty.'"[124] Vindex nominated Servius Sulpicius Galba (3 B.C.–A.D. 69), commander of the Roman forces in Spain, to succeed Nero. Galba "was also nominated by the soldiers as emperor."[125]

Nero initially made plans to put down the revolt, but "in preparing for this expedition, his first care was to provide carriages for his musical instruments and machinery to be used upon the stage; to have the hair of the concubines he carried with him dressed in the fashion of men; and to supply them with battle-axes and Amazonian bucklers."[126]

The news arrived "that the rest of the armies had declared against him," leaving Nero with no allies.[127] "Being abandoned by all without exception he [Nero] began forming plans to kill the senators, burn the city to the ground, and sail to Alexandria."[128] Eventually, even Nero's personal bodyguard deserted him, and the emperor fled to a country house just outside the city with a few associates.[129] With capture imminent, Nero decided to commit suicide. Weeping, he proclaimed "what an artist is now about to perish!"[130] Nero "drove a dagger into his throat, being assisted in the act by Epaphroditus, his secretary ... [and] he expired, with his eyes fixed and starting out of his head, to the terror of all who beheld him."[131]

## *Pliny the Elder (A.D. 23–79)*

### *Natural History*

Pliny the Elder (Gaius Plinius Secundus, A.D. 23–79) was a Roman aristocrat who spent most of his life in the service of the Roman Empire, either in administration or military duty. "Before A.D. 35 his father took him to Rome, where he was educated under his father's friend, the poet and military commander, P. Pomponius Secundus, who inspired him with a lifelong love of learning."[132] Tacitus described Pomponius Secundus as "a man of refined manners and brilliant genius."[133] Pliny was also mentored by Seneca, and under his tutelage "became a keen student of philosophy and rhetoric."[134]

Pliny wrote seven books during his lifetime. During his tenure as a cavalry commander, he authored his first work, *On Throwing the Javelin from Horseback*. The third book, *German Wars*, a history of all the wars between Rome and Germany, was a more ambitious project.[135] The work was evidently respected, because Tacitus referred to Pliny as "the historian of the German wars."[136]

All of Pliny's manuscripts have been lost with the exception of his last and most important work, *Natural History*. Pliny's *Natural History* was a monumental undertaking; it was the first encyclopedia of natural philosophy. Written in thirty-seven books, the *Natural History* covers astronomy, geography, geology, meteorology, zoology, oceanography, birds and insects, anatomy, botany, forestry, agriculture, and medicine.[137] Pliny's nephew, Pliny the Younger (c. A.D. 61–113), described it as "a work of great compass and learning, and as full of variety as nature herself."[138]

There is more. Pliny discussed the history of painting (Book 35), and sculpture in bronze (Book 34) and marble (Book 36). In his dedication to the Roman Emperor Vespasian (A.D. 9–79), Pliny stated that the *Natural History* covered 20,000 topics and was based on his perusal of 2,000 books.[139] Much of the value of *Natural History* derives from the fact that most of the original sources Pliny relied upon are now lost.

Pliny was a voracious reader and tireless worker. He would have books read to him while he ate dinner and would dictate sections of his book while relaxing in the bath. He wasted no waking moment. Pliny never married, but adopted his nephew, Pliny the Younger, as his heir. Pliny's work habits were described by his nephew in Book 3, Letter 5, of his *Epistles*.

> You will wonder how a man, so engaged as he was, could find time to compose such a number of books; and some of them upon abstruse subjects. In summer, he always began his studies as

> soon as it was night; in winter, generally at one in the morning, but never later than two, and often at midnight ... Before day-break, he used to wait upon Vespasian, who likewise chose that season to transact business. When he had finished the affairs which that emperor committed to his charge, he returned home again to his studies. After a short and light repast at noon he would frequently ... repose himself in the sun; during which time some author was read to him, from whence he made extracts and observations ... for it was a maxim of his, that, "no book was so bad, but that something might be learnt from it." When this was over, he generally went into the cold bath, and ... then reposed himself for a little while. Thus, as if it had been a new day, he immediately resumed his studies till supper-time, when a book was again read to him, upon which he would make some hasty remarks. In summer, he always rose from supper by day-light; and, in winter, as soon as it was dark.... Such was his manner of life amidst the noise and hurry of the town: but in the country his whole time was devoted to study.... In his journeys, he lost no time from his studies, but his mind, at those seasons, being disengaged from all other business, applied itself wholly to that single pursuit.... By this extraordinary application, he found time to compose the several treatises I have mentioned, besides one hundred and sixty volumes which he left me by his will, consisting of a kind of commonplace [text]; written on both sides, in a very small character; so that one might fairly reckon the number considerably more.[140]

Like many of the Greek philosophers who preceded him, Pliny was a monotheist. "Whatever God be, if there be any other God, and wherever he exists, he is all sense, all sight, all hearing, all life, all mind, and all within himself."[141] Pliny's God was not the personal deity of the Jews and Christians, but an ambiguous spiritual principle. "The power of nature is clearly proved, and is shown to be what we call God."[142]

Pliny explicitly denied polytheism, describing it as folly. "To believe that there are a number of gods, derived from the virtues and vices of man ... indicates still greater folly."[143] He also stated that it was childish to ascribe human characteristics to gods. "To suppose that marriages are contracted between the gods, and that, during so long a period, there should have been no issue from them, that some of them should be old and always grey-headed and others young and like children, some of a dark complexion, winged, lame, produced from eggs, living and dying on alternate days, is sufficiently puerile and foolish. But it is the height of impudence to imagine, that adultery takes place between them, that they have contests and quarrels, and that there are gods of theft and of various crimes."[144]

In *Decline and Fall of the Roman Empire*, Edward Gibbon described Pliny's *Natural History* as "that immense register where Pliny has deposited the discoveries, the arts, and the errors of mankind."[145] George Sarton characterized it as "at once a sepulchre of ancient science, and a cradle of medieval lore."[146]

The *Natural History* is infamous for Pliny's uncritical and credulous inclusion of every fact, fairy tale, and apocryphal story known to the ancients. These include human monstrosities such as "a race of men who have their feet turned backwards, with eight toes on each foot ... a tribe of men who have the heads of dogs ... [a] race of men ... who have only one leg, but are able to leap with surprising agility ... a tribe who are without necks, and have eyes in their shoulders ... [and] a people who have no mouths ... [who] subsist only by breathing and by the odors which they inhale through the nostrils."[147]

In addition to human curiosities, *Natural History* contains descriptions of a number of mythological and fantastic animals. Pliny wrote that the country of Ethiopia contained "horses with wings, and armed with horns," as well as an animal that had "a triple row of teeth, which fit into each other like those of a comb, the face and ears of a man, and azure eyes, is of the color of blood, has the body of a lion, and a tail ending in a sting, like that of the scorpion."[148]

Pliny claimed that a dog "will never bark at a person who carries a dog's tongue in his shoe,"[149] and that magnets placed next to certain precious stones would lose their power to attract iron. "So great is the antipathy borne by this stone to the magnet, that when placed near, it will not allow of its attracting iron."[150]

Pliny's *Natural History* also contains some implausible claims concerning stones and minerals. Pliny stated that gazing upon emeralds will refresh fatigued eyes. "Even when the vision has been fatigued with intently viewing other objects, it is refreshed by being turned upon this stone."[151] He believed that rock grew in the Earth like a living thing, and that marble quarries were thus inexhaustible. "Marble there grows in the quarries; and those who work in the quarries assure us that the wounds thus inflicted upon the mountains fill up spontaneously."[152]

If *Natural History* is reflective of general opinion, the Romans had a number of superstitious beliefs concerning women. "There is," Pliny wrote, "no limit to the marvelous powers attributed to females ... hailstorms, they say, whirlwinds, and lightning even, will be scared away by a woman uncovering her body while her monthly courses are upon her ... and out at sea, a storm may be lulled by a woman uncovering her body merely, even though not menstruating at the time."[153]

A whole host of remarkable effects were attributed to a menstruating woman and menstrual blood, most of them bad. The luster of anything colored purple would be reduced if touched by a menstruating woman, and "if a woman strips herself naked while she is menstruating, and walks round a field of wheat, the caterpillars, worms, beetles, and other vermin, will fall from off the ears of corn."[154] If "the edge of a razor," comes into contact with menstrual blood, it "will become blunted."[155] And "if the menstrual discharge coincides with an eclipse of the moon or sun, the evils resulting from it are irremediable."[156]

There were limits to Pliny's credulity. In Book 8 of *Natural History*, he rejected Greek stories of werewolves. "That men have been turned into wolves, and again restored to their original form, we must confidently look upon as untrue, unless, indeed, we are ready to believe all the tales, which, for so many ages, have been found to be fabulous.... It is really wonderful to what length the credulity of the Greeks will go! There is no falsehood, if ever so barefaced, to which some of them cannot be bound to bear testimony."[157]

Pliny's tendency to uncritically accept every fact has marred his reputation amongst historians. However, there is something to be said for Pliny's all-inclusive method of collecting and collating all the knowledge of his time. It provides us with an unfiltered look into what the ancients knew about the world instead of what Pliny himself thought. With the benefit of hindsight, it is remarkably easy to point out what should have been excluded as ridiculous, but for Pliny it was more difficult.

Some aspects of Pliny's discussion of natural wonders must have seemed to men of his age to be as incredible as the existence of unicorns, yet today we know these phenomena to be true. Pliny mentioned a stone that burned,[158] and islands that rose spontaneously out of the sea. "Land is sometimes formed ... [by] rising suddenly out of the sea."[159] The stone that burned, of course, was coal, and volcanic islands can rise spontaneously out of the sea. In 1963, the island of Surtsey was formed in the North Atlantic ocean about 33 kilometers south of Iceland. The emergence of the island was accompanied by smoke, lava flows, and a lightning storm.[160] Volcanic activity continued for four years, and the island grew to an area of 2.7 square kilometers.[161]

The fantastic stories and claims found in Pliny's *Natural History* illustrate why the ancient philosophers were skeptical of empiricism. In their time, empiricism meant anecdotal information, not systematic and controlled experimentation. Anecdotal data are notoriously unreliable.

## *Eruption of Vesuvius*

In A.D. 79, Pliny was in command of the Roman fleet at the city of Misenum, on the west side of the Bay of Naples, about seventeen miles (28 kilometers) from Mount Vesuvius. On the east side of the Bay, Vesuvius shadowed the Roman cities of Herculaneum, Pompeii, and Stabiae. Herculaneum was the closest town to Vesuvius, with Pompeii and Stabiae lying slightly farther south.

On August 24, A.D. 79, Vesuvius began a catastrophic eruption that buried the Roman cities of Pompeii and Herculaneum, erasing them from human memory for more than sixteen hundred years. Pliny the Elder observed the eruption and his scientific curiosity was aroused. He resolved to cross the bay and observe the phenomenon more closely.

Before Pliny could depart, he received a message from a concerned friend in Stabiae who begged him to come and rescue her. Pliny sailed across the bay, but underestimated the danger and died during the evacuation. Fifty-six years of age, obese, and apparently in poor health, he passed out and suffocated from noxious volcanic fumes. The events were recorded by his nephew, Pliny the Younger, in Book 6, Letter 16, of his *Epistles.*

> On the 24th of August, about one in the afternoon, my mother desired him [Pliny the Elder] to observe a cloud which appeared of a very unusual size and shape.... It was not at that distance discernible from what mountain this cloud issued, but it was found afterwards to proceed from Vesuvius. I cannot give you a more exact description of its figure, than be resembling it to that of a pine tree; for it shot up a great height in the form of a tall trunk, which spread at the top into a sort of branches.... [I]t appeared sometimes bright, and sometimes dark and spotted, as it was either more or less impregnated with earth and cinders. This uncommon appearance excited my uncle's philosophical curiosity to take a nearer view of it. He accordingly ordered a light vessel to be prepared.... As he was going out of the house with his tablets in his hand, he was met by the mariners belonging to the galleys stationed at Retina, from which they had fled in the utmost terror.... They conjured him, therefore, not to proceed and expose his life to imminent and inevitable danger. ... [I]nstead of gratifying his philosophical spirit, he resigned it to the more magnanimous principle of aiding the distressed. [H]e ordered the fleet immediately to put to sea, and went himself on board with an intention of assisting not only Retina, but the several other towns which stood thick upon that beautiful coast. Hastening to the place, therefore, from whence others fled with the utmost terror, he steered his direct course to the point of danger.... He was now so near the mountain, that the cinders, which grew thicker and hotter the more he advanced, fell into the ships, together with pumice stones, and black pieces of burning rock; they were likewise in danger, not only of being aground by the sudden retreat of the sea, but also from the vast fragments which rolled down from the mountains, and obstructed all the shore. Here he stopped, to consider whether he should return back; to which the pilot advising him, "fortune," said he, "befriends the brave; steer to Pomponianus." Pomponianus was then at Stabiae, separated by a gulf.... Pomponianus had already sent his baggage on board; for though he was not at that time in actual danger, yet ... he was determined, if it should in the least increase, to put to sea as soon as the wind should change. It was favorable, however, for carrying my uncle to Pomponianus, whom he found in the greatest consternation; and embracing him to keep up his spirits. The more to dissipate his fears, he ordered his servants ... to carry him to the baths; and ... he sat down to supper with great, or at least (what is equally heroic) with all the appearance of cheerfulness. In the mean while, the fire from Vesuvius flamed forth from several parts of the mountain with great violence ... After this he retired to rest.... The court which led to his apartment being now almost filled with stones and ashes, it would have been impossible for him, if he had continued there any longer, to have made his way out.... He got up, and joined Pomponianus, and the rest of the company.... They consulted together whether it would be most prudent to trust to the houses, which now shook from side to side with frequent and violent concussions, or flee to the open fields, where the calcined [burnt] stones and cinders, though levigated [powdered] indeed, yet fell in large showers, and threatened them with instant destruction. In this distress, they resolved for the fields, as the less dangerous situation of the two.... They went out, then, having pillows tied upon their heads with napkins; and this was their whole defense against

> the storm of stones that fell around them. ... They thought it expedient to go down farther upon the shore, in order to observe if they might safely put out to sea; but they found the waves still running extremely high and boisterous. There my uncle ... laid himself down upon a sail-cloth which was spread for him; when immediately the flames, preceded by a strong smell of sulphur, dispersed the rest of the company, and obliged him to rise. He raised himself up, with the assistance of two of his servants, and instantly fell down dead; suffocated, I conjecture, by some gross and noxious vapor.[162]

## *Galen (c. A.D. 129–200)*

### *Stature and Influence*

Galen was a Greek physician who did most of his work in Rome. He was an extremely prolific writer whose works were esteemed and influential, especially in medieval Europe. "After Aristotle, [Galen] was the author oftenest quoted in the Middle Ages, and most revered."[163] Galen and Hippocrates were the most renowned physicians in antiquity, and Galen's books "were the dominant source of all medical acquirement for more than twelve centuries."[164] In his *Ecclesiastical History*, Eusebius of Caesarea (c. A.D. 260–340), said that, "and as to Galen, he is even perhaps worshipped by some."[165] The works of the Arab physicians, Rhazes and Avicenna, were also influential in medieval Europe. But they had in turn been influenced by Galen. In the ninth century, Hunayn Ibn Ishaq (A.D. 808–873) had translated nearly the entire body of Galen's works.[166] Galen's authority in medicine persisted through the sixteenth century.[167]

### *Life and Education*

Galen was a native of Pergamum, a Greek city in present day Turkey. In the second century A.D., Pergamum was economically prosperous, and the most important city in Greek and Roman Asia. Pergamum was noted especially as a medical center. The shrine of Asclepius at Pergamum had an area exceeding ten thousand square meters. This temple was a center of pilgrimage and one of the wonders of the ancient world.[168]

Galen was born into a prominent, wealthy, and educated family.[169] His father, Nicon, was a renowned architect who owned several estates. In addition to architecture, Nicon was educated in geometry and astronomy.[170] "Nicon was a rich and learned man, skilled in the *belles lettres*, the philosophy, astronomy, geometry, and architecture of the times."[171] Both Galen and his father were Roman citizens.[172]

In his writings, Galen evidenced deep respect and admiration for his father. He described Nicon as "good-tempered, just, efficient, and benevolent."[173] But Galen was contemptuous of his mother. Her temperament was "irascible," and Galen compared her to Socrates' notoriously ill-tempered spouse, Xanthippe.[174] Galen was raised on a farm that produced crops such as peas, lentils, beans, figs, olives, and grapes. Galen's family also kept bees, cows, and horses.[175] Nicon's estates aimed at self-sufficiency, and he conducted agricultural experiments in an attempt to improve his crops.[176]

Up to age fourteen, Galen was schooled at home.[177] His father then had him educated in the philosophical systems of Plato, Aristotle, the Stoics, and the Epicureans.[178] Nicon was so concerned with his son's education that he attended the philosophical lectures with Galen.[179]

Exposure to a number of different and often conflicting philosophies taught Galen to be open minded rather than dogmatic. Many of the claims of the competing philosophical systems contradicted each other, and there seemed to be no way to resolve the discrepancies. Nicon advised his son to not hurriedly commit to the doctrine of any one school, but to weigh and test them all equitably in a search for "justice, moderation, courage, and intelligence."[180] Nicon also instructed Galen in the method of demonstrative proof utilized in geometry.[181]

Writing in later life, Galen noted that one must study the teachings of the ancients, but not accept them at face value. Everything had to be tested and subjected to the trial of experience.

> He whose purpose is to know anything better than the multitude do must far surpass all others both as regards his nature and his early training. And when he reaches early adolescence he must become possessed with an ardent love of truth, like one inspired; neither day nor night may he cease to urge and strain himself in order to learn thoroughly all that has been said by the most illustrious of the ancients. And when he has learnt this, then for a prolonged period he must test and prove it, observing what part of it is in agreement, and what in disagreement with obvious fact; thus he will choose this and turn away from that.[182]

Galen "commenced the study of medicine at the age of seventeen."[183] The decision was based upon a dream experienced by his father, in which it was revealed to Nicon that medicine was the proper occupation for his son.[184] Galen and all of his contemporaries believed in the validity of both dreams and divination.[185] An entire Hippocratic treatise was devoted to dreams and their significance. The author began by asserting, "the signs which occur in dreams will be found very valuable."[186] Galen's contemporary, the Stoic philosopher and Roman emperor, Marcus Aurelius Antoninus (A.D. 121–180), wrote that "remedies have been shown to me by dreams."[187] In his book, *On the Utility of Parts*, Galen himself said that he had been chastised by God in a dream for failing to describe the anatomy of the eye, "the divinest of our organs."[188]

Galen studied medicine in Pergamum for four years. After his father died, Galen traveled to Smyrna, Corinth, and Alexandria, taking twelve years to complete his medical training.[189] Most of Galen's medical teachers practiced in the Hippocratic tradition and were interested in anatomy.[190] While still a student at Smyrna, Galen had already begun his prolific writing.[191] Galen's education was considerably more lengthy than that of the average physician. It was also more intellectually oriented. Galen was an obsessive reader who owned an enormous personal library.[192]

In the autumn of A.D. 157, Galen returned to Pergamum to become chief physician to the gladiators.[193] A trained and skilled gladiator was an extremely valuable piece of property, and it was in the best interest of an owner to obtain the most skilled physician possible to attend to the care of his fighters.[194] In this position, Galen acquired much experience in the treatment of wounds and other serious injuries. In the second century A.D., the art of surgery was well-developed. This is evidenced by the complex array of instruments used by Roman army surgeons, many of which survive today.[195] In addition to treating wounds, Galen also attended to the health of the gladiators through the hippocratic regimen of diet and exercise.[196]

Galen emigrated to Rome in the summer of A.D. 162.[197] In the second century A.D., the Roman Empire was at, or near, the peak of its power. Rome was the center of the Mediterranean world, and a talented, ambitious, and intellectual physician would have been drawn there. It was not unusual for a Greek physician to travel to Italy. Greek medicine had arrived in Rome in 293 B.C., when the Roman government officially endorsed the cult of Asclepius.[198] That Greek medicine was known and practiced in Rome during the second century B.C. is shown by Plutarch's statement that Cato the Censor (234–149 B.C.) had "an aversion [not] only against the Greek philosophers, but the physicians also."[199]

By the third century B.C., Rome was ascendant in political and military power. But she had been seduced by the arts and culture of the Greeks. Horace (65–8 B.C.) noted, "tamed Greece to tame her victress now began, and with her arts fair Latium over-ran."[200] Physicians who came to Rome enjoyed privileges far beyond other foreigners. Around 49 B.C., Julius Caesar awarded Roman citizenship to alien doctors, and in 23 B.C. Augustus decreed that physicians in Rome were exempt from taxation.[201] Romans tended to avoid the medical profession, and in general regarded medicine as an occupation for foreigners.[202]

In Rome, Galen was assisted by many influential friends and acquaintances.[203] He quickly acquired notoriety by curing several patients of prominent power and status. Galen also achieved fame by holding public lectures and demonstrations on anatomy.[204] Galen's initial stay in Rome lasted only a few years, and returned to Pergamum c. A.D. 166,[205] possibly to avoid a plague that may have been smallpox,[206] though it is more likely that his departure resulted from an overindulgence in political intrigues, controversies, and jealousies.[207]

In Rome, Galen partly acquired his reputation by openly engaging in disputes and debates with established physicians.[208] Galen had a taste for provoking controversy and disputation. He lumped Erasistratus with Hippocrates as the "most illustrious of physicians,"[209] but once publicly criticized the work of Erasistratus just for the sake of annoying a disciple who was present.[210] This happened when Galen was thirty-four years of age. In later life, Galen tended to avoid public conflicts and demonstrations because found that these caused other physicians to envy him and call him a quack.[211]

In A.D. 169, Emperor Marcus Aurelius appointed Galen to be court physician.[212] Except for a brief return to Pergamum c. A.D. 192, Galen spent the rest of his life in Rome. In addition to his duties as court physician, Galen had a considerable practice and must have spent a large amount of time reading and writing.[213] He conducted dissections on almost a daily basis.[214] The commonly accepted dated for Galen's death is A.D. 200,[215] but he may have lived into the third century A.D., possibly to the age of eighty-seven.[216]

## *Works*

Galen was a notoriously prolific author. He began writing even as a medical student at Pergamum.[217] Galen's works "consist of nearly seven hundred books or treatises."[218] Galen wrote so many books that later in life he felt it necessary to author two books (*On His Own Books, On the Order of His Own Books*) simply devoted to listing and describing his other works.[219]

The standard edition of Galen's works in Greek and Latin was produced by Carolus Gottlob Kühn, and published in Leipzig between 1821 and 1833. It contains "22 thick volumes,"[220] that average in length a thousand pages each.[221] Very few of Galen's works have been translated into English. The body of work is so immense that scholars have perhaps found the task to be too challenging. Many of the works attributed to Galen are likely to be spurious.[222] Ancient authors may have sought exposure for their views by attributing the work to Galen. Galen himself told the amusing story that he was at a bookseller in Rome and overheard two customers disputing as to whether or not Galen was the authentic author of a work signed "Galen, the physician."[223]

## *Empiricism*

Galen's written comments and enormous number of anatomical investigations through dissection and vivisection clearly indicate that he understood and advocated the experimental method.[224] This was unusual in ancient philosophy, but commonplace for the practical arts.

In Galen's view, all theories had to be subjected to the test of experience.[225] He advocated the methods of modern science by noting that, "in believing what has been well found there are two criteria for all men, reason and experience."[226] Galen argued for empiricism by asking, "how is it we know that fire is hot? ... and how do we learn that ice is cold except through the senses?"[227] This striking example was later amplified and echoed by Roger Bacon in the thirteenth century.[228]

Galen's work, life, and influence are evidence of the progressive, cumulative, and incremental nature of science. Galen, read, wrote, and was read. Galen read voluminously, and was apparently strongly influenced by writers such as Hippocrates, Aristotle, Herophilus, and Erasistratus. His views on the value of empiricism were thus not completely original, but developed in the context of a centuries-long tradition of natural philosophy and medical science. In turn, Galen was widely read, not just by physicians, but by scholars, scientists, philosophers, and theologians in Medieval Europe. Their written comments indicate clearly and unambiguously that Galen's work was a significant influence toward the recognition that empiricism was necessary in the sciences.[229] The line of descent is thus unbroken, from Hippocrates, through Galen, to Roger Bacon, and on to the Scientific Revolution of the seventeenth century.

Galen's work also illustrates why ancient philosophers were extremely skeptical of the value of empiricism. The experiences available to them were mostly singular, anecdotal, and incidental, and consequently are notoriously unreliable. Like Hippocrates, Galen rejected superstition. He condemned "Egyptian sorceries and incantations, amulets ... philters, love-charms, dream-draughts, and imprecations."[230] But lacking the resources for systematic and controlled experimentation, Galen was credulous in his acceptance and endorsement of many folk or traditional remedies. He believed that the flesh of vipers had therapeutic value.[231] Galen recommended harvesting herbs by picking them with the left hand just before dawn.[232] As a remedy for epilepsy, Galen recommended wearing the root of a peony plant. This belief was not superstitious, but based upon observation and experience. Galen had observed an afflicted youth remain asymptomatic while wearing the root. But when the root was removed, the boy once again became afflicted.[233]

As a cure for lumbago (low back pain), Galen recommended powdering the wings of a swallow. The swallow was to be bled to death by making a cut in its leg. The bird's carcass was then skinned, roasted, and eaten in its entirety. The final step in the remedy was to rub the oil derived from the bird over the body for three days. Galen asserted that the efficacy of this remedy had often been proven by experience.[234]

Galen also proscribed the eating of fruit because his father had once fallen ill after consuming a quantity of fresh fruit.[235] Thus Galen's works, like others, are a monument to the dangers of anecdotal empiricism.

## *Anatomy and Dissections*

Galen was not the first to investigate animal physiology through dissection. His renowned predecessors at Alexandria, Herophilus and Erasistratus, began the scientific study of anatomy and physiology through careful and systematic dissections. Galen had access to their writings, but none of the works of Herophilus or Erasistratus are extant today.

Herophilus flourished during the last third of the fourth century B.C. He conducted medical research at Alexandria under the patronage of Ptolemy I and II.[236] Tertullian (c. A.D. 155–220) described Herophilus as a "butcher who cut up no end of persons, in order to investigate the secrets of nature."[237]

Celsus claimed that Herophilus and Erasistratus, with the sanction of the Ptolemaic kings, also conducted human vivisections on condemned criminals. "Herophilus and Erasistratus ... dissected such criminals alive, as were delivered over to them from the prisons by royal sanction."[238] In ancient Greece, dissection of the human body was taboo. But it is possible that the standard was relaxed in Alexandria. Egyptians had been opening up human corpses for thousands of years to remove organs and place them in canopic jars.[239]

Through dissection, Herophilus studied the anatomy of the human eye, brain, liver, nervous system, and genital organs.[240] By tracing the origin of the nerves to the brain, he corrected Aristotle's mistaken supposition that the source of the human intellect was the heart.[241]

Erasistratus, born c. 304 B.C., was a younger colleague of Herophilus.[242] To determine if mass was consumed by the metabolism of a bird, Erasistratus recommended that the animal be kept in a cage but not fed. Excrements were carefully collected and weighed. The bird was weighed at the beginning and end of the experimental period to determine if any substance had been given off.[243]

Erasistratus was fond of using mechanical analogies. He compared the grinding action of the stomach to that of a grain mill, and envisaged the human body as a machine that could be explained in mechanical terms.[244] He may have been influenced by Strato of Lampsacus (c. 340–270 B.C.) to adopt a mechanistic outlook and experimental approach.[245]

Erasistratus became convinced that only veins carried human blood. He thought the arteries carried a *pneuma*, or invisible life force. The pneuma of the physicians was analogous to the pneuma of the Stoic philosophers, a fundamental substance that permeated the cosmos.[246] As the cosmic pneuma was present in the macrocosm, the microcosm of the human body required a pneuma to sustain the vital activity that separated the living and non-living worlds.

The idea that human arteries contained pneuma, not blood, originated with Praxagoras of Cos (born c. 340 B.C.). Praxagoras observed that the arteries of animals sacrificed for study apparently had air in their arteries. This was because Aristotle had recommended that animals dissected for anatomical studies be killed by strangulation. Strangulation empties blood from the arteries and the left side of the heart.[247]

Pneuma was related to air, but was a type of refined air or spirit that carried a vital force. There was also more than one type of pneuma. Erasistratus maintained that the human brain transformed the vital pneuma of the arteries into a psychic pneuma that could be transported by nerves.[248] To explain why opened human arteries were found to contain blood, not pneuma, Erasistratus invoked the principle of the vacuum. When an artery was opened, the pneuma escaped instantaneously, leaving a vacuum. The vacuum was filled by blood rushing in from veins. Erasistratus thus recognized that capillary vessels, or *anastomoses*, must exist, although they were too small to be detected directly.[249] The idea of a vital pneuma obtained from air is not so different from the modern recognition that air is a heterogeneous mixture of gasses, of which only one, oxygen, provides energy for the human metabolism. In this respect, Erasistratus was essentially correct in his view that respiration filled the arteries with a vital life force.[250] Erasistratus must have also dissected animals, because he observed that the number of convolutions in the human brain exceeded those in other animals. From this observation he inferred that the presence of convolutions in the cerebellum was proportional to intellectual development.[251]

In second century Rome, the dissection of human corpses was not possible, but Galen conducted many dissections and experiments on animals. He preferred to use monkeys because of their close human resemblance, but also dissected pigs, sheep, and goats.[252] Galen advised that a trip to Alexandria was indispensable for the education of a medical students, because it was possible to learn anatomy there through autopsies on human corpses. Galen also described how he took advantage of chance opportunities to examine human skeletons. In one such case, he was able to examine the "skeleton of a robber" that no one would bury.[253]

Galen authored at least three books on anatomy and dissection, the most important of which was *Anatomical Procedures*. In his writings, Galen emphasized repeatedly that a physician could not be trained by reading books. A doctor had to obtain a firsthand and immediate knowledge of animal bodies through dissection. Theoretical knowledge was also necessary. If properly conjoined, the theoretical and applied were mutually reinforcing and beneficial.[254]

Through dissection and experimental investigations into anatomy, Galen confirmed the existence of capillaries by observing that when the blood was drained from an animal's body by opening a vein, the arteries were also emptied.[255] "If you will kill an animal by cutting through a number of its large arteries, you will find the veins becoming empty along with the arteries; now, this could never occur if there were not anastomoses [capillaries] between them."[256]

Galen demonstrated that the arteries indeed contained both blood [and pneuma, as he believed], not pneuma alone. He argued that when an artery was initially opened, no pneuma was seen to escape. Because pneuma was related to ordinary air, presumably its escape could not be instantaneous. Galen also confirmed the presence of blood in the arteries experimentally. He isolated an artery by tying it off with ligatures and isolating it from any possible connection with veins. When the artery was then opened, it invariably contained blood, not pneuma.[257] Although Galen understood that blood circulated through the arteries, he did not understand the proper function of arteries and veins. He believed that arteries and veins delivered different types of blood to different organs. The lungs were nourished by arteries, the liver by veins alone.[258]

Galen showed that nerves were responsible for vocalizing through a public demonstration. He stopped the squealing of a pig instantly by severing the nerves leading to the larynx.[259] Galen also studied the digestive system by conducting vivisections on animals. In *On the Natural Faculties*, he testified that "I have personally, on countless occasions, divided the peritoneum of a still living animal and have always found all the intestines contracting peristaltically upon their contents."[260]

In his book, *On Anatomical Procedure*, Galen outlined the principle that the parts of the body must be in sympathy with each other. An animal was a coherent whole. The implication was that the whole of the greater animal could be inferred from a lesser part.[261] This principle was later applied by the French anatomist and paleontologist, Georges Cuvier (A.D. 1769–1832) in his studies of Pleistocene fauna. Speaking in 1798, Cuvier claimed "after inspecting a single bone, one can often determine the class, and sometimes even the genus of the animal to which it belonged."[262]

In *On the Natural Faculties*, Galen described an experimental proof of the fact that urine could only flow from the kidneys to the bladder, and never backward.

> The method of demonstration is as follows. One has to divide the peritoneum in front of the ureters, then secure these with ligatures, and next, having bandaged up the animal, let him go (for he will not continue to urinate). After this one loosens the external bandages and shows the bladder empty and the ureters quite full and distended — in fact almost on the point of rupturing; on removing the ligature from them, one then plainly sees the bladder becoming filled with urine. When this has been made quite clear, then, before the animal urinates, one has to tie a ligature round his penis and then to squeeze the bladder all over; still nothing goes back through the ureters to the kidneys. Here, then, it becomes obvious that not only in a dead animal, but in one which is still living, the ureters are prevented from receiving back the urine from the bladder.[263]

## *Medical Schools and Theories*

In Galen's time, Rome offered "a veritable marketplace of medicine."[264] Competing medical sects or schools included the Hippocratic, Dogmatic, Empiric, Methodic, Pneumatic, and Eclectic.[265] Despite the fact that the Romans were infamous for lacking originality in science and philosophy, the chronicle of medical debates showed that there was an active intellectual interest in at least medicine during the second century in Rome. Intellectuals in Rome met regularly at "the Temple of Peace," a place Galen described as "the usual meeting-place of all people who practiced the logical arts."[266]

Writing in the first century A.D., Galen's predecessor, Celsus, had broadly divided physicians into "theorists" and "empirics."[267] "The theorists say, that a knowledge of the occult or containing causes of disease, of the evident causes, of the natural functions, and lastly of the internal parts, are all indispensable prerequisites to practice ... for their belief is, that no man can know how to treat disease, who is ignorant of its origin ... nor do they deny the necessity of experiment, but they contend that without theory there is no avenue to experiment."[268]

> On the other hand, those who attach themselves to experience, and from that circumstance style themselves empirics, admit indeed the necessity of evident causes; but contend that research after occult causes and the natural functions are superfluous, because in their nature incomprehensible.... [T]herefore one's reliance should be servilely placed on no man's argument, on no man's authority. [E]xpectations of cure [must be grounded] on well-tried matters of fact; that is to say, like as in all other arts, upon those things which experience shall have taught us in the treatment of disease: for it is practice and not controversy that makes the husbandman and the pilot.[269]

Galen concurred with Celsus, dividing the medical schools broadly into two main divisions, the rationalists and the empiricists. The empiricists were defined by Galen as those "that make experience alone their starting point," while the rationalists were those "who start from reason."[270]

The existence of different schools did not imply that Hippocratic methods had been so much superseded as elaborated. All schools generally held the Hippocratic writings in high regard.[271] Being of an independent mindset, Galen ascribed to no dogmatic sect. He was eclectic, and saw dogmatism as an indicator of intellectual weakness. In *On the Natural Faculties*, he exclaimed "those who are enslaved to their sects are not merely devoid of all sound knowledge, but they will not even stop to learn!"[272] Galen encouraged his readers to think for themselves.[273]

Galen considered the Hippocratic tradition to be the most sound,[274] but freely criticized Hippocrates when he considered it appropriate. In *On the Natural Faculties*, Galen observed that "the extent of exactitude and truth in the doctrines of Hippocrates may be gauged ... from the various subjects of natural research themselves."[275]

From Hippocrates and Aristotle, Galen derived the theory that the four primary qualities in nature were hot, cold, moist, and dry.[276] In *On the Natural Faculties*, he explained, "the principles of heat, cold, dryness and moisture act upon and are acted upon by one another, the hot principle being the most active, and the cold coming next to it in power; all this was stated in the first place by Hippocrates and secondly by Aristotle."[277] Galen made it clear that he regarded these qualities or principles as primary when he asserted that "everything throughout nature is governed ... [by] the warm, cold, dry, and moist [principles]."[278] According to Aristotle, even the four elements could be resolved into these four primary principles.[279]

It followed that the four humors were in deficit or excess as a result of imbalances in the four primary qualities. For example, black bile was "cold and dry,"[280] while "yellow bile ... [resulted from] disproportionate heat."[281]

Excesses or deficiencies in the four humors were not only regarded as the cause of disease, but were also considered to determine four primary human temperaments: sanguine, choleric, melancholic, and phlegmatic.[282] The theory of the four temperaments dated at least back to the time of Hippocrates. In the third book of *Epidemics*, Hippocrates had described individuals of different temperaments as exhibiting different symptoms.[283]

Galen supposed the existence of a vital spirit or *pneuma* in the human body. The Greek word "pneuma" meant "both breath and wind."[284] But in this context, it meant a vital, spiritual substance. The theory of a spiritual pneuma originated perhaps with the presocratic natural philosopher, Diogenes of Apollonia (fl. 430–440 B.C.). Diogenes claimed air was the fundamental substance that permeated the cosmos, and identified it with both intelligence and God. Diogenes argued that it was apparent that air was the source of life, because when air was taken away from animals they died. "Men and all other animals live upon air by breathing it, and this is their soul and their intelligence, as will be clearly shown in this work; while, when this is taken away, they die, and their intelligence fails."[285]

The theory of pneuma was invoked by Aristotle to explain the vitality of living things. The presence of an element of this nature was necessary to explain the obvious and dramatic break between the living and non-living worlds. The four elements were inorganic. So there had to be a fifth substance that provided the vitality present in living things.[286] In *De Generatione Animalium* (*On the Generation of Animals*), Aristotle proposed that pneuma was the means by which "the faculties of the soul are transmitted from parent to offspring."[287]

The human pneuma was analogous to the cosmic pneuma of the Stoics, a "primitive substance," that constituted "the totality of all existence; out of it the whole visible universe proceeds, hereafter to be again resolved into it."[288] The Stoic pneuma "permeates and gives life to all things and connects them together in one organic whole."[289]

Aristotle's theory was adopted by Erasistratus,[290] and elaborated by Galen and others. In the first century B.C. there developed a school of pneumatists who placed emphasis on pneuma as the most important factor in human health.[291]

According to Galen, the three fundamental organs in the human body, the heart, brain, and liver, were fed or dominated by three different types of pneuma, the physical, vital, and psychic. Air was taken in through the process of respiration and the various pneuma extracted or produced within the body.[292] The liver was controlled by the physical pneuma and the heart by the vital pneuma. The brain transformed the vital pneuma into a psychic pneuma that was distributed by the nervous system.[293]

The pneumatic theory persisted until more mechanistic views of the human body prevailed in the seventeenth century.[294]

## *Teleology*

Like Aristotle, Galen was a monotheist and teleologist who saw purpose in nature. Galen believed in God, and this view likely made his writings more acceptable to medieval Europeans.

But Galen's God was not the personified deity of Judaism or Christianity. It was the God of the Greek philosophers, an impersonal and supreme spiritual principle that was the source of order in the cosmos. In *On the Utility of Parts*, Galen referred to his God as "the power that made us," a power that was "wise, powerful, and good."[295]

Galen's religious and philosophical views were more in line with the Stoics, and he opposed the views of the atomists and Epicureans. Epicurus (341–270 B.C.) had disingenuously conceded the existence of gods, but maintained that they took no interest in men or the universe. The Epicureans admitted no design in nature.[296]

On the contrary, Galen was a strong advocate of the *Design Argument*, the proposition that the appearance of design and order in the universe is evidence for the existence of God. "Nature," he argued, "does everything artistically, and nothing in vain."[297]

In his treatise, *The Use of the Various Parts of the Body*, Galen argued that both animal and human bodies showed evidence of perfect design for their necessary functions. Because man was the only intelligent animal, God had not equipped him with claws, hoofs, or horns, but with the hand. "To man, the only animal that partakes of divine intelligence, the Creator has given, in lieu of every other weapon or organ of defense, that instrument the hand — an instrument applicable to every act and occasion, as well of peace as of war."[298]

Galen found a striking example of design in the arrangement of human teeth. "Did chance dispose the teeth themselves in their present order? Which, if it were any other than it is, what would be the consequence? If, for instance, the incisors and canine teeth had occupied the back part of the mouth, and the molar or grinding teeth had occupied the front, what use could we have made of either? Shall we, then, admire the skill of him who disposes a chorus of thirty-two men in just order? And can we deny the skill of the Creator in disposing the same number of teeth in an order so convenient, so necessary even, for our existence?"[299]

## Chapter 5

# Hellenistic Philosophy in Rome

The most popular philosophies in Rome, and throughout the Hellenistic Mediterranean, from about 200 B.C. through A.D. 300, were Epicureanism and Stoicism. Both philosophies were pragmatic, in that they were designed to instruct their followers on how to live an ethical life. But there were disagreements on points of doctrine.

The more influential of the two, especially in Rome, was Stoicism. "Stoicism [did not] achieve its crowning triumph until it was brought to Rome, where the grave earnestness of the national character could appreciate its doctrine, and where for two centuries or more it was the creed, if not the philosophy, of all the best of the Romans."[1] Stoicism was also a source for many of the doctrines popular in Medieval European thought. These included the Doctrine of the Macrocosm and Microcosm, Divine Providence, and the Design Argument.

After the third century A.D., the popularity of both Epicureanism and Stoicism faded. The major trend of philosophic thought was Neoplatonism.

## *Epicureanism (c. 300 B.C.–A.D. 300)*

### *Life of Epicurus (341–270 B.C.)*

Epicurus (341–270 B.C.) was born in Samos.[2] His father, Neocles, was an Athenian colonist. Neocles was an elementary schoolteacher, so the family must have been of middle class status.[3] Epicurus reportedly began the study of philosophy at the age of 12 or 14.[4] "Even as a schoolboy he is said to have given proofs of an inquiring mind. When reading in Hesiod how all things had their origin in Chaos, he puzzled the master by asking, 'whence came Chaos?'"[5]

In 323 B.C., at the age of eighteen, Epicurus returned to Athens "to be enrolled as a citizen and to undergo that training in military duties which the constitution assigned to youths between the ages of eighteen and twenty."[6] After the death of Alexander the Great, the Athenian colonists in Samos were evicted, and Epicurus joined his family in the Ionian city of Colophon.[7] In 310 B.C., at the age of thirty-two, Epicurus was teaching philosophy at Mitylene, a city on the island of Lesbos.[8] Around 306 B.C., Epicurus returned to Athens, and founded a philosophical school that became known as *The Garden*.[9] The Garden was unusual in that both women and slaves could be admitted as students.[10]

"In the character of Epicurus the conspicuous traits are sympathy, generosity, and sweet reasonableness ... [but] no man was ever more vilely slandered or more cruelly misunderstood."[11] The attacks on Epicurus' character may have been prompted by his habit of savagely criticizing other philosophers, even those from whom he derived or inherited principles of his own philosophy. Epicurus called Democritus "the babbler," and said that Aristotle was "a profligate who, after squandering his patrimony, joined the army and sold drugs."[12] These attacks seem inconsistent with sympathetic characterizations of Epicurus, but Diogenes Laërtius assures us that "there are plenty of witnesses of the unsurpassable kindness of the man [Epicurus] to every body; both his own country which honored him with brazen statutes, and his friends who were so numerous that they could not be contained in whole cities."[13]

Epicurus reportedly authored more than three hundred books, all of which have been lost.[14] The extant original sources consist of letters to Herodotus, Pythocles, and Menoeceus, a short summary of Epicurus' primary doctrines, and some miscellaneous fragments.[15] The letters and summary of Epicurean maxims were preserved by Diogenes Laërtius.[16] The most important secondary source is the work *De Rerum Natura* (*On the Nature of Things*) by the Epicurean Roman poet, Titus Lucretius Carus (c. 99–55 B.C.). *De Rerum Natura* is believed to be based on a lost work of Epicurus that was an extended epitome of Epicurean doctrines.[17]

In the modern sense, an epicure is a gourmet, or a person devoted to sensual pleasures. But Epicurus was "no epicure,"[18] and Epicureans were not hedonists. Epicurus lived frugally, on bread and water.[19] He and his students "all lived in the most simple and economical manner."[20] In *Letter to Menoeceus*, Epicurus maintained that "prudence is something more valuable than even philosophy."[21]

Epicurus taught in The Garden at Athens for more than thirty years, and died at seventy years of age from a kidney stone.[22]

## *Epicurean Doctrines*

Epicureanism, like Stoicism, was a pragmatic philosophy whose purpose was to provide an individual with a set of ethical guidelines that would enable them to live a happy life. In the view of Epicurus, "philosophy ... is a life and not merely a doctrine."[23] Epicureanism was "promulgated first and foremost as a rule of life, a means of escape from human misery."[24]

The goal of the Epicurean was to obtain happiness. In *Letter to Menoeceus*, Epicurus explained, "if happiness is present, we have everything, and when it is absent, we do everything with a view to possess it."[25] Happiness comes from pleasure, unhappiness from pain.[26] "The only absolute good is pleasure, after which all living things strive; the only absolute evil is pain, which all avoid."[27]

In *Letter to Menoeceus*, Epicurus affirmed that "pleasure is the beginning and end of living happily."[28] But he then went on to explain that to an Epicurean, pleasure meant an ethical and spiritual contentment, not an uninhibited and hedonistic indulgence in sensual gratification:

> When ... we say that pleasure is a chief good, we are not speaking of the pleasures of the debauched man, or those which lie in sensual enjoyment, as some think who are ignorant, and who do not entertain our opinions, or else interpret them perversely; but we mean the freedom of the body from pain, and of the soul from confusion. For it is not continued drinkings and revels, or the enjoyment of female society, or feasts of fish and other such things, as a costly table supplies, that make life pleasant, but sober contemplation, which examines into the reasons for all choice and

> avoidance, and which puts to flight the vain opinions from which the greater part of the confusion arises which troubles the soul.[29]

Indeed, Epicureanism was the antithesis of hedonism. Luxury was forbidden.[30] Happiness was to be obtained, not by indulging appetites, but by suppressing them. "The limitation of desire is seen to involve habituation to an almost ascetic bodily discipline, in order that the wise man may become self-sufficing, that is, independent of external things."[31] "What Epicurus meant by pleasure was not what most people meant by it, but something very different — a tranquil and comfortable state of mind and body."[32]

Friendship was viewed as an indispensable adjunct to happiness. One of Epicurus' maxims was "he who desires to live tranquilly without having anything to fear from other men, ought to make himself friends."[33] But Epicurean friendship was not the altruistic brotherhood of Christianity. A man sought friendship because it was to his benefit. "No one," said Epicurus, "loves another except for his own interest."[34]

The contented life also necessitated the study of natural philosophy. This was done, not for the sake of knowledge alone, but for the pragmatic end of personal happiness. In *Letter to Pythocles*, Epicurus explained "knowledge of the heavenly phenomena ... has no other aim but that freedom from anxiety, and that calmness which is derived from a firm belief."[35] The Epicurean studied nature "because, if we are to be happy, we must be released from mental trouble, above all from groundless fears, more particularly the terrors of superstition, the fear of the gods, and the dread of death."[36]

The Epicurean epistemological criteria for determining truth were perception, preconceptions, and feelings.[37] Perception refers to positive knowledge obtained through the senses. Perception has primacy; it "is the immediate and ultimately the sole guarantee of truth."[38] The Epicureans endorsed logical reasoning, but subordinated it to observation and sense perception. "The reason cannot pronounce on the senses; for ... all reasoning has the senses for its foundation."[39] To the ancient criticism that the senses deceive, the Epicureans rebuttal was to argue that there was no such thing as a false perception, only a faulty interpretation of a true perception. "The senses cannot be deceived. There can be no such thing, properly speaking, as sense-illusion or hallucination. The mistake lies in the misinterpretation of our sensations. What we suppose that we perceive is too often our own mental presupposition, our own over-hasty inference from what we actually do perceive."[40]

What the Epicureans termed "preconceptions," were mental archetypes that were the basis of definitions. There could be no communication without universally-accepted definitions. "By preconception, the Epicureans mean ... a general idea which exists in us."[41] "At the same moment that we utter the word man, we conceive the figure of a man, in virtue of a preconception which we owe to the preceding operations of the senses.... [We] could not give names to things, if we had not a preliminary notion of what the things were."[42]

The third criterion of truth, feeling, referred to sensations or perceptions that led to either pleasure or pain.[43] As the entire point of Epicureanism was to seek pleasure and avoid pain, feeling was an indispensable criterion for determining appropriate actions.

Though Epicurus advocated the investigation of nature, he was cautious about drawing definite conclusions from the limited data available to him in areas such as astronomy. In *Letter to Pythocles*, he pointed out that several possible explanations were plausible for phenomena such as the source of the Moon's light, and that it would be a mistake to insist that one theory was correct, or to engage in endless speculations. "One ... [should not] obstinately adopt an exclusive mode of explanation; and that, for wanting of knowing what is possible for a man to explain, and what is inaccessible to his intelligence; one does not throw one's self into interminable speculations."[44] *Letter to Pythocles* contains an extended discussion wherein Epicurus listed a number of natural phenomena for which multiple explanations were possible. These included eclipses, the formation of clouds, thunder, lightning, earthquakes, hail, snow, rainbows, comets, and meteors.[45]

Because Epicurus saw no point in speculating about the causes of phenomena that could not be investigated, at least one historian has concluded that the Epicureans were unscientific, and had "an essentially negative attitude towards research into particular natural phenomena."[46] But it would be more correct to view Epicurus' attitude as an anticipation of Chamberlin's doctrine of multiple working hypotheses. The American geologist, Thomas Chrowder Chamberlin (1843–1928), defined this as an effort "to bring up into view every rational explanation of new phenomena, and to develop every tenable hypothesis respecting their cause and history."[47]

For hundreds of years, natural philosophers had engaged in endless speculation on questions that were essentially unresolvable with the methods and instrumentation available to them. Thus "experiment being impossible, [Epicurus] was content to take up an attitude of suspense, excluding no possibility, but waiting for further evidence. This modest attitude, it is maintained, is more becoming to the true man of science than over-hasty speculation, which jumps to conclusions."[48]

The Epicureans were atomists, and Democritus was clearly an important influence on Epicurus' thought. Everything in the cosmos was reducible to atoms and the void. These combined to form the elements and composite bodies. "The atoms which form the bodies, these full elements from which the combined bodies come, and into which they resolve themselves, assume an incalculable variety of forms."[49] These atoms come in different shapes. The number of shapes is large, but finite. But the quantity of atoms having a particular shape is infinite.[50] Atoms are also unchangeable, solid, and indestructible.[51]

Human souls exist, but are material and perishable. "The soul is a bodily substance composed of slight particles, diffused over all the members of the body, and presenting a great analogy to a sort of spirit.... [W]hen ... the body is dissolved, then the soul too is dissolved."[52]

Epicureans considered the cosmos to be infinite. In *Letter to Herodotus*, Epicurus plainly stated, "the universe is infinite."[53] Other worlds exist, and a world is defined to be "a collection of things embraced by the heaven, containing the stars, the earth, and all visible objects.... [S]uch worlds are infinite in number."[54]

Because "nothing can come of that which does not exist,"[55] worlds originate from seeds. "Production of a world may be explained thus: seeds suitably appropriated to such an end may emanate either from one or from several worlds, or from the space that separates them; they flow towards a particular point where they become collected together and organized; after that, other germs come to unite them together in such a way as to form a durable whole, a basis, a nucleus to which all successive additions unite themselves."[56]

Epicureans were also materialists who saw matter as a self-sufficient explanation for the ordering of the universe. They rejected teleology and any suggestion that the world had been created by a deity. In *De Rerum Natura*, Lucretius explained, "this world was made naturally, and without design, and the seeds of things of their own accord jostling together by variety of motions, rashly sometimes, in vain often, and to no purpose, at length suddenly agreed and united, and became the beginning of mighty productions, of the Earth, the sea, and the heavens, and the whole animal creation."[57]

Epicurus professed a belief in the gods, but these deities were abstract spiritual beings that never interacted with, or took an interest in, the affairs of men. In *Letter to Menoeceus*, Epicurus affirmed the existence of gods, but decried the common tendency to ascribe human characteristics to them. "There are gods; for our knowledge of them is indistinct. But they are not of the character which people in general attribute to them ... the assertions of the many about the gods are not anticipations, but false opinions ... the greatest evils which befall wicked men, and the benefits which are conferred on the good, are all attributed to the gods; for they connect all their ideas of them with a comparison of human virtues, and everything which is different from human qualities, they regard as incompatible with the divine nature."[58]

Although Epicurus warned against attributing human characteristics to the gods, strangely enough, he considered them to have human shapes, but also to be invisible to the senses. "In the judgment of Epicurus all the gods are anthropomorphites, or have the shape of men; but they are perceptible only by reason, for their nature admits of no other manner of being apprehended, their parts being so small and fine that they give no corporeal representations."[59]

Epicurus would not entertain supernatural explanations for natural phenomena. In his discussion of possible explanations for astronomical phenomena, he cautioned "above all things let us beware of making the deity interpose here."[60] The Epicurean gods were entirely detached from the workings of the universe. "The whole nature of the gods must spend an immortality in softest peace, removed from our affairs, and separated by distance infinite."[61] Epicureans insisted upon naturalism. "Things are formed without the help and trouble of the gods."[62] "The heavens, the Earth, the Sun, the Moon, the sea, and all other beings ... are equally bound by the general laws of nature ... if you rightly apprehend, nature will appear free in her operations, wholly from under the power of domineering deities, and to act all things voluntarily, and of herself, without the assistance of gods."[63]

In *De Rerum Natura*, Lucretius condemned religion. "Mankind, in wretched bondage held, lay groveling on the ground, galled with the yoke of what is called religion."[64] "A vain fear of the gods torments men in this life, and terrifies them with all the ills that fortune thinks fit to lay upon them."[65]

Democritus was a determinist. "Everything that happens, happens of necessity."[66] But Epicurus believed in free will. In *Letter to Menoeceus*, Epicurus wrote "our own will is free."[67] Epicureans also rejected divination. Diogenes Laërtius wrote that Epicurus "discards divination ... and he says divination has no existence."[68]

Although many of the Ionian natural philosophers were active in civic affairs, the Epicureans shunned politics "as the ruin and confusion of true happiness."[69] "The wise man ... will not be fond of frequenting assemblies."[70]

The Epicureans also had a disdain for much of the conventional Greek education. The dialectic that Socrates and Plato regarded as essential, "they wholly reject[ed] as superfluous."[71] "Epicurus regarded with indifference the ordinary routine education of the day in grammar, rhetoric, dialectic, and music, and for mere erudition he had a hearty contempt. The only study necessary for a philosopher was the study of nature, or what we now call natural science, and this must be cultivated, not for its own sake, but merely as the indispensable means to a happy life. Unless and until we have learned the natural causes of phenomena, we are at the mercy of superstition, fears, and terrors."[72]

Epicureans had no fear of death. In *Letter to Menoeceus*, Epicurus explained, "the most formidable of all evils, death, is nothing to us, since, when we exist, death is not present to us; and when death is present, then we have no existence."[73]

## *Stoicism (c. 300 B.C.–A.D. 200)*

### *The Stoic Philosophers*

The founder of Stoicism was Zeno of Citium (c. 335–263 B.C.). Zeno was a native of Cyprus, and may have been an ethnic Phoenician, as Cyprus was populated by both Greeks and Phoenicians. Zeno was commonly referred to as "the Phoenician." However, Zeno spent most of life and did all of his philosophic work in Athens.

Diogenes Laërtius relates different versions of how Zeno came to settle in Athens: "He was shipwrecked close to the Piraeus [a city in Attica]; and when he had made his way from the coast as far as Athens, he sat down by a bookseller's stall, being now about thirty years of age. And as he took up the second book of Xenophon's *Memorabilia* and began to read it, he was delighted with it, and asked where such men as were described in that book lived."[74] The life of Socrates must have been a significant influence on Zeno, and "it is in Stoicism ... that we find the Socratic tradition most faithfully represented."[75]

In a more mundane version, Zeno was motivated to come to Athens and study philosophy by reading books his father brought back from the city. "His father Innaseas often came to Athens, as he was a merchant, and ... he used to bring back many of the books of the Socratic philosophers, to Zeno, while he was still only a boy ... and that in consequence of this he [Zeno] went to Athens."[76]

Before starting his own school, Zeno studied under a number of philosophers in Athens. These included Polemo, who was Plato's heir as head of the Academy, Stilpo, and Crates.[77] The strongest influence on Zeno was exerted by the Cynics.[78] But Zeno proved "too modest for the shamelessnesss of the Cynics."[79]

After studying for years with a number of other philosophers, Zeno founded his own school around 300 B.C..[80] Zeno's students became known as Stoics, because he chose to teach on the *stoa*, a colonnade "on the north side of the market-place at Athens."[81]

Diogenes Laërtius described Zeno as possessing "a very accommodating temper."[82] Zeno also had intellectual curiosity. "He was a man of a very investigating spirit, and one who inquired very minutely into everything."[83] In emulation of Socrates, Zeno's lifestyle was frugal, verging on ascetic. "He [Zeno] was a person of great powers of abstinence and endurance; and of very simple habits, living on food which required no fire to dress it, and wearing a thin cloak."[84] Zeno regarded "vanity ... [as] the most unbecoming of all things, and especially so in the young."[85] The following poem was said to be a description of Zeno.

> The cold of winter, and the ceaseless rain,
> Come powerless against him; weak is the dart
> Of the fierce summer sun, or fell disease,
> To bend that iron frame. He stands apart,
> In nought resembling the vast common crowd;
> But, patient and unwearied, night and day,
> Clings to his studies and philosophy.[86]

Zeno was "very concise in his speeches,"[87] and disliked excessive loquaciousness. "When a youth was asking him questions with a pertinacity unsuited to his age, [Zeno] led him to a looking-glass and bade him look at himself, and then asked him whether such questions appeared suitable to the face he saw there."[88] On another occasion, "a young man was talking a great deal of nonsense, and [Zeno] said to him, 'this is the reason why we have two ears and only one mouth, that we may hear more and speak less.'"[89] At a meeting with ambassadors from one of the Ptolemaic kings, Zeno remained entirely silent. Puzzled, the ambassadors asked Zeno what "report they were to make of him to the king."[90] Zeno replied, "tell the king that there was a man in the room who knew how to hold his tongue."[91]

Having apprehended one of his slaves in the act of theft, Zeno decided to inflict punishment by scourging. The slave, having some understanding of Zeno's doctrine of predetermination, sought to avoid the whipping by stating "it was fated that I should steal."[92] "Yes," Zeno agreed, but it was also fated "that you should be beaten."[93]

Zeno was evidently well regarded in Athens, because toward the end of his life the Athenians resolved "to present him with a golden crown ... on account of his virtue and temperance, and to build him a tomb ... at the public expense."[94]

Zeno died in 263 B.C., and was succeeded by Cleanthes (c. 330–230 B.C.) as the head of the Stoic School.[95] "Attaching himself to Zeno, he [Cleanthes] devoted himself to philosophy in a most noble manner; and he adhered to the same doctrines as his master."[96] Diogenes Laërtius described Cleanthes as "very industrious; but he was not well endowed by nature, and was very slow in his intellect."[97]

It seems strange that a philosopher may be described as stupid, and it is evident that not everyone agrees with Diogenes Laërtius' description of Cleanthes as dim-witted. Cleanthes "certainly was not the merely docile and receptive intelligence he is sometimes represented as being."[98] Claims that Cleanthes were dull are contradicted by the fact that Cleanthes succeeded Zeno. Cleanthes "became so eminent, that, though Zeno had many other disciples of high reputation, he [Cleanthes] succeeded him [Zeno] as the president of his School."[99]

More significant contributions to Stoicism were made by Cleanthes' student, and the third head of the Stoic School, Chrysippus (c. 280–205 B.C.). "Chrysippus addressed himself to the congenial task of assimilating, developing, [and] systematizing the doctrines bequeathed to him, and, above all, securing them in their stereotyped and final form ... Chrysippus made the Stoic system what it was."[100]

In contrast to his description of Cleanthes as dull, Diogenes Laërtius portrayed Chrysippus as "a man of very great natural ability, and of great acuteness in every way ... he was industrious beyond all other men ... for he wrote more than seven hundred and five books."[101]

Diogenes Laërtius attributed to Chrysippus the vice of arrogance. "He [Chrysippus] appears to have been a man of exceeding arrogance ... he had so high an opinion of himself, that once, when a man asked him, 'to whom shall I entrust my son?' he said 'to me, for if I thought that there was any one better than myself, I would have gone to him to teach me philosophy.'"[102]

Chrysippus was also excoriated for "having written a great deal that is very shameful and indecent ... in his treatise on polity, he allows people to marry their mothers, or their daughters, or their sons ... and in the third book of his treatise on justice, he devotes a thousand lines to bidding people devour even the dead."[103]

Although he inherited the philosophy of Zeno and Cleanthes, Chrysippus was an independent thinker. Asked why he did not attend "the lectures of Ariston, who was drawing a great crowd ... he said 'if I had attended to the multitude I should not have been a philosopher.'"[104] Diogenes Laërtius stated that Chrysippus "in many points ... dissented from Zeno, and also from Cleanthes."[105] Perhaps this simply illustrates the lack of perfect concordance in any system of thought.

Diogenes Laërtius related the unlikely and odd story that Chrysippus "died of a fit of intemperate laughter ... [after] seeing his ass eating figs.... [Chrysippus] laughed so violently that he died."[106] By the time of Chrysippus' death in 205 B.C., "the structure of Stoic doctrine was complete."[107]

## *Stoic Doctrines*

Like Epicureanism, Stoicism was a pragmatic philosophy whose practice was designed to enable a person to attain an idealized good. The Epicurean good was pleasure, but the Stoic good was virtue. The Stoics "regarded virtue as the only good, the one thing in life worth striving for."[108] The Stoics acknowledged that Epicureans valued virtue and did not seek pleasure "without virtue."[109] But they believed that virtue should take primacy over pleasure. Seneca argued, "why then for pleasure, say I, before virtue?"[110] The idea that virtue is the only requisite for happiness can be traced to Plato's *Republic*.[111]

Stoicism was "a practical philosophy, a rallying-point for strong and noble spirits contending against odds."[112] Philosophy was studied, not as an end in itself, or an idle intellectual preoccupation, but as a means to an end. "We study philosophy in order to live and act. Conduct is the one thing of supreme importance."[113] Virtue was an unobtainable ideal that an individual progressed toward, "the wise man is an ideal."[114] "It [virtue] is only to be attained by continual toil, effort, and self-discipline."[115] Stoicism was not a philosophy to be studied in books, it was a way of life to be put into practice.

Stoics defined virtue as being in harmony with the universe.[116] "The virtue of the happy man, his even flow of life, is realized only when in all the actions he does his individual genius is in harmony with the will of the ruler of the universe."[117]

The pursuit of virtue required "active participation in public life."[118] So the Stoics differed from the Epicureans, in that they considered involvement in government to be a duty, at least so far as it was consistent with a virtuous life. Thus a Stoic would properly eschew partaking in a corrupt administration.

The primary epistemological criterion of truth in Stoicism was sense perception. "The Stoics say that what the senses represent is true."[119] Knowing this, Ptolemy Philopator attempted to demonstrate to the Stoic philosopher, Sphaerus [c. 285–210 B.C.], that information obtained through the senses was unreliable. "The king [Ptolemy IV], wishing to refute him [Sphaerus], ordered some pomegranates of wax to be set before him; and when Sphaerus was deceived by them, the king shouted that he had given his assent to a false perception. But Sphaerus answered very neatly, that he had not given his assent to the fact that they were pomegranates, but to the fact that it was probable that they might be pomegranates. And that a perception which could be comprehended differed from one that was only probable."[120]

Stoics divided philosophy into logic, ethics, and physics (natural philosophy). In order to attain virtue by aligning oneself with the cosmos, it was necessary to employ physics or natural philosophy to understand the universe. Logic was the tool applied in correct thinking, "to form right judgments and make a right use of the data of sense."[121] Ethics depends on logic and physics, but was the most important of the trio, because it served as a guide to actions appropriate to a virtuous life.[122]

Stoic cosmology was cyclic. "When the present cycle has run its course all matter will be absorbed once more into primary substance or deity and the world be consumed in a general conflagration."[123] In the beginning, there is only a fundamental and "primitive substance or *pneuma* ... this is the totality of all existence; out of it the whole visible universe proceeds, hereafter to be again resolved into it."[124] Stoics believed the cosmic cycle of creation and destruction repeated indefinitely. The cosmos had been created and perished innumerable times in the past, and would continue to do so in the future.

After creation, the four elements of fire, air, water, and earth appeared. Thus the Stoics did not regard the four elements as being truly elemental. The four elements were capable of transforming into each other. "We stoics are satisfied that the earth is interchangeable in its elements."[125] "The Stoics are of [the] opinion that matter is changeable, mutable, [and] convertible."[126] But of the four elements, the Stoics followed Heraclitus in awarding primacy to fire. "We Stoics ... say that it is fire that lays hold upon the world and changes all things into its own nature."[127] Fire is the material manifestation of the divine element in nature.[128]

Like the Epicureans, the Stoics admitted the existence of a vacuum, but confined its existence to outside the world, or the visible universe. "It is often said by Chrysippus, that there is without the world an infinite vacuum."[129] "The Stoics [say] that within the compass of the world there is no vacuum, but beyond it the vacuum is infinite."[130]

After creation, the cosmos divides itself into active and a passive elements. The passive element is inanimate matter. The active element is a vital principle or *Logos* that permeates the world. The Logos is divine reason or intelligence that manifests in coherence and order. The term was introduced by Heraclitus, who said "all things happen according to this Logos."[131] An earlier influence may have been Thales, who "said that the mind in the universe is god, and the all is endowed with soul and is full of spirits."[132] "Logos is not only a sovereign ordinance which nature invariably obeys, but also the divine reason, immanent in nature and man, which possesses intelligence and thinks—nay, is itself intelligence."[133] Logos is referred to in the *Bible* as the *Word*. "In the beginning was the Word, and the Word was with God, and the Word was God."[134]

The Stoics were monotheists, in the sense that they conceived of a single, supreme deity. However, their God was not an anthropomorphic or personal God, but an ambiguous spiritual principle. Seneca defined God as "the universal intelligence."[135] In a longer passage, Seneca discoursed on the nature of the Deity.

> If you prefer to call him fate, you will not be wrong. He it is on whom depend all things, from whom proceed all causes of causes. If you prefer to call him providence, you will still be right; for he it is by whose counsel provision is made for the world that it may pursue its orderly course and unfold the drama of its being. If you prefer to call him nature, you will make no mistake; for it is he from whom all things derive being, and by whose breath we live. If you prefer to call him the world, you will not be in error; for he is everything that you can see, he is wholly infused in all his parts, self-sustained through inherent power.[136]

Epictetus (c. 55–A.D. 135) asked, "What then is the nature of God? Flesh? Certainly not. An estate in land? By no means. Fame? No. Is it intelligence, knowledge, right reason? Yes."[137] Plutarch claimed that the Stoics derived their conception of God by observing the presence of beauty and order in the world. "The Stoics thus define the essence of a God. It is a spirit intellectual and fiery, which acknowledges no shape, but is continually changed into what it pleases, and assimilates itself to all things. The knowledge of this Deity they first received from the pulchritude of those things which so visibly appeared to us; for they concluded that nothing beauteous could casually or fortuitously be formed, but that it was framed from the art of a great understanding that produced the world."[138]

Stoics believed "God therefore is the world, the stars, the earth, and (highest of all) the supreme mind in the heavens."[139] "To the question, what is God? Stoicism rejoins, What is God not?"[140]

Although the Stoics believed in a supreme deity, they also admitted the existence of lesser deities, spiritual beings intermediate between man and God. "The Stoics believe the same as we do concerning the daemons, and that amongst the great company of gods which are commonly believed, there is but one who is eternal and immortal; all the rest, having been born in time, shall end by death."[141]

The Stoic belief in an ordered universe permeated with, and controlled by, a divine intelligence, led to a number of corollaries. These included the Doctrine of the Macrocosm and Microcosm, fate and predestination, divination, teleology, providence, and design.

The Stoics believed in the existence of the soul, but like the Epicureans, maintained that it was corporeal.[142] As the Logos permeated the soul of the cosmos, the human soul by analogy, was diffused throughout the human body. The Epicureans believed that human souls perished after the death of the physical body, but the Stoics thought "the souls of the unlearned and ignorant descend to the coagumentation [joining] of earthly things, but the learned and vigorous endure till the general fire."[143]

Because the human soul was a manifestation of the Logos, the Stoics believed that "man is the only rational creature on earth and the possession of reason stamps him with the divine image."[144] By analogy, the microcosm of man was therefore a reflection of the macrocosm of the cosmos. "The soul of the world fills and penetrates it: in like manner the human soul pervades and breathes all through the body, informing and guiding it, stamping the man with his essential character of rational."[145]

The manifestation of Logos in nature was the appearance of design. In *De Natura Deorum* (*Of the Nature of the Gods*), Marcus Tullius Cicero (106–43 B.C.) made the *watchmaker analogy*, the argument that the contrivance of the universe resembles a clock made by human hands. "Nature therefore cannot be void of reason, if art can bring nothing to perfection without it, and if the works of nature exceed those of art. When you view an image or a picture, you imagine it is wrought by art; when you behold afar off a ship under sail, you judge it is steered by reason and art; when you see a dial or water-clock, you believe the hours are showed by art, and not by chance; can you then imagine that the universe, which contains all arts and artificers, can be void of reason and understanding?"[146]

If the world was made by design and with purpose, it followed that the causal order of events was governed by divine providence. Stoics believed that "the universe and all its parts are ordered and administrated by divine providence and that all events subserve the highest end, the welfare and advantage of rational beings ... everything had been ordained by perfect reason for the general good; everything, therefore, happens in the best way possible."[147]

Roman Emperor, and Stoic philosopher, Marcus Aurelius Antoninus (A.D. 121–180) explained that what appeared to be the result of chance or fortune was in fact the preordained result of providence. "All that is from the gods is full of providence. That which is from fortune is not separated from nature or without an interweaving and involution with the things which are ordered by providence. From thence all things flow; and there is besides necessity, and that which is for the advantage of the whole universe, of which thou art a part."[148] As the universe was ordered by divine providence, the Stoics believed that "we are all working together to one end, some with knowledge and design, and others without knowing what they do."[149]

Stoics thought that all events were linked by a chain of causation, thus they believed in predestination and fate. The Stoic doctrine of fate was inherited from, or influenced by, Heraclitus, who said that "all things happen according to fate and that fate itself is necessity."[150]

The doctrine of fate and necessity meant that "for all that is or happens there is an immediate cause or antecedent."[151] Thus "in the unalterable succession of causes and effects every event is necessary and predestined."[152] In *Natural Questions*, Seneca maintained that "the order of events is rolled on by the eternal succession of fate, whose first law it is to abide by its decrees ... fate ... is the binding necessity of all events and actions, a necessity no force can break."[153] According to Plutarch, Chrysippus said "that fate is the reason of the world ... it is that law whereby providence rules and administers every thing that is in the world ... it is a chain of causes, that is, an order and connection of causes which cannot be resisted."[154]

There is an apparent contradiction in Stoicism, in that Stoics believed both in predestination and free will. The contradiction is resolved if we understand the chain of causes and effects that order nature is the result and manifestation of the Logos that permeates the cosmos. Fate is the working of the divine intelligence in the macrocosm. Because man is a microcosm, his intelligence is a part, albeit a small part, of the Logos. Thus an individual human being by exerting their free will may have some effect on fate, but this must be contrasted with the larger effect of the universal Logos. Plutarch explained, "the Stoics ... say that ... fate is the ordered complication of causes, in which there is an intexture [blending] of those things which proceed from our own determination, so that some things are to be attributed to fate, others not."

The Stoics endorsed divination as a logical corollary to their belief in a universe governed by providence in which all events were interrelated. They believed there was a "mutual coherence and interconnection between all the parts in the whole universe. Omens and portents are thus produced in sympathy with those events of which they are precursors and indications, so that by natural aptitude or acquired art the connection between them may be empirically observed and noted."[155]

Seneca explained that divination was made possible by fate. "The roll of fate is unfolded; it sends ahead in all directions intimations of what is to follow, which are in part familiar, in part unknown to us. Everything that happens is a sign of something that is going to happen: mere chance occurrences uncontrolled by any rational principle do not admit of the application of divination ... as a matter of fact, there is no living creature whose movement or meeting with us does not foretell something."[156]

The problem of evil is the apparent irreconcilability of the existence of God, an omnipotent and good Being, with the existence of evil and suffering in the world. If God exists, why does he permit evil? The Stoic answer to this problem was to point out that the existence of evil was indispensable for the existence of good. Good could not exist without evil. "Chrysippus ... observes that nothing can be more absurd or foolish ... [than to] think that there can be good, without the existence of evil. For as good is contrary to evil, and it is necessary that both should exist, opposite to each other, and as it were dependent upon mutual and opposite exertions, so ... no opposing thing [can] exist, without its particular opposite. For how could there be a sense of justice, if there were no injustice?"[157]

Stoics were notorious for being indifferent and resigned to the course of events in life. The common definition of "stoic" is indifference to pleasure or pain. Stoic indifference is the logical outcome of their entire belief system. Because the goal of life was the pursuit of virtue, pleasure or pain was irrelevant. The doctrine of predestination implied that resistance to the course of life was futile. "Resignation to the course of destiny, submission to the divinely appointed order of the world is the proper attitude for man."[158]

Seneca maintained that it was man's duty to submit to the will of God. "We should not only submit to God, but assent to him, and obey him out of duty, even if there were no necessity. All those terrible appearances that make us groan and tremble are but the tribute of life; we are neither to wish, nor to ask, nor to hope to escape them."[159] If the pursuit of virtue meant suffering or pain, then the more painful, the better. "In suffering for virtue, it is not the torment but the cause, that we are to consider; and the more pain, the more renown."[160]

Marcus Aurelius explained, "I am a part of the whole which is governed by nature ... by remembering then, that I am a part of such a whole, I shall be content with everything that happens."[161] We are to accept everything that happens, because "that which happens to every man is fixed in a manner for him suitably to his destiny."[162] Annoyances are to be recognized as an inevitable and necessary part of existence. "When thou art offended with any man's shameless conduct, immediately ask thyself, Is it possible, then, that shameless men should not be in the world? It is not possible. Do not, then, require what is impossible."[163]

Stoics did not avoid emotions, they only condemned those passions incompatible with the calm and reasoned pursuit of virtue. "The Stoic temper does not imply absolute freedom from all emotion, but only from irrational mental storms. The sage is not hard and unfeeling, like a block of marble."[164] Epictetus explained, "whatever then we shall discover to be at the same time affectionate and also consistent with reason, this we confidently declare to be right and good."[165]

## *Neoplatonism (c. A.D. 250–550)*

### *Plotinus (A.D. 204–270) and Porphyry (c. A.D. 233–304)*

In the third century A.D., Stoicism and Epicureanism withered. The dominant pagan philosophy for the next three hundred years was Neoplatonism. Neoplatonism was a revival of Platonic and Pythagorean philosophy, with the added influence of Aristotle and Stoicism. "It was eclectic, but so were most of the later philosophies, and combined the systems of Plato, Aristotle, and the Stoics under the aegis of Pythagoras."[166]

As the central element in Neoplatonism is ecstatic religious communion with God, it is more properly regarded as a religion than a philosophy. The rise of Neoplatonism in the Mediterranean world signified the inward turn of the Western mind, and adumbrated the onset of the Dark Age in Europe.

The founder of Neoplatonism was Plotinus (A.D. 204–270). Plotinus "was born of Roman parents at Lycopolis in Egypt."[167] He began the study of philosophy at the age of twenty. At first, Plotinus was dissatisfied with all of his instructors. He finally came under the tutelage of Ammonius Saccas, and "followed Ammonius continuously" to the age of thirty-nine.[168]

Because Ammonius Saccas was Plotinus' teacher, some interpreters attribute the origin of Neoplatonism to Ammonius. But almost nothing is known of Ammonius, not even his birth and death dates. He left no writings. According to Eusebius, Ammonius "had been educated among Christians by his parents ... [but] when he began to exercise his own understanding, and apply himself to philosophy, he immediately changed his views."[169] Plotinus is regarded as the founder of Neoplatonism, because he was the first to set down the Neoplatonic doctrines in writing and form an influential school with an intellectual lineage.

At age forty, Plotinus "settled in Rome," where he lived until his death at age sixty-six.[170] Plotinus' student and biographer, Porphyry (c. A.D. 233–304), described his master's teaching style as informal. "He [Plotinus] was entirely free from all the inflated pomp of the professor: his lectures had the air of conversation, and he never forced upon his hearers the severely logical substructure of his thesis."[171]

Porphyry's assessment of his teacher was favorable, if not worshipful. Plotinus "was gentle.... [A]fter spending twenty-six entire years in Rome, acting, too, as arbiter in many differences, he had never made an enemy of any citizen."[172] Porphyry said that Plotinus had a "remarkable" insight into human character, and related the story of how Plotinus was able to discover a thief simply by lining up all the suspects and staring into their eyes. Plotinus' judgment was verified when the thief revealed the location of the stolen item while being whipped.[173]

Porphyry did condemn Plotinus for poor writing. Plotinus wrote spontaneously, and refused to rewrite. "Plotinus could not bear to go back on his work even for one re-reading; and indeed the condition of his sight would scarcely allow it; his handwriting was slovenly; his misjoined his words; he cared nothing about spelling; his one concern was for the idea."[174]

Upon his death, Plotinus left fifty-four books. These, Porphyry edited and organized into "six sets of nine," the six *Enneads*.[175] In compiling Plotinus' work, Porphyry revised and edited it. "Such revision was necessary," he argued, because of Plotinus' disorganized and sloppy writing style.[176]

Porphyry also authored an influential introduction to Aristotle's *Categories*, the *Isagoge*, that was translated into Latin by Boethius (c. A.D. 480–525). The *Isagoge* became "a cornerstone of the early medieval knowledge of logic."[177]

Following Plotinus and Porphyry, the most important Neoplatonists were Iamblichus (c. A.D. 250–330), and Proclus (A.D. 410–485).[178] At the beginning of the sixth century A.D., the Academy in Athens taught Neoplatonic doctrines, but it was closed in A.D. 529 by order of the Emperor Justinian (A.D. 482–565).[179] Neoplatonism, like every other school of ancient philosophy, was obscured by and partially absorbed into Christianity.

## *God, Nous, and Soul*

There were three primary elements in Neoplatonic philosophy: the One, *nous*, and the soul. These are not separate entities, but states of consciousness. The One is God, the supreme reality, unknowable, inexpressible, and ineffable. The One "is the source of all life, and therefore absolute causality and the only real existence."[180]

In *Against the Gnostics*, Plotinus identified the One as being identical with Plato's conception of the *Good*. "The nature of the Good is simple and the first; for every thing which is not the first is not simple; and since it has nothing in itself, but is one alone, and the nature of what is called the One is the same with the Good."[181]

The One cannot be described, only experienced. It is above thought and reason, completely transcendental and ineffable. "It has no attributes of any kind,"[182] because to ascribe an attribute to it would diminish its unity.

Immediately below the One in consciousness is *nous*. *Nous* is a Greek word that is usually translated as intellect, mind, reason, or intellectual principle. It is "at once being and thought, ideal world and idea."[183]

Soul is the lowest of the three elements. It is the movement of soul that creates the illusion of time and space, elements that have no real existence in either the One or *nous*. The soul "stands between the nous and the phenomenal world, is permeated and illuminated by the former, but is also in contact with the latter."[184] Like the One and *nous*, soul is immaterial.

Matter, or the material world, is not counted as an element, because it has no real existence. The world revealed to us by our senses is a transitory illusion. "Matter is ... the dark principle ... destitute of form and idea, it is evil."[185]

The Neoplatonic cosmos was thus composed of hierarchical states of consciousness, with the One being the highest, and matter being the lowest. "The totality of being may thus be conceived as a series of concentric circles, fading away towards the verge of non-existence, the force of the original Being in the outermost circle being a vanishing quantity."[186] The world of multiplicity was created from the One by a process of "successive emanations such that each principle emanated from one above it in the scale of being."[187] Neoplatonism was pantheistic. *Pantheism* is the "belief or philosophical theory that God is immanent in or identical with the universe."[188]

Neoplatonists rejected materialism and atomism. A materialistic doctrine could not explain the cosmos, because matter lacked intellect and order. "Matter will not give form to itself, nor insert soul in itself.... [T]he universe also would rapidly perish if all things were bodies ... or rather, nothing would be generated, but all things would stop in matter, as there would not be any thing to invest it with for ... for what order is there, or reason or intellect, in a pneumatic substance, which is in want of order from soul?"[189]

Because Neoplatonism regarded sensory information as illusory, and the material world at best to be neutral, at worst, evil, it was hardly a philosophy that would promote the development of positive science, empiricism, or the experimental method. Truth was to be found by looking within, not without.

## *Soul Travel*

The goal of the Neoplatonic philosopher was to obtain ecstatic communion with the One by means of soul travel.* Through contemplation, "the soul must retrace its steps back to the supreme God."[190] By soul travel, we leave the illusory world of matter behind. We rise even above intellect, to finally merge our souls with the One, the universal soul, the indescribable and ultimate reality. "The last stage is reached when, in the highest tension and concentration, beholding in silence and utter forgetfulness of all things, it [soul] is able as it were to lose itself. Then it may see God, the fountain of life, the source of being, the origin of all good, the root of the soul. In that moment it [soul] enjoys the highest indescribable bliss; it is as it were swallowed up of divinity, bathed in the light of eternity."[191]

An individual presumably prepared themselves for soul travel by abandoning worldly and sensual pleasures and taking up a frugal if not ascetic lifestyle. Porphyry related that one of Plotinus' students, the Roman Senator, Rogatianus, "advanced to such detachment

*Soul travel was revived in the twentieth century by the American author Paul Twitchell (1908?–1971) who founded a religious movement named Eckankar. In his book *The Tiger's Fang*, Twitchell chronicled the journey of his soul to the level of complete omniscience and omnipresence. He then described advancing several levels higher.

from political ambitions that he gave up all his property, dismissed all his slaves, [and] renounced every dignity ... this new regime of abstinence and abnegation restored his health."[192] Plotinus regarded Rogatianus as an exemplary model.

In *The Soul's Descent Into Body* (*Ennead 4.8*), Plotinus described his ascent to the One. "Many times it has happened: lifted out of the body into myself; becoming external to all other things and self-encentered; beholding a marvelous beauty; then, more than ever, assured of community with the loftiest order; enacting the noblest life, acquiring identity with the divine; stationing within It by having attained that activity; poised above whatsoever within the intellectual is less than the Supreme."[193]

But ecstatic communion with the One was a rare event, even for the master. Porphyry related that in the years he spent with Plotinus, his teacher ascended to the One only four times, and that he, Porphyry, obtained union only once, at the age of sixty-eight.[194] It is not difficult to see why Neoplatonism eventually was overwhelmed by Christianity. Christianity offered salvation to every man through a simple act of faith. In Neoplatonism salvation was achieved through the mystical experience, a rare event for sages and out of the reach of the average man.

Although Neoplatonism is commonly described as a philosophy, it would be more correct to call it a religion. Philosophy is based on reason, the deployment of the intellect. But the ecstatic communion that was the central element in Neoplatonism is revelation, an entirely different way of knowing, and the traditional fount of religion.

# Conclusion

Science began in Greece in the sixth century B.C. with the invocation of naturalism, the explanation and interpretation of phenomena in terms of natural law rather than the interdiction of supernatural forces or beings. The concomitant corollary to naturalism was uniformity, the supposition that nature behaves in an unvarying manner throughout both time and space.

It is doubtful if science and philosophy could have originated in ancient Greece if Greece were not an open society that permitted intellectual freedom and tolerated individual differences. Students were not only permitted, but encouraged to criticize their teachers and authorities. Innovation was welcomed. Intellectual freedom in Greece may have been the natural partner to political freedom. A democratic system invariably is more open and free than an autocratic one. The multitude necessarily possesses a diverse set of opinions and beliefs, and their participation in government therefore requires toleration of individual differences.

Arguably, Greek contributions to Western Civilization have never been equaled. In addition to their contributions in politics, art, and literature, the Greeks invented science, philosophy, and logic. Thales, Pythagoras and their colleagues found mathematics as a set of empirical rules and turned it into an exact science. In philosophy, physics, chemistry, medicine, and mathematics, Greek knowledge was ascendant for nearly two thousand years. As late as the sixteenth century, Aristotle was still regarded as the preeminent authority in philosophy and physics. The theories of the four humors and the four elements persisted into the eighteenth century. And Euclid's *Elements* was widely used as a mathematics textbook until the middle of the nineteenth century.

Perhaps most important of all, the Greeks invented the principle of demonstration, the concept that it was possible to establish a method by which demonstrable truth could be established. Opinion, on which everyone disagreed, was replaced by demonstration, the construction of a logical proof with which everyone was compelled to agree.

Ironically it was the seductive appeal of mathematical proof that proved to be a major impediment to the development of an experimental method and modern science. Pure deductive logic without constant recourse to empirical input invariably results in unreliable knowledge. The Greeks tended to disparage empiricism, because the empirical data available to them consisted entirely of anecdotal data, a type of information that is notoriously unreliable. Modern science passes phenomenological data through the filter of repeatability. But this methodology depends upon an efficient means of mass communication through a printing press or its electronic equivalent. The Greeks did not have this technology, nor did they have any conception that it might be advantageous.

The Romans dominated the Mediterranean world for nearly a thousand years, but made few significant or original contributions to science or philosophy. The most popular philosophies in Rome, Epicureanism and Stoicism, were both Greek. Roman genius manifested itself in law, civil administration, and engineering works. The system of aqueducts that supplied Rome with water was an engineering marvel, and Roman bridges and roads were built to endure.

Neither the Greeks or the Romans hit upon the idea of a universal human society based upon a respect for the basic dignity and worth of the individual. Greek and Roman societies were socially stratified and their economies were based on slavery. Roman rule was based on brute force, not the spontaneous cooperation of peoples possessing different belief systems.

By the fifth century A.D., the ancient systems were spent. The intellectual momentum provided by the Presocratic Enlightenment was exhausted. In the Middle East and Europe, the human mind turned inward. For the next millennium most serious thinkers were theologians, and all the important social and cultural milestones were religious. In the fourth century, Christianity became the official religion of the Roman Empire and began to spread through northern and western Europe. In the seventh century, Islam rapidly expanded, reaching as far north and west as France in the eighth century. No significant advances were made in rational philosophy or the sciences for nearly a thousand years. But the spread of Christianity and Islam united diverse tribes, peoples, and cultures under common creeds. And Christianity's ethic of universal human brotherhood encouraged productive cooperation and discouraged conflict.

While the sciences stagnated, the unwritten narrative of crafts and technology continued to evolve. Neither the Greeks nor the Romans had much appreciation for the technical arts. But as early as the eighth century, Europeans were developing new technologies and an entirely new attitude toward man's relationship with the natural world. For the first time, philosophers began to obtain an appreciation of the necessity of linking theory and practice. This new and progressive attitude would culminate in the Scientific and Industrial Revolutions.

# Chapter Notes

## *Preface*

1. Aristotle, 1906, *The Nicomachean Ethics of Aristotle*, Book 6, Chapter 3, Tenth Edition, translated by F. H. Peters, Kegan Paul, Trench, Trübner & Co., Ltd., London, p. 185 (1139b).
2. Herodotus, 1910, *The History of Herodotus*, Book 6, Chapter 117, translated by George Rawlinson (1812–1902), vol. 2 (first published in 1858), J. M. Dent & Sons, New York, p. 104.
3. Butterfield, H., 1931, *The Whig Interpretation of History*, G. Bell and Sons, London.
4. *Encyclopædia Britannica*, Eleventh Edition, 1911, Whig and Tory, vol. 28, Encyclopædia Britannica Company, New York, p. 588.
5. Butterfield, H., 1965, *The Whig Interpretation of History* (first published in 1931 by G. Bell and Sons in London), W. W. Norton, New York, p. 16.
6. Ibid., p. 12.
7. Harrison, E., 1987, Whigs, Prigs and Historians of Science, *Nature*, vol. 329, p. 213–214.
8. Butterfield, H., 1965, *The Whig Interpretation of History* (first published in 1931 by G. Bell and Sons in London), W. W. Norton, New York, p. 75.
9. Sarton, G., 1924, Review of "A History of Magic and Experimental Science During the First Thirteen Centuries of Our Era," *Isis*, no. 1, vol. 6, p. 78.
10. Sarton, G., 1927, *Introduction to the History of Science*, vol. 1, Carnegie Institution of Washington, Washington, DC, p. 6.
11. Sarton, G., 1924, Review of "A History of Magic and Experimental Science During the First Thirteen Centuries of Our Era," *Isis*, no. 1, vol. 6, p. 82.
12. al-Ghazali, 1997, *The Incoherence of the Philosophers*, translated by Michael E. Marmura, Brigham Young Press, Provo, Utah, p. 171.
13. Drake, S., 1978, *Galileo at Work, His Scientific Biography* (first published by the University of Chicago Press in 1978), Dover, New York, p. 377.
14. Turnbull, H. W. (Editor), 1959, *The Correspondence of Isaac Newton*, vol 1, Cambridge University Press, London, p. 416.
15. Drake, S., 1978, *Galileo at Work, His Scientific Biography* (first published by the University of Chicago Press in 1978), Dover, New York, p. 377.
16. Butterfield, H., 1965, *The Whig Interpretation of History* (first published in 1931 by G. Bell and Sons in London): W.W. Norton, New York, p. 105.

## *Introduction*

1. Sarton, G., 1927, *Introduction to the History of Science*, vol. 1, Carnegie Institution of Washington, Washington, DC, p. 4.
2. *Oxford English Dictionary Online*, Second Edition, 1989.
3. Deming, D., 2008, Design, Science and Naturalism, *Earth Science Reviews*, vol. 90, p. 60–63.
4. *Oxford English Dictionary*, Second Edition, 1989.
5. Sarton, G., 1927, *Introduction to the History of Science*, vol. 1, Carnegie Institution of Washington, Washington, DC, p. 4.
6. Saint Teresa of Avila, 1916, *The Life of St. Teresa of Jesus*, Fifth Edition, translated by David Lewis, Benziger Brothers, New York, p. 266–267.
7. Sarton, G., 1924, Review of "A History of Magic and Experimental Science During the First Thirteen Centuries of Our Era," *Isis*, no. 1, vol. 6, p. 78.
8. Wade, N., 2006, *Before the Dawn*, Penguin Press, New York, p. 14.
9. Semaw, S., 2000, The World's Oldest Stone Artefacts from Gona, Ethiopa, *Journal of Archaeological Science*, vol. 27, p. 1197.
10. Wade, N., 2006, *Before the Dawn*, Penguin Press, New York, p. 20.
11. Hodges, H., 1970, *Technology in the Ancient World*, Marboro Books, London, 1992 edition, p. 23.
12. Wade, N., 2006, *Before the Dawn*, Penguin Press, New York, p. 36.
13. Darwin, C., 1897, *The Descent of Man and Selection in Relation to Sex*, D. Appleton, New York, p. 86.
14. Wade, N., 2006, *Before the Dawn*, Penguin Press, New York, p. 32.
15. Hillman, G., et al., 2001, New Evidence of Lateglacial cereal cultivation at Abu Hureyra on the Euphrates, *The Holocene*, vol. 11, no. 4, p. 383.
16. Akkermans, P. M. M. G., and Schwartz, G. M., 2003, *The Archeology of Syria*, Cambridge University Press, p. 45.
17. Singer, C., 1949, *A Short History of Science to the Nineteenth Century*, Clarendon Press, p. 4.
18. Proclus, 1792, *The Philosophical and Mathematical Commentaries of Proclus, on the First Book of Euclid's Elements*, vol. 1, printed for the author, London, p. 98–99.
19. Hesiod, 1920, *Works and Days* (lines 609–617), in *Hesiod, the Homeric Hymns and Homerica*, with an English translation by Hugh G. Evelyn-White, London, William Heinemann, p. 49.
20. Dreyer, J. L. E., 1953, *A History of Astronomy from Thales to Kepler* (first published in 1906 as *History of the Planetary Systems from Thales to Kepler*), Dover, New York, p. 1.
21. Clagett, M., 1994, *Greek Science in Antiquity*, Barnes and Noble, New York (reprint of 1955 edition), p. 13.
22. Edwards, C. (translator), 1904, *The Hammurabi Code and the Sinaitic Legislation*, Watts & Co., London, p. 23.
23. Ibid., p. 37.
24. Sarton, G., 1960, *A History of Science, Ancient Science Through the Golden Age of Greece*, Harvard University Press, Cambridge, p. 87–88.
25. Frazer, J. G., 1922, *The Golden Bough* (abridged edition), Macmillan, New York, p. 54–55.
26. Crump, T., 2001, *A Brief History of Science as Seen Through the Development of Scientific Instruments*, Carroll & Graf, New York, p. 12.
27. Davisson, W. I., and Harper, J. E., 1972, *European Economic History, Volume 1, The Acient World*, Appleton-Century-Crofts, New York, p. 51.
28. Harden, D. B., 1956, Glass and Glazes, in *A History of Technology*, vol. 2, edited by Charles Singer, Oxford University Press, London, p. 311.
29. Neuburger, A., 1930, *The Technical Arts and Sciences of the Ancients*, Macmillan, New York, p. 153.

30. Crowfoot, G. M., 1954, Textiles, Basketry, and Mats, in *A History of Technology*, vol. 1, edited by Charles Singer, Oxford University Press, London, p. 413.
31. Neuburger, A., 1930, *The Technical Arts and Sciences of the Ancients*, Macmillan, New York, p. 165.
32. Crowfoot, G. M., 1954, Textiles, Basketry, and Mats, in *A History of Technology*, vol. 1, edited by Charles Singer, Oxford University Press, London, p. 424.
33. Ibid.
34. Ibid., p. 434.
35. Ibid., p. 435.
36. Neuburger, A., 1930, *The Technical Arts and Sciences of the Ancients*, Macmillan, New York, p. 8.
37. Bromehead, C. N., 1954, Mining and Quarrying, in *A History of Technology*, vol. 1, edited by Charles Singer, Oxford University Press, London, p. 563.
38. Herodotus, 1910, *The History of Herodotus*, Book 3, Chapter 23, translated by George Rawlinson (1812–1902), vol. 1 (first published in 1858), J. M. Dent & Sons, New York, p. 221.
39. Diodorus Siculus, 1814, *The Historical Library of Diodorus the Sicilian*, Book 3, Chapter 1, vol. 1, translated by G. Booth, W. M'Dowall for J. Davis, London, p. 158.
40. Ibid., p. 158–159.
41. Ibid., p. 159–160.
42. Strabo, 1856, *The Geography of Strabo*, Book 11, Chapter 2, Paragraph 19, vol. 2, translated by H. C. Hamilton and W. Falconer, Henry G. Bohn, London, p. 229.
43. Neuburger, A., 1930, *The Technical Arts and Sciences of the Ancients*, Macmillan, New York, p. 12.
44. Homer, 1922, *The Iliad of Homer*, Book 19, lines 363–396, translated by Andrew Lang, Walter Leaf, and Ernest Myers, Revised Edition, Macmillan and Co., London, p. 397.
45. Strabo, 1856, *The Geography of Strabo*, Book 3, Chapter 2, Paragraph 8, vol. 1, translated by H. C. Hamilton and W. Falconer, Henry G. Bohn, London, p. 219.
46. Ibid., p. 220.
47. Derry, T. K., and Williams, T. I., 1993, *A Short History of Technology, From the Earliest Times to A.D. 1900* (reprint of 1960 edition published by Oxford University Press), Dover, New York, p. 117.
48. Neuburger, A., 1930, *The Technical Arts and Sciences of the Ancients*, Macmillan, New York, p. 13.
49. Bromehead, C. N., 1954, Mining and Quarrying, in *A History of Technology*, vol. 1, edited by Charles Singer, Oxford University Press, London, p. 563.
50. *Encyclopædia Britannica*, 1972, "Techonology, History of," vol. 21, William Benton, Chicago, p. 750C.
51. Herodotus, 1910, *The History of Herodotus*, Book 2, Chapter 109, translated by George Rawlinson (1812–1902), vol. 1 (first published in 1858), J. M. Dent & Sons, New York, p. 166.
52. Lindberg, D. C., 1992, *The Beginnings of Western Science*, University of Chicago Press, Chicago, p. 12.
53. Plato, 1937, Phaedrus, in, *The Dialogues of Plato*, translated into English by Benjamin Jowett (1817–1893), vol. 1, Random House, New York, p. 278 (274–275).
54. Ibid.
55. McMurtrie, D. C., 1943, *The Book: The Story of Printing and Bookmaking*, Oxford University Press, New York, p. 5.
56. Clagett, M., 1994, *Greek Science in Antiquity*, Barnes and Noble, New York (reprint of 1955 edition), p. 21.

## Chapter 1

1. Grant, E., 2007, *A History of Natural Philosophy*, Cambridge University Press, Cambridge, p. 2.
2. Burnet, J., 1920, *Early Greek Philosophy*, Third Edition, A. & C. Black, London, p. v.
3. Burnet, J., 1920, *Greek Philosophy, Part I, Thales to Plato*, Macmillan & Co., Limited, London, p. 4.
4. Sarton, G., 1936, The Unity and Diversity of the Mediterranean World, *Osiris*, vol. 2, p. 406–463.
5. Burnet, J., 1920, *Greek Philosophy, Part I, Thales to Plato*, Macmillan & Co., Limited, London, p. 4–5.
6. Sarton, G., 1960, *A History of Science, Ancient Science Through the Golden Age of Greece*, Harvard University Press, Cambridge, p. 37.
7. Ibid., p. 40.
8. Zeller, E., 1908, *Outlines of the History of Greek Philosophy*, translated by Sarah Frances Alleyne and Evelyn Abbott, Henry Holt, New York, p. 23.
9. Sarton, G., 1960, *A History of Science, Ancient Science Through the Golden Age of Greece*, Harvard University Press, Cambridge, p. 21–22.
10. Clagett, M., 1994, *Greek Science in Antiquity*, Barnes and Noble, New York (reprint of 1955 edition), p. 13.
11. Reinach, S., 1904, *The Story of Art Throughout the Ages, an Illustrated Record*, Charles Scribner's Sons, New York, p. 19.
12. Ibid., p. 17.
13. Lloyd, G. E. R., 1979, *Magic, Reason and Experience*, Cambridge University Press, Cambridge, p. 14.
14. Popper, K., 1998, *The World of Parmenides, Essays on the Presocratic Enlightenment*, Routlege, New York, p. 20.
15. Ibid., p. 22.
16. Plato, 1955, *The Republic*, Book 1, translated by Henry Desmond Pritchard Lee, Penguin, Baltimore, Maryland, p. 64 (336).
17. Plato, 1937, Timaeus, in *The Dialogues of Plato*, translated into English by Benjamin Jowett (1817–1893), vol. 2, Random House, New York, p. 34 (54).
18. Muller, H. J., 1961, *Freedom in the Ancient World*, Harper & Brothers, New York, p. 168.
19. Dalberg-Acton, J. E. E., 1907, Freedom in Antiquity, in *The History of Freedom and Other Essays*, edited by John Neville Figgis and Reginald Vere Laurence, Macmillan and Co., Limited, London, p. 11.
20. Ibid., p. 10.
21. Lewes, G. H., 1864, *Aristotle, a Chapter from the History of Science*, Smith, Elder, and Co., London, p. 44.
22. Grote, G., 1875, *Plato and the Other Companions of Sokrates*, vol. 1, John Murray, London, p. 2.
23. Needham, J., 1956, *Science and Civilisation in China*, vol. 2, Cambridge University Press, London, p. 582.
24. Strabo, 1856, *The Geography of Strabo*, Book 8, Chapter 7, Paragraph 1, vol. 2, translated by H. C. Hamilton and W. Falconer, Henry G. Bohn, London, p. 67–68.
25. Herodotus, 1910, *The History of Herodotus*, Book 1, Chapter 142, translated by George Rawlinson (1812–1902), vol. 1 (first published in 1858), J. M. Dent & Sons, New York, p. 74.
26. Ibid., Book 1, Chapter 146, p. 76.
27. Burnet, J., 1920, *Early Greek Philosophy*, Third Edition, London, Adam & Charles Black, p. 2–3.
28. Grant, E., 2007, *A History of Natural Philosophy*, Cambridge University Press, Cambridge, p. 7.
29. Herodotus, 1910, *The History of Herodotus*, Book 5, Chapter 28, translated by George Rawlinson (1812–1902), vol. 2 (first published in 1858), J. M. Dent & Sons, New York, p. 11.
30. Pliny the Elder, 1855, *The Natural History of Pliny*, Book 5, Chapter 31, translated by John Bostock and H. T. Riley, vol. 1, Henry G. Bohn, London, p. 466–467.
31. Burnet, J., 1920, *Early Greek Philosophy*, Third Edition, London, Adam & Charles Black, p. 13–14.
32. Hesiod, 1920, *Hesiod, the Homeric Hymns and Homerica*, with an English translation by Hugh G. Evelyn-White, London, William Heinemann.
33. Turner, W., 1903, *History of Philosophy*, Ginn and Company, Boston, p. 31.
34. Burnet, J., 1920, *Early Greek Philosophy*, Third Edition, London, Adam & Charles Black, p. 80.
35. Haskins, C. H., 1957, *The Renaissance of the 12th Century* (first published in 1927 by Harvard University Press, Cambridge, Massachusetts), Meridian Books, New York, p. 64.
36. Sarton, G., 1960, *A History of Science, Ancient Science Through the Golden Age of Greece*, Harvard University Press, Cambridge, p. 162.
37. Fairbanks, A. 1898, *The First Philosophers of Greece*, Kegan Paul, Trench, Trübner & Co., Ltd., London, p. 1.

38. Popper, K., 1998, *The World of Parmenides, Essays on the Presocratic Enlightenment*, Routlege, New York, p. 23.
39. Burnet, J., 1920, *Early Greek Philosophy*, Third Edition, A. & C. Black, Ltd., London, p. 46.
40. Heath, T., 1981, *A History of Greek Mathematics*, vol. 1 (first published in 1921 by the Clarendon Press, Oxford), Dover, New York, p. 128–140.
41. Proclus, 1792, *The Philosophical and Mathematical Commentaries of Proclus, on the First Book of Euclid's Elements*, vol 1, printed for the author, London, p. 99.
42. Plutarch, 1718, Banquet of the Seven Wisemen (translated by Roger Davis), in *Plutarch's Morals: in Five Volumes, translated from the Greek, by Several Hands*, Fifth Edition, vol. 2, W. Taylor, London, p. 3.
43. Fairbanks, A., 1898, *The First Philosophers of Greece*, Kegan Paul, Trench, Trübner & Co., Ltd., London, p. 2.
44. Hubbert, M. K., 1963, Are We Retrogressing in Science? *Geological Society of America Bulletin*, vol. 74, p. 376.
45. Burnet, J., 1920, *Early Greek Philosophy*, Third Edition, A. & C. Black, Ltd., London, p. 48–49.
46. Aristotle, 1939, *On the Heavens* (*De Caelo*), Book 2, Chapter 13, translated by W. K. C. Guthrie, William Heinemann, London, p. 225 (294a).
47. Ibid., Book 2, Chapter 13, p. 227 (294b).
48. Grote, G., 1888, *Plato and the Other Companions of Sokrates*, New Edition, vol. 1, John Murray, London, p. 4.
49. Fairbanks, A. 1898, *The First Philosophers of Greece*, Kegan Paul, Trench, Trübner & Co., Ltd., London, p. 5.
50. Apuleius, L., 1909, *The Apologia and Florida of Apuleius of Madaura*, translated by Harold Edgeworth Butler (1878–1951), Florida, Chapter 18, Oxford at the Clarendon Press, p. 204–205.
51. Plato, 1937, Theaetetus, in *The Dialogues of Plato*, translated into English by Benjamin Jowett (1817–1893), vol. 2, Random House, New York, p. 176 (174).
52. Herodotus, 1910, *The History of Herodotus*, Book 1, Chapter 74, translated by George Rawlinson (1812–1902), vol 1 (first published in 1858), J. M. Dent & Sons, New York, p. 37.
53. Fairbanks, A. 1898, *The First Philosophers of Greece*, Kegan Paul, Trench, Trübner & Co., Ltd., London, p. 6–7.
54. Ibid., p. 6.
55. Plutarch, 1870, *Of Those Sentiments Concerning Nature with which Philosophers were Delighted*, Book 1, Chapter 7, translated by John Dowel, in *Plutarch's Morals, Translated from the Greek by Several Hands, Corrected and Revised by William W. Goodwin*, vol. 3, Little, Brown, and Company, Boston, p. 121.
56. Aristotle, 1885, *The Politics of Aristotle*, Book 1, Chapter 11, vol. 1, translated by Benjamin Jowett (1817–1893), Oxford at the Clarendon Press, London, p. 21 (1259a).
57. Ibid.
58. Laërtius, Diogenes, 1905, *The Lives and Opinions of Eminent Philosophers*, Book 1, translated by C. D. Yonge, London, George Bell and Sons, p. 19.
59. Ibid., p. 20.
60. Strabo, 1856, *The Geography of Strabo*, Book 14, Chapter 1, Paragraph 7, vol. 3, translated by H. C. Hamilton and W. Falconer, Henry G. Bohn, London, p. 5.
61. Laërtius, Diogenes, 1905, *The Lives and Opinions of Eminent Philosophers*, Book 2, translated by C. D. Yonge, London, George Bell and Sons, p. 57.
62. Herodotus, 1910, *The History of Herodotus*, Book 2, Chapter 109, translated by George Rawlinson (1812–1902), vol 1, (first published in 1858), J. M. Dent & Sons, New York, p. 166.
63. Fairbanks, A. 1898, *The First Philosophers of Greece*, Kegan Paul, Trench, Trübner & Co., Ltd., London, p. 2.
64. Burnet, J., 1920, *Early Greek Philosophy*, Third Edition, London, Adam & Charles Black, p. 65.
65. Popper, K., 1998, *The World of Parmenides, Essays on the Presocratic Enlightenment*, Routlege, New York, p. 9.
66. Fairbanks, A. 1898, *The First Philosophers of Greece*, Kegan Paul, Trench, Trübner & Co., Ltd., London, p. 13.
67. Dreyer, J. L. E., 1953, *A History of Astronomy from Thales to Kepler* (first published in 1906 as *History of the Planetary Systems from Thales to Kepler*), Dover, New York, p. 14.
68. Burnet, J., 1920, *Early Greek Philosophy*, Third Edition, London, Adam & Charles Black, p. 66–67.
69. Ibid., p. 52.
70. Ibid., p. 53.
71. Ibid., p. 58.
72. Cicero, Marcus Tullius, 1775, *Of the Nature of the Gods, in Three Books*, Book 1, T. Davies, London, p. 21.
73. Burnet, J., 1920, *Early Greek Philosophy*, Third Edition, London, Adam & Charles Black, p. 68.
74. Ibid., p. 67.
75. Pliny the Elder, 1855, *The Natural History of Pliny*, Book 2, Chapter 81, translated by John Bostock and H. T. Riley, vol. 1, Henry G. Bohn, London, p. 112.
76. Laërtius, Diogenes, 1905, *The Lives and Opinions of Eminent Philosophers*, Book 2, translated by C. D. Yonge, London, George Bell and Sons, p. 57.
77. Strabo, 1856, *The Geography of Strabo*, Book 1, Chapter 1, Paragraph 1, translated by H. C. Hamilton and W. Falconer, vol. 1, Henry G. Bohn, London, p. 1.
78. Burnet, J., 1920, *Early Greek Philosophy*, Third Edition, London, Adam & Charles Black, p. 70–71.
79. Laërtius, Diogenes, 1905, *The Lives and Opinions of Eminent Philosophers*, Book 2, translated by C. D. Yonge, London, George Bell and Sons, p. 57.
80. Kirk, G. S., and Raven, J. E., 1957, *The Presocratic Philsophers*, Cambridge University Press, London, p. 186.
81. Laërtius, Diogenes, 1905, *The Lives and Opinions of Eminent Philosophers*, Book 9, translated by C. D. Yonge, London, George Bell and Sons, p. 376.
82. Ibid., p. 377–378.
83. Fairbanks, A. 1898, *The First Philosophers of Greece*, Kegan Paul, Trench, Trübner & Co., Ltd., London, p. 25.
84. Ibid., p. 39.
85. Freeman, K., 1966, *Ancilla to the Pre-Socratic Philosophers, a Complete Translation of the Fragments in Diels, Fragmente der Vorsokratiker*, Harvard University Press, Cambridge, p. 26.
86. Fairbanks, A. 1898, *The First Philosophers of Greece*, Kegan Paul, Trench, Trübner & Co., Ltd., London, p. 27, 37.
87. Ibid., p. 35.
88. Ibid., p. 57.
89. Popper, K., 1998, *The World of Parmenides, Essays on the Presocratic Enlightenment*, Routlege, New York, p. 19.
90. Fairbanks, A. 1898, *The First Philosophers of Greece*, Kegan Paul, Trench, Trübner & Co., Ltd., London, p. 35.
91. Bakewell, C. M., 1907, *Source Book in Ancient Philosophy*, Charles Scribner's Sons, New York, p. 31.
92. Fairbanks, A. 1898, *The First Philosophers of Greece*, Kegan Paul, Trench, Trübner & Co., Ltd., London, p. 37.
93. Ibid., p. 39.
94. Ibid., p. 59.
95. Ibid., p. 31.
96. Ibid., p. 61.
97. Hicks, R. D., 1910, *Stoic and Epicurean*, Charles Scribner's Sons, New York, p. 11.
98. Fairbanks, A. 1898, *The First Philosophers of Greece*, Kegan Paul, Trench, Trübner & Co., Ltd., London, p. 57.
99. Ibid., p. 58.
100. Laërtius, Diogenes, 1905, *The Lives and Opinions of Eminent Philosophers*, Book 9, translated by C. D. Yonge, London, George Bell and Sons, p. 378–379.
101. Kirk, G. S., and Raven, J. E., 1957, *The Presocratic Philsophers*, Cambridge University Press, London, p. 203.
102. Fairbanks, A. 1898, *The First Philosophers of Greece*, Kegan Paul, Trench, Trübner & Co., Ltd., London, p. 62.
103. Ibid., p. 45.
104. Bakewell, C. M., 1907, *Source Book in Ancient Philosophy*, Charles Scribner's Sons, New York, p. 34.
105. Fairbanks, A. 1898, *The First Philosophers of Greece*, Kegan Paul, Trench, Trübner & Co., Ltd., London, p. 47.
106. Ibid., p. 49.
107. Ibid., p. 35.
108. Burnet, J., 1920, *Early Greek Philosophy*, Third Edition, London, Adam & Charles Black, p. 134.

109. Lloyd, G. E. R., 1970, *Early Greek Science: Thales to Aristotle*, W. W. Norton & Co., New York, p. 10.
110. Laërtius, Diogenes, 1905, *The Lives and Opinions of Eminent Philosophers*, Book 9, translated by C. D. Yonge, London, George Bell and Sons, p. 377.
111. Bailey, C., 1928, *The Greek Atomists and Epicurus*, Oxford at the Clarendon Press, London, p. 66.
112. Burnet, J., 1920, *Greek Philosophy, Part 1, Thales to Plato*, Macmillan, London, p. 94.
113. Laërtius, Diogenes, 1905, *The Lives and Opinions of Eminent Philosophers*, Book 9, translated by C. D. Yonge, London, George Bell and Sons, p. 390.
114. Ibid., p. 390.
115. Heath, T., 1981, *A History of Greek Mathematics, Volume I, From Thales to Euclid* (first published in 1921 by the Clarendon Press, Oxford), Dover, New York, p. 176.
116. Laërtius, Diogenes, 1905, *The Lives and Opinions of Eminent Philosophers*, Book 9, translated by C. D. Yonge, London, George Bell and Sons, p. 391.
117. Ibid., p. 176.
118. Bailey, C., 1928, *The Greek Atomists and Epicurus*, Oxford at the Clarendon Press, London, p. 112.
119. Horace, 1748, *The Satires, Epistles, and Art of Poetry of Horace Translated into English Prose*, Third Edition, Epistle 1, Book 2, printed for Joseph Davidson, London, p, 337.
120. Julian, 1784, *Select Works of the Emperor Julian, translated from the Greek by John Duncombe*, vol. 2, J. Nichols, London, p. 95–96.
121. Bailey, C., 1928, *The Greek Atomists and Epicurus*: Oxford at the Clarendon Press, London, p. 76.
122. Aristotle, 1941, *On Generation and Corruption*, Book 1, Chapter 8, translated by Harold H. Joachim, in *Basic Works of Aristotle*, Random House, New York, p. 499 (325b).
123. Aristotle, 1941, *Physics*, Book 1, Chapter 5, translated by R. P. Hardie and R. K. Gaye, in *Basic Works of Aristotle*, Random House, New York, p. 226 (188a).
124. Aristotle, 1941, *Metaphysics*, Book 1, Chapter 4, translated by W. D. Ross, in *Basic Works of Aristotle*, Random House, New York, p. 697 (985b).
125. Aristotle, 1941, *De Caelo* (*On the Heavens*), Book 3, Chapter 4, translated by J. L. Stocks, in *Basic Works of Aristotle*, Random House, New York, p. 446 (303a).
126. Laërtius, Diogenes, 1905, *The Lives and Opinions of Eminent Philosophers*, Book 9, translated by C. D. Yonge, London, George Bell and Sons, p. 391.
127. Bakewell, C. M., 1907, *Source Book in Ancient Philosophy*, Charles Scribner's Sons, New York, p. 60.
128. Burnet, J., 1920, *Greek Philosophy, Part 1, Thales to Plato*, Macmillan, London, p. 196.
129. Bailey, C., 1928, *The Greek Atomists and Epicurus*, Oxford at the Clarendon Press, London, p. 122.
130. Bakewell, C. M., 1907, *Source Book in Ancient Philosophy*, Charles Scribner's Sons, New York, p. 57.
131. Burnet, J., 1920, *Greek Philosophy, Part 1, Thales to Plato*, Macmillan, London, p. 347.
132. Hippolytus, 1921, *Philosophumena, or the Refutation of All Heresies*, vol. 1, translated by F. Legge, Macmillan, New York, p. 48–49.
133. Bakewell, C. M., 1907, *Source Book in Ancient Philosophy*, Charles Scribner's Sons, New York, p. 61–63.
134. Laërtius, Diogenes, 1905, *The Lives and Opinions of Eminent Philosophers*, Book 9, translated by C. D. Yonge, London, George Bell and Sons, p. 392.
135. Ibid., p. 394.
136. Herodotus, 1910, *The History of Herodotus*, Book 3, Chapter 39, translated by George Rawlinson (1812–1902), vol. 1 (first published in 1858), J. M. Dent & Sons, New York, p. 230.
137. Grote, G., 1899, *A History of Greece*, vol. 3, Harper & Brothers, New York, p. 454.
138. Aristotle, 1885, *The Politics of Aristotle*, vol. 1, Book 5, Chapter 11, translated by Benjamin Jowett (1817–1893), Oxford at the Clarendon Press, London, p. 178 (1313b).
139. Herodotus, 1910, *The History of Herodotus*, Book 3, Chapter 60, translated by George Rawlinson (1812–1902), vol. 1 (first published in 1858), J. M. Dent & Sons, New York, p. 239–240.
140. Goodfield, J., 1964, The Tunnel of Eupalinus, *Scientific American*, vol. 210, no. 6, p. 104–112.
141. Apostol, T. M., 2004, The Tunnel of Samos, *Engineering & Science*, vol. 67, no. 1, p. 31.
142. Burns, A., 1971, The Tunnel of Eupalinus and the Tunnel Problem of Hero of Alexandria, *Isis*, vol. 62, no. 2, p. 176.
143. Ibid., p. 174–175.
144. Herodotus, 1910, *The History of Herodotus*, Book 3, Chapters 54, 56, translated by George Rawlinson (1812–1902), vol. 1 (first published in 1858), J. M. Dent & Sons, New York, p. 237.
145. Ibid., Book 3, Chapter 44, p. 232.
146. Ibid., Book 3, Chapter 40, p. 230–231.
147. Ibid., Book 3, Chapter 41, p. 231.
148. Ibid., Book 3, Chapter 42, p. 231.
149. Ibid., Book 3, Chapter 43, p. 231.
150. Ibid.
151. Ibid., Book 3, Chapter 122, p. 268.
152. Ibid., Book 3, Chapter 123, p. 269.
153. Ibid., Book 3, Chapter 125, p. 269.
154. Kirk, G. S., and Raven, J. E., 1957, *The Presocratic Philsophers*, Cambridge University Press, London, p. 228.
155. Ibid., p. 221.
156. Iamblichus, 1986, *Life of Pythagoras*, Chapter 2, translated by Thomas Taylor (1758–1835; reprinted from the edition published in 1818 by J. M. Watkins in London), Inner Traditions, Rochester, Vermont, p. 3.
157. Ibid., p. 5.
158. Ibid., p. 6.
159. Burnet, J., 1920, *Greek Philosophy, Part I, Thales to Plato*, Macmillan & Co., Limited, London, p. 39.
160. Laërtius, Diogenes, 1905, *The Lives and Opinions of Eminent Philosophers*, Book 8, translated by C. D. Yonge, London, George Bell and Sons, p. 338–339.
161. Iamblichus, 1986, *Life of Pythagoras*, Chapter 4, translated by Thomas Taylor (1758–1835; reprinted from the edition published in 1818 by J. M. Watkins in London), Inner Traditions, Rochester, Vermont, p. 9.
162. *Encyclopædia Britannica*, 1972, "Cambysses," vol. 4, William Benton, Chicago, p. 695.
163. Iamblichus, 1986, *Life of Pythagoras*, Chapter 4, translated by Thomas Taylor (1758–1835; reprinted from the edition published in 1818 by J. M. Watkins in London), Inner Traditions, Rochester, Vermont, p. 9.
164. Ibid., Chapter 5, p. 12.
165. Ibid.
166. Heath, T., 1913, *Aristarchus of Samos, the Ancient Copernicus*, Oxford at the Clarendon Press, London, p. 47.
167. Burnet, J., 1920, *Greek Philosophy, Part I, Thales to Plato*, Macmillan & Co., Limited, London, p. 37.
168. Kirk, G. S., and Raven, J. E., 1957, *The Presocratic Philsophers*, Cambridge University Press, London, p. 220.
169. Strabo, 1856, *The Geography of Strabo*, Book 14, Chapter 1, Paragraph 16, vol. 3, translated by H. C. Hamilton and W. Falconer, Henry G. Bohn, London, p. 9.
170. Iamblichus, 1986, *Life of Pythagoras*, Chapter 12, translated by Thomas Taylor (1758–1835; reprinted from the edition published in 1818 by J. M. Watkins in London), Inner Traditions, Rochester, Vermont, p. 28.
171. Burnet, J., 1920, *Greek Philosophy, Part I, Thales to Plato*, Macmillan & Co., Limited, London, p. 42.
172. Laërtius, Diogenes, 1905, *The Lives and Opinions of Eminent Philosophers*, Book 8, translated by C. D. Yonge, London, George Bell and Sons, p. 341.
173. Iamblichus, 1986, *Life of Pythagoras*, Chapter 12, translated by Thomas Taylor (1758–1835; reprinted from the edition published in 1818 by J. M. Watkins in London), Inner Traditions, Rochester, Vermont, p. 28.
174. Laërtius, Diogenes, 1905, *The Lives and Opinions of Eminent Philosophers*, Book 8, translated by C. D. Yonge, London, George Bell and Sons, p. 342.
175. Kirk, G. S., and Raven, J. E., 1957, *The Presocratic Philsophers*, Cambridge University Press, London, p. 221.
176. Burnet, J., 1920, *Greek Philosophy, Part I, Thales to Plato*, Macmillan & Co., Limited, London, p. 52.

177. Aristotle, 1941, *Metaphysics*, Book 1, Chapter 5, translated by W. D. Ross, in *Basic Works of Aristotle*, Random House, New York, p. 698 (986a).
178. Because Pythagoras left no writings, virtually all of the stories and accounts of his life are uncertain. In *Early Greek Philospty* (Fourth Edition, 1930, p. 84) John Burnet wrote, "it is not easy to give any account of Pythagoras that can claim to be regarded as historical."
179. Iamblichus, 1986, *Life of Pythagoras*, Chapter 26, translated by Thomas Taylor (1758–1835; reprinted from the edition published in 1818 by J. M. Watkins in London), Inner Traditions, Rochester, Vermont, p. 62.
180. Koestler, A., 1963, *The Sleepwalkers*, Grosset & Dunlap, New York, p. 28.
181. Aristotle, 1941, *Metaphysics*, Book 1, Chapter 5, translated by W. D. Ross, in *Basic Works of Aristotle*, Random House, New York, p. 698 (985b–986a).
182. Proclus, 1792, *The Philosophical and Mathematical Commentaries of Proclus, on the First Book of Euclid's Elements*, vol. 1, printed for the author, London, p. 99.
183. Heath, T., 1981, *A History of Greek Mathematics: Volume I, From Thales to Euclid* (first published in 1921 by the Clarendon Press, Oxford), Dover, New York, p. 144.
184. Ibid., p. 158–162.
185. Allman, G. J., 1889, *Greek Geometry from Thales to Euclid*, Longmans, Green, & Co., London, p. 50.
186. Laërtius, Diogenes, 1905, *The Lives and Opinions of Eminent Philosophers*, Book 8, translated by C. D. Yonge, London, George Bell and Sons, p. 358.
187. Dreyer, J. L. E., 1953, *A History of Astronomy from Thales to Kepler* (first published in 1906 as *History of the Planetary Systems from Thales to Kepler*), Dover, New York, p. 37–38.
188. Heath, T., 1913, *Aristarchus of Samos, the Ancient Copernicus*, Oxford at the Clarendon Press, London, p. 48–49.
189. Aristotle, 1939, *On the Heavens* (*De Caelo*), Book 2, Chapter 13, translated by W. K. C. Guthrie, William Heinemann, London, p. 217 (293a).
190. Heath, T., 1913, *Aristarchus of Samos, the Ancient Copernicus*, Oxford at the Clarendon Press, London, p. 269.
191. Ibid., p. 106.
192. Iamblichus, 1986, *Life of Pythagoras*, Chapter 15, translated by Thomas Taylor (1758–1835; reprinted from the edition published in 1818 by J. M. Watkins in London), Inner Traditions, Rochester, Vermont, p. 32–34.
193. Ibid., p. 72.
194. Heath, T., 1913, *Aristarchus of Samos, the Ancient Copernicus*, Oxford at the Clarendon Press, London, p. 105 (translation of *De Caelo*, Book 2, Chapter 9, p. 290b)
195. Aristotle, 1939, *On the Heavens* (*De Caelo*), Book 2, Chapter 9, translated by W. K. C. Guthrie, William Heinemann, London, p. 195 (291a).
196. Ibid.
197. The proof is given in Lloyd, G. E. R., 1970, *Early Greek Science: Thales to Aristotle*, W. W. Norton & Co., New York, p. 35; also, Heath, T. L., 1956, *The Thirteen Books of Euclid's Elements*, Second Edition, vol. 3 (first published in 1908 by Cambridge University Press), Dover, New York, p. 2.
198. Heath, T. L., 1956, *The Thirteen Books of Euclid's Elements*, Second Edition, vol. 1 (first published in 1908 by Cambridge University Press), Dover, New York, p. 411.
199. Iamblichus, 1986, *Life of Pythagoras*, Chapter 14, translated by Thomas Taylor (1758–1835; reprinted from the edition published in 1818 by J. M. Watkins in London), Inner Traditions, Rochester, Vermont, p. 31.
200. Laërtius, Diogenes, 1905, *The Lives and Opinions of Eminent Philosophers*, Book 8, translated by C. D. Yonge, London, George Bell and Sons, p. 339.
201. Kirk, G. S., and Raven, J. E., 1957, *The Presocratic Philsophers*, Cambridge University Press, London, p. 222.
202. Herodotus, 1910, *The History of Herodotus*, Book 2, Chapter 123, translated by George Rawlinson (1812–1902), vol. 1 (first published in 1858), J. M. Dent & Sons, New York, p. 177.
203. Iamblichus, 1986, *Life of Pythagoras*, Chapter 15, translated by Thomas Taylor (1758–1835; reprinted from the edition published in 1818 by J. M. Watkins in London), Inner Traditions, Rochester, Vermont, p. 31–32.
204. Porphyry, 1987, The Life of Pythagoras, in *The Pythagorean Sourcebook and Library*, translated by Kenneth Sylvan Guthrie (1871–1940), Phanes Press, Grand Rapids, Michigan, p. 130.
205. Laërtius, Diogenes, 1905, *The Lives and Opinions of Eminent Philosophers*, Book 8, translated by C. D. Yonge, London, George Bell and Sons, p. 345–346.
206. Ibid., p. 346.
207. Iamblichus, 1986, *Life of Pythagoras*, Chapter 16, translated by Thomas Taylor (1758–1835; reprinted from the edition published in 1818 by J. M. Watkins in London), Inner Traditions, Rochester, Vermont, p. 36.
208. Laërtius, Diogenes, 1905, *The Lives and Opinions of Eminent Philosophers*, Book 8, translated by C. D. Yonge, London, George Bell and Sons, p. 356.
209. Iamblichus, 1986, *Life of Pythagoras*, Chapter 35, translated by Thomas Taylor (1758–1835; reprinted from the edition published in 1818 by J. M. Watkins in London), Inner Traditions, Rochester, Vermont, p. 127–128.
210. Shuckburgh, E. S., 1889, *The Histories of Polybius*, footnote to Book 2, Chapter 39, Macmillan, London, p. 135.
211. Polybius, 1889, *The Histories of Polybius*, Book 2, Chapter 39, vol. 1, translated by Evelyn S. Shuckburgh, Macmillan and Co., London, p. 135.
212. Herodotus, 1910, *The History of Herodotus*, Book 1, Chapter 167, translated by George Rawlinson (1812–1902), vol. 1 (first published in 1858), J. M. Dent & Sons, New York, p. 85; also, Grote, G., 1899, *A History of Greece*, vol. 3, p. 421.
213. Burnet, J., 1920, *Early Greek Philosophy*, Third Edition, London, Adam & Charles Black, p. 119.
214. Laërtius, Diogenes, 1905, *The Lives and Opinions of Eminent Philosophers*, Book 9, translated by C. D. Yonge, London, George Bell and Sons, p. 384.
215. Plato, 1937, Parmenides, in *The Dialogues of Plato*, translated into English by Benjamin Jowett (1817–1893), vol. 2, Random House, New York, p. 88 (127).
216. Laërtius, Diogenes, 1905, *The Lives and Opinions of Eminent Philosophers*, Book 9, translated by C. D. Yonge, London, George Bell and Sons, p. 384.
217. Heath, T., 1913, *Aristarchus of Samos, the Ancient Copernicus*, Oxford at the Clarendon Press, London, p. 64.
218. Ibid., p. 66.
219. Vlastos, G., 1975, *Plato's Universe*, University of Washington Press, Seattle, p. 45.
220. Heath, T., 1913, *Aristarchus of Samos, the Ancient Copernicus*, Oxford at the Clarendon Press, London, p. 66.
221. Ibid., p. 75.
222. Burnet, J., 1920, *Early Greek Philosophy*, Third Edition, London, Adam & Charles Black, p. 177.
223. Kirk, G. S., and Raven, J. E., 1957, *The Presocratic Philsophers*, Cambridge University Press, London, p. 265.
224. Burnet, J., 1920, *Early Greek Philosophy*, Third Edition, London, Adam & Charles Black, p. 174.
225. Ibid., p. 181.
226. Ibid., p. 182.
227. Laërtius, Diogenes, 1905, *The Lives and Opinions of Eminent Philosophers*, Book 9, translated by C. D. Yonge, London, George Bell and Sons, p. 385.
228. Ibid., p. 386.
229. Burnet, J., 1920, *Early Greek Philosophy*, Third Edition, London, Adam & Charles Black, p. 316.
230. Ibid., p. 318.
231. Ibid.
232. Plato, 1937, Theaetetus, in *The Dialogues of Plato*, translated into English by Benjamin Jowett (1817–1893), vol. 2, Random House, New York, p. 183 (181).
233. Aristotle, 1941, *De Generatione et Corruptione* (*On Generation and Corruption*), Book 1, Chapter 8, translated by Harold H. Joachim, in *Basic Works of Aristotle*, Random House, New York, p. 497–498 (325a).
234. Plato, 1937, Sophist, in *The Dialogues of Plato*, translated into English by Benjamin Jowett (1817–1893), vol. 2, Random House, New York, p. 249 (242).

235. Farrington, B., 1961, *Greek Science*, Penguin Books, Baltimore, Maryland, p. 56.
236. Thorndike, L., 1923, *A History of Magic and Experimental Science*, vol. 1, Columbia University Press, New York, p. 140.
237. Grote, G., 1899, *A History of Greece*, vol. 8, Harper & Brothers, New York, p. 405.
238. Ibid., p. 406.
239. Mourelatos, A. P. D., 2008, Empedocles, in *Complete Dictionary of Scientific Biography*, edited by Charles Gillispie, vol. 4, Cengage Learning, New York, p. 367.
240. Laërtius, Diogenes, 1905, *The Lives and Opinions of Eminent Philosophers*, Book 8, translated by C. D. Yonge, London, George Bell and Sons, p. 360.
241. Iamblichus, 1986, *Life of Pythagoras*, Chapter 36, translated by Thomas Taylor (1758–1835; reprinted from the edition published in 1818 by J. M. Watkins in London), Inner Traditions, Rochester, Vermont, p. 137.
242. Sarton, G., 1960, *A History of Science, Ancient Science Through the Golden Age of Greece*, Harvard University Press, Cambridge, p.246.
243. Burnet, J., 1920, *Early Greek Philosophy*, Third Edition, London, Adam & Charles Black, p. 199.
244. Fairbanks, A. 1898, *The First Philosophers of Greece*, Kegan Paul, Trench, Trübner & Co., Ltd., London, p. 223.
245. Kirk, G. S., and Raven, J. E., 1957, *The Presocratic Philsophers*, Cambridge University Press, London, p. 322.
246. Freeman, K., 1966, *Ancilla to the Pre-Socratic Philosophers, a Complete Translation of the Fragments in Diels, Fragmente der Vorsokratiker*, Harvard University Press, Cambridge, p. 65.
247. Bakewell, C. M., 1907, *Source Book in Ancient Philosophy*, Charles Scribner's Sons, New York, p. 46.
248. Laërtius, Diogenes, 1905, *The Lives and Opinions of Eminent Philosophers*, Book 8, translated by C. D. Yonge, London, George Bell and Sons, p. 364.
249. Fairbanks, A. 1898, *The First Philosophers of Greece*, Kegan Paul, Trench, Trübner & Co., Ltd., London, p. 218.
250. Bakewell, C. M., 1907, *Source Book in Ancient Philosophy*, Charles Scribner's Sons, New York, p. 44.
251. Fairbanks, A. 1898, *The First Philosophers of Greece*, Kegan Paul, Trench, Trübner & Co., Ltd., London, p. 216.
252. Burnet, J., 1920, *Early Greek Philosophy*, Third Edition, London, Adam & Charles Black, p. 208.
253. Ibid., p. 219.
254. Ibid., p. 27.
255. Kirk, G. S., and Raven, J. E., 1957, *The Presocratic Philsophers*, Cambridge University Press, London, p. 342.
256. Farrington, B., 1961, *Greek Science*, Penguin Books, Baltimore, Maryland, p. 58.
257. Fairbanks, A. 1898, *The First Philosophers of Greece*, Kegan Paul, Trench, Trübner & Co., Ltd., London, p. 219.
258. Burnet, J., 1920, *Early Greek Philosophy*, Third Edition, London, Adam & Charles Black, p. 200.
259. Ibid., p. 201.
260. Coxe, J. R., 1846, Introduction, in *The Writings of Hippocrates and Galen*, Lindsay and Blakiston, Philadelphia, p. 22. This Introduction appears to be Coxe's translation of *Historie de la Medecine* by Daniel Le Clerc, published at A la Haye (The Hague) by Chez Isaac van der Kloot in 1729 (first edition in 1696).
261. Clarke, J., 1910, *Physical Science in the Time of Nero, Being a Translation of the Quaestiones Naturales of Seneca*, Book 3, Chapter 24, Macmillan, London, p. 136.
262. Heath, T., 1913, *Aristarchus of Samos, the Ancient Copernicus*, Oxford at the Clarendon Press, London, p. 89.
263. Dreyer, J. L. E., 1953, *A History of Astronomy from Thales to Kepler* (first published in 1906 as *History of the Planetary Systems from Thales to Kepler*), Dover, New York, p. 26, 28.
264. Aristotle, 1941, *De Caelo* (*On the Heavens*), Book 2, Chapter 13, translated by John Leofric Stocks, in *Basic Works of Aristotle*, Random House, New York, p. 431.
265. Heath, T., 1913, *Aristarchus of Samos, the Ancient Copernicus*, Oxford at the Clarendon Press, London, p. 89.
266. Ibid., p. 90.
267. Ibid., p. 89.
268. Ibid., p. 89.
269. Laërtius, Diogenes, 1905, *The Lives and Opinions of Eminent Philosophers*, Book 8, translated by C. D. Yonge, London, George Bell and Sons, p. 365–366.
270. Nutton, V., 2004, *Ancient Medicine*, Routledge, London, p. 87, 152–153.
271. Joly, R., 2008, Hippocrates of Cos, in *Complete Dictionary of Scientific Biography*, edited by Charles Gillispie, vol. 6, Cengage Learning, New York, p. 419.
272. Plato, 1937, Protagoras, in *The Dialogues of Plato*, translated into English by Benjamin Jowett (1817–1893), vol. 1, Random House, New York, p. 83 (311).
273. Plato, 1937, Phaedrus, in *The Dialogues of Plato*, translated into English by Benjamin Jowett (1817–1893), vol. 1, Random House, New York, p. 273 (270).
274. Aristotle, 1885, *The Politics of Aristotle*, vol. 1, Book 7, Chapter 4, translated by Benjamin Jowett (1817–1893), Oxford at the Clarendon Press, London, p. 214 (1326a).
275. Strabo, 1857, *The Geography of Strabo*, Book 14, Chapter 2, Paragraph 19, vol. 3, translated by H. C. Hamilton and W. Falconer, Henry G. Bohn, London, p. 36.
276. Pliny the Elder, 1856, *The Natural History of Pliny*, translated by John Bostock and H. T. Riley, book 26, chapter 6, vol. 5, Henry G. Bohn, London, p. 156.
277. Ibid., book 29, chapter 2, vol. 5, p. 371.
278. Withington, E. T., 1921, The Asclepiadae and the Priests of Asclepius, in *Studies in the History and Method of Science*, vol. 2, edited by Charles Singer, Oxford at the Clarendon Press, London, p. 194.
279. Coxe, J. R., 1846, Introduction, in *The Writings of Hippocrates and Galen*, Lindsay and Blakiston, Philadelphia, p. 20. This Introduction appears to be Coxe's translation of *Historie de la Medecine* by Daniel Le Clerc, published at A la Haye (The Hague) by Chez Isaac van der Kloot in 1729 (first edition in 1696).
280. Homer, 1920, To Asclepius, Hymn 16, in *Hesiod, the Homeric Hymns and Homerica*, with an English translation by Hugh G. Evelyn-White, London, William Heinemann, p. 441.
281. Grote, G., 1899, *A History of Greece*, vol. 1, Harper & Brothers, New York, p. 164–165.
282. Ibid., p. 165.
283. Homer, 1922, *The Iliad of Homer*, Book 2, Line 730, revised edition, translated by Andrew Lang, Walter Leaf, and Ernest Myers, Macmillan, London, p. 43–44.
284. Bryce, T. R., 2002, The Trojan War: is There Truth Behind the Legend? *Near Eastern Archaeology*, no. 3, vol. 65, p. 185.
285. Grote, G., 1899, *A History of Greece*, vol. 1, Harper & Brothers, New York, p. 167.
286. Ibid., p. 168.
287. Adams, F., 1849, Preliminary Discourse, in *The Genuine Works of Hippocrates*, vol. 1, Sydenham Society, London, p. 5.
288. Ibid.
289. Nutton, V., 2004, *Ancient Medicine*, Routledge, London, p. 110.
290. Withington, E. T., 1921, The Asclepiadae and the Priests of Asclepius, in *Studies in the History and Method of Science*, vol. 2, edited by Charles Singer, Oxford at the Clarendon Press, London, p. 192–205.
291. Adams, F., 1849, Preliminary Discourse, in *The Genuine Works of Hippocrates*, vol. 1, Sydenham Society, London, p. 8.
292. Plato, 1937, Theaetetus, in *The Dialogues of Plato*, translated into English by Benjamin Jowett (1817–1893), vol. 2, Random House, New York, p. 151 (150).
293. Ibid., p. 151 (149).
294. Kudlien, F., 1968, Early Greek Primitive Medicine, *Clio Medica*, vol. 3, p. 310.
295. Ibid., p. 312.
296. Ibid., p. 315–316.
297. Homer, 1922, *The Iliad of Homer*, Book 1, Lines 48–52, revised edition, translated by Andrew Lang, Walter Leaf, and Ernest Myers, Macmillan, London, p. 2.

298. *Encyclopædia Britannica*, Eleventh Edition, 1910, Hippocrates, vol. 13, Encyclopædia Britannica Company, New York, p. 517–518.
299. Ibid., p. 517.
300. Nutton, V., 2004, *Ancient Medicine*, Routledge, London, p. 113–114.
301. Adams, F., 1849, Preliminary Discourse, in *The Genuine Works of Hippocrates*, vol. 1, Sydenham Society, London, p. 12.
302. Ibid.
303. Thucydides, 1942, *The Peloponnesian War*, Book 2, Chapter 47, translated by Benjamin Jowett (1817–1893), in *The Greek Historians*, vol. 1, Random House, New York, p. 653.
304. Ibid., Book 2, Chapter 50, p. 654–655.
305. Ibid., Book 2, Chapter 51–52, p. 655.
306. Grote, G., 1899, *A History of Greece*, vol. 5, Harper & Brothers, New York, p. 78.
307. Ibid., p. 79.
308. Joly, R., 2008, Hippocrates of Cos, in *Complete Dictionary of Scientific Biography*, edited by Charles Gillispie, vol. 6, Cengage Learning, New York, p. 419.
309. Ibid.
310. Hippocrates, 1950, *The Medical Works of Hippocrates*, translated by John Chadwick and W. N. Mann, Charles C. Thomas, Springfield, Illinois, p. vii; also, Hippocrates, 1849, *The Genuine Works of Hippocrates*, vol. 1, translated by Francis Adams, Sydenham Society, London, p. ix–x.
311. Nutton, V., 2004, *Ancient Medicine*, Routledge, London, p. 44.
312. Ibid., p. 71.
313. Hippocrates, 1886, *The Genuine Works of Hippocrates*, vol. 1, translated by Francis Adams, William Wood and Company, New York, p. 300.
314. Hippocrates, 1950, Aphorisms, in *The Medical Works of Hippocrates*, translated by John Chadwick and W. N. Mann, Charles C. Thomas, Springfield, Illinois, p. 148.
315. Hippocrates, 1846, The Law of Hippocrates, in *The Writings of Hippocrates and Galen*, translated by John Redman Coxe, Lindsay and Blakiston, Philadelphia, p. 45.
316. Hippocrates, 1846, The Art of Medicine in Former Times (On Ancient Medicine), in *The Writings of Hippocrates and Galen*, translated by John Redman Coxe, Lindsay and Blakiston, Philadelphia, p. 56.
317. Herodotus, 1910, *The History of Herodotus*, Book 1, Chapter 197, translated by George Rawlinson (1812–1902), vol. 1 (first published in 1858), J. M. Dent & Sons, New York, p. 101.
318. Withington, E. T., 1921, The Asclepiadae and the Priests of Asclepius, in *Studies in the History and Method of Science*, vol. 2, edited by Charles Singer, Oxford at the Clarendon Press, London, p. 196–197.
319. Grote, G., 1899, *A History of Greece*, vol. 1, Harper & Brothers, New York, p. 168.
320. Coxe, J. R., 1846, Introduction, in *The Writings of Hippocrates and Galen*, translated by John Redman Coxe, Lindsay and Blakiston, Philadelphia, p. 23. This Introduction appears to be Coxe's translation of *Historie de la Medecine* by Daniel Le Clerc, published at A la Haye (The Hague) by Chez Isaac van der Kloot in 1729 (first edition in 1696).
321. Hippocrates, 1950, Epidemics, Books 1, 2, 3, in *The Medical Works of Hippocrates*, translated by John Chadwick and W. N. Mann, Charles C. Thomas, Springfield, Illinois, p. 29–80.
322. Hippocrates, 1886, *The Genuine Works of Hippocrates*, vol. 2, translated by Francis Adams, William Wood and Company, New York, p. 192.
323. Adams, F., 1849, Preliminary Discourse, in *The Genuine Works of Hippocrates*, vol. 1, Sydenham Society, London, p. 17–18.
324. Coxe, J. R., 1846, Introduction, in *The Writings of Hippocrates and Galen*, translated by John Redman Coxe, Lindsay and Blakiston, Philadelphia, p. 22. This Introduction appears to be Coxe's translation of *Historie de la Medecine* by Daniel Le Clerc, published at A la Haye (The Hague) by Chez Isaac van der Kloot in 1729 (first edition in 1696).
325. Celsus, A. C., 1831, *A Translation of the Eight Books of Aul. Corn. Celsus on Medicine*, Book 1, Preface, Second Edition, revised by G. F. Collier, Simpkin and Marshall, London, p. 2.
326. Hippocrates, 1846, On the Art of Medicine, in *The Writings of Hippocrates and Galen*, translated by John Redman Coxe, Lindsay and Blakiston, Philadelphia, p. 54.
327. Kudlien, F., 1968, Early Greek Primitive Medicine, *Clio Medica*, vol. 3, p. 311.
328. Hippocrates, 1846, The Art of Medicine in Former Times (On Ancient Medicine), in *The Writings of Hippocrates and Galen*, translated by John Redman Coxe, Lindsay and Blakiston, Philadelphia, p. 55.
329. Ibid., p. 56.
330. Ibid.
331. Hippocrates, 1846, On the Nature of Man, in *The Writings of Hippocrates and Galen*, translated by John Redman Coxe, Lindsay and Blakiston, Philadelphia, p. 148.
332. Hippocrates, 1846, On the Art of Medicine, in *The Writings of Hippocrates and Galen*, translated by John Redman Coxe, Lindsay and Blakiston, Philadelphia, p. 49.
333. Hippocrates, 1886, *The Genuine Works of Hippocrates*, vol. 2, translated by Francis Adams, William Wood and Company, New York, p. 334.
334. Ibid., p. 338.
335. Ibid., p. 335.
336. Hippocrates, 1849, Airs, Waters, and Places, in *The Genuine Works of Hippocrates*, vol. 1, translated by Francis Adams, Sydenham Society, London, p. 216.
337. Ibid.
338. Nutton, V., 2004, *Ancient Medicine*, Routledge, London, p. 22–22.
339. Porter, R., 2001, Medical Science, in *The Cambridge Illustrated History of Medicine*, edited by Roy Porter, Cambridge University Press, New York, p. 184.
340. Nutton, V., 2004, *Ancient Medicine*, Routledge, London, p. 96.
341. Hippocrates, 1849, On Airs, Waters, and Places, in *The Genuine Works of Hippocrates*, vol. 1, translated by Francis Adams, Sydenham Society, London, p. 190.
342. Ibid., p. 191–195.
343. Ibid., p. 195.
344. Ibid., p. 197.
345. Ibid., p. 191.
346. Ibid., p. 202.
347. Ibid., p. 205.
348. Ibid., p. 210.
349. Vitruvius, 1960, *The Ten Books on Architecture*, Book 6, Chapter 1, Paragraph 10, translated by Morris Hicky Morgan (first published by Harvard University Press in 1914), Dover, New York, p. 173.
350. Coxe, J. R., 1846, Introduction, in *The Writings of Hippocrates and Galen*, Lindsay and Blakiston, Philadelphia, p. 26. This Introduction appears to be Coxe's translation of *Historie de la Medecine* by Daniel Le Clerc, published at A la Haye (The Hague) by Chez Isaac van der Kloot in 1729 (first edition in 1696).
351. Nutton, V., 2004, *Ancient Medicine*, Routledge, London, p. 79.
352. Hippocrates, 1846, On the Nature of Man, in *The Writings of Hippocrates and Galen*, translated by John Redman Coxe, Lindsay and Blakiston, Philadelphia, p. 149.
353. Nutton, V., 2004, *Ancient Medicine*, Routledge, London, p. 78.
354. Hippocrates, 1846, On the Nature of Man, in *The Writings of Hippocrates and Galen*, translated by John Redman Coxe, Lindsay and Blakiston, Philadelphia, p. 152.
355. Nutton, V., 2004, *Ancient Medicine*, Routledge, London, p. 80.
356. Ibid., p. 26.
357. Hippocrates, 1846, On the Nature of Man, in *The Writings of Hippocrates and Galen*, translated by John Redman Coxe, Lindsay and Blakiston, Philadelphia, p. 154.
358. Ibid., p. 153.
359. Hippocrates, 1950, A Regimen for Health, in *The Med-*

*ical Works of Hippocrates*, translated by John Chadwick and W. N. Mann, Charles C. Thomas, Springfield, Illinois, p. 215.

360. Ibid., p. 217.

361. Hippocrates, 1849, On Ancient Medicine, in *The Genuine Works of Hippocrates*, vol. 1, translated by Francis Adams, Sydenham Society, London, p. 164.

362. Plato, 1937, *The Republic*, Book 3, in *The Dialogues of Plato*, translated into English by Benjamin Jowett (1817–1893), vol. 1, Random House, New York, p. 670 (406).

363. Nutton, V., 2004, *Ancient Medicine*, Routledge, London, p. 97.

364. Hippocrates, 1950, A Regimen for Health, in *The Medical Works of Hippocrates*, translated by John Chadwick and W. N. Mann, Charles C. Thomas, Springfield, Illinois, p. 218.

365. Celsus, A. C., 1831, *A Translation of the Eight Books of Aul. Corn. Celsus on Medicine*, Book 1, Preface, Second Edition, revised by G. F. Collier, Simpkin and Marshall, London, p. 1.

366. Nutton, V., 2004, *Ancient Medicine*, Routledge, London, p. 174–175.

367. Coxe, J. R., 1846, Introduction, in *The Writings of Hippocrates and Galen*, Lindsay and Blakiston, Philadelphia, p. 27. This Introduction appears to be Coxe's translation of *Historie de la Medecine* by Daniel Le Clerc, published at A la Haye (The Hague) by Chez Isaac van der Kloot in 1729 (first edition in 1696). Also, Adams, F., 1849, Preliminary Discourse, in *The Genuine Works of Hippocrates*, vol. 1, Sydenham Society, London, p. 20.

368. Coxe, J. R., 1846, Introduction, in *The Writings of Hippocrates and Galen*, Lindsay and Blakiston, Philadelphia, p. 23. This Introduction appears to be Coxe's translation of *Historie de la Medecine* by Daniel Le Clerc, published at A la Haye (The Hague) by Chez Isaac van der Kloot in 1729 (first edition in 1696). Also, Adams, F., 1849, Preliminary Discourse, in *The Genuine Works of Hippocrates*, vol. 1, Sydenham Society, London, p. 20.

369. Adams, F., 1849, Preliminary Discourse, in *The Genuine Works of Hippocrates*, vol. 1, Sydenham Society, London, p. 20.

370. *Encyclopædia Britannica*, Eleventh Edition, 1911, Medicine, vol. 18, Encyclopædia Britannica Company, New York, p. 42.

371. Adams, F., 1849, Preliminary Discourse, in *The Genuine Works of Hippocrates*, vol. 1, Sydenham Society, London, p. 20–21.

372. Creasy, E. S., 1908, *The Fifteen Decisive Battles of the World from Marathon to Waterloo*, Everyman's Library Edition, E. P. Dutton, New York, p. 14–15.

373. *Encyclopædia Britannica*, Eleventh Edition, 1911, Satrap, vol. 24, Encyclopædia Britannica Company, New York, p. 230.

374. Creasy, E. S., 1908, *The Fifteen Decisive Battles of the World from Marathon to Waterloo*, Everyman's Library Edition, E. P. Dutton, New York, p. 12.

375. Whittaker, T., 1911, *Priests, Philosophers and Prophets*, Adam and Charles Black, London, p. 33.

376. Burnet, J., 1908, *Early Greek Philosophy*, Second Edition, Adam and Charles Black, London, p. 87

377. Creasy, E. S., 1908, *The Fifteen Decisive Battles of the World from Marathon to Waterloo*, Everyman's Library Edition, E. P. Dutton, New York, p. 13.

378. Herodotus, 1910, *The History of Herodotus*, Book 5, Chapter 98, translated by George Rawlinson (1812–1902), vol. 2 (first published in 1858), J. M. Dent & Sons, New York, p. 46.

379. Ibid., Book 5, Chapter 99, p. 47.

380. Ibid., Book 5, Chapters 100, 101, 102, p. 47–48.

381. Ibid., Book 5, Chapter 124, p. 55.

382. Ibid., Book 6, Chapters 18, 19, p. 63.

383. Grote, G., 1899, *A History of Greece*, vol. 3, Harper & Brothers, New York, p. 518.

384. Herodotus, 1910, *The History of Herodotus*, Book 5, Chapter 105, translated by George Rawlinson (1812–1902), vol. 2 (first published in 1858), J. M. Dent & Sons, New York, p. 49.

385. Ibid., Book 6, Chapter 48, p. 74.

386. Ibid., Book 7, Chapter 133, p. 168.

387. Ibid., Book 6, Chapter 101, p. 97.

388. Creasy, E. S., 1908, *The Fifteen Decisive Battles of the World from Marathon to Waterloo*, Everyman's Library Edition, E. P. Dutton, New York, p. 19–20.

389. Herodotus, 1910, *The History of Herodotus*, Book 6, Chapter 106, translated by George Rawlinson (1812–1902), vol. 2 (first published in 1858), J. M. Dent & Sons, New York, p. 99.

390. Grote, G., 1899, *A History of Greece*, vol. 4, Harper & Brothers, New York, p. 33–34.

391. Keegan, J., 1993, *A History of Warfare*, Alfred A. Knopf, New York, p. 253.

392. Creasy, E. S., 1908, *The Fifteen Decisive Battles of the World from Marathon to Waterloo*, Everyman's Library Edition, E. P. Dutton, New York, p. 5.

393. Ibid., p. 9. Creasy's source is Herodotus, Book 6, Chapter 109.

394. Herodotus, 1910, *The History of Herodotus*, Book 6, Chapter 112, translated by George Rawlinson (1812–1902), vol. 2 (first published in 1858), J. M. Dent & Sons, New York, p. 102–103.

395. Ibid., Book 6, Chapter 117, p. 104.

396. Keegan, J., 1993, *A History of Warfare*, Alfred A. Knopf, New York, p. 253.

397. Creasy, E. S., 1908, *The Fifteen Decisive Battles of the World from Marathon to Waterloo*, Everyman's Library Edition, E. P. Dutton, New York, p. 32–33.

398. Grote, G., 1899, *A History of Greece*, vol. 2, Harper & Brothers, New York, p. 283–284.

399. Thucydides, 1942, *The Peloponnesian War*, Book 1, Chapter 10, translated by Benjamin Jowett (1817–1893), in *The Greek Historians*, vol. 1, Random House, New York, p. 571.

400. Herodotus, 1910, *The History of Herodotus*, Book 1, Chapter 65, translated by George Rawlinson (1812–1902), vol. 1 (first published in 1858), J. M. Dent & Sons, New York, p. 30.

401. Xenophon, 1898, *The Government of Lacedaemon*, Chapter 1, Paragraph 1, in *Xenophon's Minor Works*, translated by J. S. Watson, George Bell & Sons, London, p. 204.

402. Plutarch, 1952, Lycurgus, in *The Lives of the Noble Grecians and Romans*, translated by John Dryden (1631–1700), *Great Books of the Western World*, vol. 14, William Benton, Chicago, p. 33.

403. Ibid., p. 42.

404. Grote, G., 1899, *A History of Greece*, vol. 2, Harper & Brothers, New York, p. 278.

405. Aristotle, 1885, *The Politics of Aristotle*, vol. 1, Book 2, Chapter 6, translated by Benjamin Jowett (1817–1893), Oxford at the Clarendon Press, London, p. 41 (1266a).

406. Grote, G., 1899, *A History of Greece*, vol. 2, Harper & Brothers, New York, p. 270.

407. Ibid.

408. Ibid., p. 274–275.

409. Thucydides, 1942, *The Peloponnesian War*, Book 1, Chapter 18, translated by Benjamin Jowett (1817–1893), in *The Greek Historians*, vol. 1, Random House, New York, p. 574.

410. Grote, G., 1899, *A History of Greece*, vol. 2, Harper & Brothers, New York, p. 270.

411. Plutarch, 1952, Lycurgus, in *The Lives of the Noble Grecians and Romans*, translated by John Dryden (1631–1700), *Great Books of the Western World*, vol. 14, William Benton, Chicago, p. 38.

412. Xenophon, 1898, *The Government of Lacedaemon*, Chapter 8, Paragraph 1, in *Xenophon's Minor Works*, translated by J. S. Watson, George Bell & Sons, London, p. 216.

413. Ibid., Chapter 10, Paragraph 4, p. 220.

414. Grote, G., 1899, *A History of Greece*, vol. 2, Harper & Brothers, New York, p. 308.

415. Plutarch, 1952, Lycurgus, in *The Lives of the Noble Grecians and Romans*, translated by John Dryden (1631–1700), *Great Books of the Western World*, vol. 14, William Benton, Chicago, p. 36.

416. Ibid.

417. Grote, G., 1899, *A History of Greece*, vol. 2, Harper & Brothers, New York, p. 310.

418. Plutarch, 1952, Lycurgus, in *The Lives of the Noble Gre-

*cians and Romans*, translated by John Dryden (1631–1700), *Great Books of the Western World*, vol. 14, William Benton, Chicago, p. 36–37.

419. Xenophon, 1898, *The Government of Lacedaemon*, Chapter 5, Paragraph 4, in *Xenophon's Minor Works*, translated by J. S. Watson, George Bell & Sons, London, p. 213.

420. Plato, 1937, *The Republic*, Book 5, in *The Dialogues of Plato*, translated into English by Benjamin Jowett (1817–1893), vol. 1, Random House, New York, p. 713 (452).

421. Ibid., Book 5, p. 713–714 (452).

422. Plutarch, 1952, Lycurgus, in *The Lives of the Noble Grecians and Romans*, translated by John Dryden (1631–1700), *Great Books of the Western World*, vol. 14, William Benton, Chicago, p. 39.

423. Grote, G., 1899, *A History of Greece*, vol. 2, Harper & Brothers, New York, p. 302.

424. Aristotle, 1885, *The Politics of Aristotle*, vol. 1, Book 2, Chapter 9, translated by Benjamin Jowett (1817–1893), Oxford at the Clarendon Press, London, p. 53 (1270a).

425. Ibid., p. 52 (1269b).

426. Plutarch, 1952, Lycurgus, in *The Lives of the Noble Grecians and Romans*, translated by John Dryden (1631–1700), *Great Books of the Western World*, vol. 14, William Benton, Chicago, p. 39.

427. Grote, G., 1899, *A History of Greece*, vol. 2, Harper & Brothers, New York, p. 305.

428. Plutarch, 1952, *Agesilaus, in The Lives of the Noble Grecians and Romans*, translated by John Dryden (1631–1700), *Great Books of the Western World*, vol. 14, William Benton, Chicago, p. 493–494. A similar account is found in Xenophon's *Hellenica*, Book 6, Chapter 4.

429. Ibid., p. 40.

430. Ibid., p. 40.

431. Ibid., p. 44.

432. Ibid., p. 41.

433. Ibid., p. 41.

434. Ibid., p. 41.

435. Thucydides, 1942, *The Peloponnesian War*, Book 4, Chapter 126, translated by Benjamin Jowett (1817–1893), in *The Greek Historians*, vol. 1, Random House, New York, p. 796.

436. Xenophon, 1898, *The Government of Lacedaemon*, Chapter 2, Paragraph 7, in *Xenophon's Minor Works*, translated by J. S. Watson, George Bell & Sons, London, p. 208.

437. Plutarch, 1952, Lycurgus, in *The Lives of the Noble Grecians and Romans*, translated by John Dryden (1631–1700), *Great Books of the Western World*, vol. 14, William Benton, Chicago, p. 42.

438. Plato, 1937, Protagoras, in *The Dialogues of Plato*, translated into English by Benjamin Jowett (1817–1893), vol. 1, Random House, New York, p. 112 (343).

439. Ibid., p. 111–112 (342–343).

440. Dover, K. J., 1989, *Greek Homosexuality, Updated and With a New Postscript*, Harvard University Press, Cambridge, p. 1.

441. Xenophon, 1898, *The Government of Lacedaemon*, Chapter 2, Paragraphs 13–14, in *Xenophon's Minor Works*, translated by J. S. Watson, George Bell & Sons, London, p. 209.

442. Plato, 1937, Laws, in *The Dialogues of Plato*, translated into English by Benjamin Jowett (1817–1893), vol. 2, Random House, New York, p. 418 (636).

443. Michell, H., 1952, *Sparta*, Cambridge University Press, London, p. 168.

444. Keegan, J., 1993, *A History of Warfare*, Knopf, New York, p. 250.

445. Plutarch, 1952, Lycurgus, in *The Lives of the Noble Grecians and Romans*, translated by John Dryden (1631–1700), *Great Books of the Western World*, vol. 14, William Benton, Chicago, p. 44.

446. Durant, W., 1939, *The Life of Greece*, Simon and Schuster, New York, p. 87.

447. Herodotus, 1910, *The History of Herodotus*, Book 7, Chapter 8, translated by George Rawlinson (1812–1902), vol. 2 (first published in 1858), J. M. Dent & Sons, New York, p. 119.

448. Ibid., Book 7, Chapter 8, p. 119–120.

449. Ibid., Book 7, Chapter 20–21, p. 129.

450. Ibid., Book 7, Chapter 186, p. 193.

451. Maurice, F., 1930, The Size of the Army of Xerxes in the Invasion of Greece 480 B.C., *The Journal of Hellenic Studies*, vol. 50, part 2, p. 210–235.

452. Herodotus, 1910, *The History of Herodotus*, Book 7, Chapter 147, translated by George Rawlinson (1812–1902), vol. 2 (first published in 1858), J. M. Dent & Sons, New York, p. 175.

453. Ibid., Book 7, Chapter 102, p. 157.

454. Ibid., Book 7, Chapters 175, 177, p. 189–190.

455. Plutarch, 1870, Laconic Apophthegms, or Remarkable Sayings of the Spartans, in *Plutarch's Morals, Translated from the Greek by Several Hands, Corrected and Revised by William W. Goodwin*, vol. 1, Little, Brown, and Company, Boston, p. 418.

456. Herodotus, 1910, *The History of Herodotus*, Book 7, Chapter 210, translated by George Rawlinson (1812–1902), vol. 2 (first published in 1858), J. M. Dent & Sons, New York, p. 202.

457. Plutarch, 1870, Laconic Apophthegms, or Remarkable Sayings of the Spartans, in *Plutarch's Morals, Translated from the Greek by Several Hands, Corrected and Revised by William W. Goodwin*, vol. 1, Little, Brown, and Company, Boston, p. 418.

458. Ibid.

459. Herodotus, 1910, *The History of Herodotus*, Book 7, Chapter 210, translated by George Rawlinson (1812–1902), vol. 2 (first published in 1858), J. M. Dent & Sons, New York, p. 202–203.

460. Grote, G., 1899, *A History of Greece*, vol. 4, Harper & Brothers, New York, p. 185.

461. Herodotus, 1910, *The History of Herodotus*, Book 7, Chapter 211, translated by George Rawlinson (1812–1902), vol. 2 (first published in 1858), J. M. Dent & Sons, New York, p. 203.

462. Plutarch, 1870, Laconic Apophthegms, or Remarkable Sayings of the Spartans, in *Plutarch's Morals, Translated from the Greek by Several Hands, Corrected and Revised by William W. Goodwin*, vol. 1, Little, Brown, and Company, Boston, p. 418.

463. Herodotus, 1910, *The History of Herodotus*, Book 7, Chapter 83, translated by George Rawlinson (1812–1902), vol. 2 (first published in 1858), J. M. Dent & Sons, New York, p. 151.

464. Grote, G., 1899, *A History of Greece*, vol. 4, Harper & Brothers, New York, p. 187–188.

465. Plutarch, 1870, Laconic Apophthegms, or Remarkable Sayings of the Spartans, in *Plutarch's Morals, Translated from the Greek by Several Hands, Corrected and Revised by William W. Goodwin*, vol. 1, Little, Brown, and Company, Boston, p. 418.

466. Grote, G., 1899, *A History of Greece*, vol. 4, Harper & Brothers, New York, p. 191–192.

467. Ibid., p. 190.

468. Ibid., p. 200.

469. Herodotus, 1910, *The History of Herodotus*, Book 7, Chapter 228, translated by George Rawlinson (1812–1902), vol. 2 (first published in 1858), J. M. Dent & Sons, New York, p. 209.

470. Ibid., Book 8, Chapter 86, p. 246.

471. Grote, G., 1899, *A History of Greece*, vol. 4, Harper & Brothers, New York, p. 229.

472. Herodotus, 1910, *The History of Herodotus*, Book 9, Chapters 62–63, translated by George Rawlinson (1812–1902), vol. 2 (first published in 1858), J. M. Dent & Sons, New York, p. 303.

473. Grote, G., 1899, *A History of Greece*, vol. 4, Harper & Brothers, New York, p. 507.

474. Popper, K., 1966, *The Open Society and Its Enemies*, vol. 1, Fifth Edition, Routledge, London, p. 129.

475. Burnet, J., 1920, *Greek Philosophy, Part I, Thales to Plato*, Macmillan & Co., Limited, London, p. 128.

476. Grote, G., 1899, *A History of Greece*, vol. 7, Harper & Brothers, New York, p. 84.

477. May, H., 2000, *On Socrates*, Wadsworth, Belmont, California, p. 22–23.

478. Burnet, J., 1920, *Greek Philosophy, Part I, Thales to Plato*, Macmillan & Co., Limited, London, p. 179.

479. Grote, G., 1899, *A History of Greece*, vol. 7, Harper & Brothers, New York, p. 86.

480. Ibid., p. 82.

481. Grote, G., 1899, *A History of Greece*, vol. 5, Harper & Brothers, New York, p. 311.
482. Strabo, 1856, *The Geography of Strabo*, Book 9, Chapter 2, Paragraph 7, vol. 2, translated by H. C. Hamilton and W. Falconer, Henry G. Bohn, London, p. 95.
483. Athenaeus, 1854, *The Deipnosophists, or Banquet of the Learned*, Book 5, Chapter 55, translated by C. D. Yonge, vol. 1, Henry G. Bohn, London, p. 343.
484. Grote, G., 1899, *A History of Greece*, vol. 7, Harper & Brothers, New York, p. 83.
485. Plato, 1937, Symposium, in *The Dialogues of Plato*, translated into English by Benjamin Jowett (1817–1893), vol. 1, Random House, New York, p. 338 (215).
486. Plato, 1937, Apology, in *The Dialogues of Plato*, translated into English by Benjamin Jowett (1817–1893), vol. 1, Random House, New York, p. 414 (31).
487. Xenophon, 1894, *Memorabilia of Socrates*, Book 1, Chapter 2, Paragraph 1, in *The Anabasis, or Expedition of Cyrus, and the Memorabilia of Socrates*, translated by J. S. Watson, Harper & Brothers, New York, p. 355.
488. Ibid., Book 1, Chapter 6, Paragraph 7, p. 381.
489. Ibid., Book 3, Chapter 12, Paragraph 4, p. 462.
490. Xenophon, 1898, *The Banquet*, Chapter 2, Paragraph 10, in *Xenophon's Minor Works*, translated by J. S. Watson, George Bell & Sons, London, p. 156.
491. Laërtius, Diogenes, 1905, *The Lives and Opinions of Eminent Philosophers*, Book 2, translated by C. D. Yonge, London, George Bell and Sons, p. 71.
492. Ibid.
493. Xenophon, 1898, *The Banquet*, Chapter 2, Paragraph 10, in *Xenophon's Minor Works*, translated by J. S. Watson, George Bell & Sons, London, p. 156.
494. Plutarch, 1952, Marcus Cato, in *The Lives of the Noble Grecians and Romans*, translated by John Dryden (1631–1700), *Great Books of the Western World*, vol. 14, William Benton, Chicago, p. 286.
495. Xenophon, 1894, *Memorabilia of Socrates*, Book 1, Chapter 1, Paragraph 10, in *The Anabasis, or Expedition of Cyrus, and the Memorabilia of Socrates*, translated by J. S. Watson, Harper & Brothers, New York, p. 352.
496. Laërtius, Diogenes, 1905, *The Lives and Opinions of Eminent Philosophers*, Book 2, translated by C. D. Yonge, London, George Bell and Sons, p. 64.
497. Ibid., p. 62.
498. Xenophon, 1894, *Memorabilia of Socrates*, Book 1, Chapter 1, Paragraph 15, in *The Anabasis, or Expedition of Cyrus, and the Memorabilia of Socrates*, translated by J. S. Watson, Harper & Brothers, New York, p. 353.
499. Ibid., Book 1, Chapter 1, Paragraph 13, p. 352.
500. Plato, 1937, Apology, in *The Dialogues of Plato*, translated into English by Benjamin Jowett (1817–1893), vol. 1, Random House, New York, p. 403 (19).
501. Grote, G., 1899, *A History of Greece*, vol. 7, Harper & Brothers, New York, p. 96.
502. Cicero, 1899, *Cicero's Tusculan Disputations*, Book 5, Paragraph 4, translated by C. D. Yonge, Harper & Brothers, New York, p. 166.
503. Xenophon, 1898, Fragments of Letters of Xenophon, translated by John Selby Watson (1804–1884), in *Xenophon's Minor Works*, George Bell & Sons, London, p. 375.
504. Xenophon, 1894, *Memorabilia of Socrates*, Book 1, Chapter 1, Paragraph 16, in *The Anabasis, or Expedition of Cyrus, and the Memorabilia of Socrates*, translated by J. S. Watson, Harper & Brothers, New York, p. 353.
505. Grote, G., 1899, *A History of Greece*, vol. 7, Harper & Brothers, New York, p. 102.
506. See discussion by George Sarton in Sarton, G., 1960, *A History of Science, Ancient Science Through the Golden Age of Greece*, Harvard University Press, Cambridge, p. 271–272.
507. Sarton, G., 1960, *A History of Science, Ancient Science Through the Golden Age of Greece*, Harvard University Press, Cambridge, p. 272.
508. Grote, G., 1899, *A History of Greece*, vol. 7, Harper & Brothers, New York, p. 126–127.
509. Aristotle, 1941, *Metaphysics*, Book 1, Chapter 6, translated by W. D. Ross, in *Basic Works of Aristotle*, Random House, New York, p. 700–701 (987b).
510. Grote, G., 1899, *A History of Greece*, vol. 2, Harper & Brothers, New York, p. 179.
511. Hamilton, E., 1953, *Mythology*, Mentor, New York, p. 30.
512. Ibid., p. 256.
513. Plato, 1937, Apology, in *The Dialogues of Plato*, translated into English by Benjamin Jowett (1817–1893), vol. 1, Random House, New York, p. 404 (21).
514. Ibid.
515. Grote, G., 1899, *A History of Greece*, vol. 7, Harper & Brothers, New York, p. 141.
516. Plato, 1937, Apology, in *The Dialogues of Plato*, translated into English by Benjamin Jowett (1817–1893), vol. 1, Random House, New York, p. 420 (38).
517. Grote, G., 1899, *A History of Greece*, vol. 7, Harper & Brothers, New York, p. 167.
518. Plato, 1937, Euthyphro, in *The Dialogues of Plato*, translated into English by Benjamin Jowett (1817–1893), vol. 1, Random House, New York, p. 385 (4).
519. Ibid., p. 386.
520. Ibid., p. 387.
521. Ibid., p. 388.
522. Ibid., p. 387.
523. Ibid., p. 389.
524. Plato, 1937, Meno, in *The Dialogues of Plato*, translated into English by Benjamin Jowett (1817–1893), vol. 1, Random House, New York, p. 359 (80).
525. Laërtius, Diogenes, 1905, *The Lives and Opinions of Eminent Philosophers*, Book 2, translated by C. D. Yonge, London, George Bell and Sons, p. 65.
526. Plato, 1937, Apology, in *The Dialogues of Plato*, translated into English by Benjamin Jowett (1817–1893), vol. 1, Random House, New York, p. 405 (22).
527. Grote, G., 1899, *A History of Greece*, vol. 7, Harper & Brothers, New York, p. 134.
528. Ibid., p. 135.
529. Thucydides, 1942, *The Peloponnesian War*, Book 1, Chapter 23, translated by Benjamin Jowett (1817–1893), in *The Greek Historians*, vol. 1, Random House, New York, p. 577.
530. Dalberg-Acton, J. E. E., 1907, Freedom in Antiquity, in *The History of Freedom and Other Essays*, edited by John Neville Figgis and Reginald Vere Laurence, Macmillan and Co., Limited, London, p. 12.
531. Thucydides, 1942, *The Peloponnesian War*, Book 1, Chapter 23, translated by Benjamin Jowett (1817–1893), in *The Greek Historians*, vol. 1, Random House, New York, p. 577.
532. Grote, G., 1899, *A History of Greece*, vol. 6, Harper & Brothers, New York, p. 450.
533. Xenophon, 1942, *Hellenica*, Book 2, Chapter 2, translated by Henry G. Dakyns (first published in 1890), in *The Greek Historians*, vol. 2, Random House, New York, p. 36.
534. Grote, G., 1899, *A History of Greece*, vol. 6, Harper & Brothers, New York, p. 458.
535. Xenophon, 1942, *Hellenica*, Book 2, Chapter 3, translated by Henry G. Dakyns (first published in 1890), in *The Greek Historians*, vol. 2, Random House, New York, p. 40.
536. Grote, G., 1899, *A History of Greece*, vol. 6, Harper & Brothers, New York, p. 464.
537. Ibid., p. 498.
538. Ibid., p. 464.
539. Xenophon, 1894, *Memorabilia of Socrates*, Book 4, Chapter 2, Paragraph 2, in *The Anabasis, or Expedition of Cyrus, and the Memorabilia of Socrates*, translated by J. S. Watson, Harper & Brothers, New York, p. 470–471.
540. Plato, 1937, *The Republic*, Book 8, in *The Dialogues of Plato*, translated into English by Benjamin Jowett (1817–1893), vol. 1, Random House, New York, p. 820–822 (562–564).
541. Madison, J., 1952, Federalist Paper 10, in *American State Papers, Great Books of the Western World*, vol. 43, William Benton, Chicago, p. 51.
542. Burnet, J., 1920, *Greek Philosophy, Part 1, Thales to Plato*, Macmillan, London, p. 183.

543. Plato, 1937, Apology, in *The Dialogues of Plato*, translated into English by Benjamin Jowett (1817–1893), vol. 1, Random House, New York, p. 408–409 (25).
544. Ibid., p. 409.
545. Ibid., p. 412.
546. Ibid., p. 413–414..
547. Ibid., p. 417.
548. Laërtius, Diogenes, 1905, *The Lives and Opinions of Eminent Philosophers*, translated by C. D. Yonge, London, George Bell and Sons, p. 72.
549. Plato, 1937, Apology, in *The Dialogues of Plato*, translated into English by Benjamin Jowett (1817–1893), vol. 1, Random House, New York, p. 419 (37).
550. Plato, 1937, Phaedo, in *The Dialogues of Plato*, translated into English by Benjamin Jowett (1817–1893), vol. 1, Random House, New York, p. 500–501 (117–118).
551. Grote, G., 1899, *A History of Greece*, vol. 7, Harper & Brothers, New York, p. 143.
552. Thucydides, 1942, *The Peloponnesian War*, Book 2, Chapter 37, translated by Benjamin Jowett (1817–1893), in *The Greek Historians*, vol. 1, Random House, New York, p. 648.
553. Grote, G., 1899, *A History of Greece*, vol. 7, Harper & Brothers, New York, p. 172.
554. Ibid.
555. Shields, C., 2006, Learning about Plato from Aristotle, in *A Companion to Plato*, edited by Hugh H. Benson, Blackwell, Malden, Massachusetts, p. 403.
556. Whitehead, A. N., 1929, *Process and Reality*, Cambridge University Press, Cambridge, p. 39.
557. Laërtius, Diogenes, 1905, *The Lives and Opinions of Eminent Philosophers*, Book 3, translated by C. D. Yonge, London, George Bell and Sons, p. 113.
558. Ibid.
559. Burnet, J., 1920, *Greek Philosophy, Part 1, Thales to Plato*, Macmillan, London, p. 207.
560. Xenophon, 1894, *Memorabilia of Socrates*, Book 3, Chapter 7, in *The Anabasis, or Expedition of Cyrus, and the Memorabilia of Socrates*, translated by J. S. Watson, Harper & Brothers, New York, p. 444–446.
561. Grote, G., 1885, *Plato, and the Other Companions of Sokrates*, vol. 1, John Murray, London, p. 247–248.
562. Laërtius, Diogenes, 1905, *The Lives and Opinions of Eminent Philosophers*, Book 3, translated by C. D. Yonge, London, George Bell and Sons, p. 114.
563. Grote, G., 1885, *Plato, and the Other Companions of Sokrates*, vol. 1, John Murray, London, p. 248.
564. Ibid., p. 249–250.
565. Laërtius, Diogenes, 1905, *The Lives and Opinions of Eminent Philosophers*, Book 3, translated by C. D. Yonge, London, George Bell and Sons, p. 115.
566. Grote, G., 1899, *A History of Greece*, vol. 6, Harper & Brothers, New York, p. 502.
567. Laërtius, Diogenes, 1905, *The Lives and Opinions of Eminent Philosophers*, Book 3, translated by C. D. Yonge, London, George Bell and Sons, p. 115.
568. Plato, 1937, Apology, in *The Dialogues of Plato*, translated into English by Benjamin Jowett (1817–1893), vol. 1, Random House, New York, p. 416 (34).
569. Plato, 1937, Phaedo, in *The Dialogues of Plato*, translated into English by Benjamin Jowett (1817–1893), vol. 1, Random House, New York, p. 442 (59).
570. Aristotle, 1941, *Metaphysics*, Book 1, Chapter 6, translated by W. D. Ross, in *Basic Works of Aristotle*, Random House, New York, p. 700–701 (987b).
571. Plato, 1891, Epistle 7, in *The Works of Plato*, Translated from the Text of Stallbaum, translated by George Burges, vol. 4, G. Bell, London, p. 500.
572. Ibid., p. 502.
573. Plato, 1937, *The Republic*, Book 5, in *The Dialogues of Plato*, translated into English by Benjamin Jowett (1817–1893), vol. 1, Random House, New York, p. 737 (473).
574. Grote, G., 1885, *Plato, and the Other Companions of Sokrates*, vol. 1, John Murray, London, p. 253.
575. Laërtius, Diogenes, 1905, *The Lives and Opinions of Eminent Philosophers*, Book 3, translated by C. D. Yonge, London, George Bell and Sons, p. 115.
576. Boas, G., 1948, Fact and Legend in the Biography of Plato, *The Philosophical Review*, vol. 57, no. 5, p. 444.
577. Laërtius, Diogenes, 1905, *The Lives and Opinions of Eminent Philosophers*, Book 3, translated by C. D. Yonge, London, George Bell and Sons, p. 120.
578. Ibid.
579. Grote, G., 1885, *Plato, and the Other Companions of Sokrates*, vol. 1, John Murray, London, p. 254.
580. Heath, T., 1981, *A History of Greek Mathematics*, vol. 1 (first published in 1921 by the Clarendon Press, Oxford), Dover, New York, p. 24.
581. Grote, G., 1885, *Plato, and the Other Companions of Sokrates*, vol. 1, John Murray, London, p. 256–257.
582. Grote, G., 1888, *Plato and the Other Companions of Sokrates*, New Edition, vol. 1, John Murray, London, p. vi–vii.
583. Sarton, G., 1960, *A History of Science, Ancient Science Through the Golden Age of Greece*, Harvard University Press, Cambridge, p. 408.
584. Ferrier, J. F., 1854, *Institutes of Metaphysic*, William Blackwood, Edinburgh and London, p. 2.
585. Khashaba, D. R., 2002, Excursions into the Dialogues of Plato, *The Examined Life Online Philosophy Journal*, vol. 3, no. 12.
586. Grote, G., 1888, *Plato and the Other Companions of Sokrates*, New Edition, vol. 1, John Murray, London, p. vii.
587. Ibid.
588. Burnet, J., 1920, *Greek Philosophy, Part I, Thales to Plato*, Macmillan & Co., Limited, London, p. 221.
589. Ibid., p. 221–222.
590. Burnet, J., 1920, *Greek Philosophy, Part I, Thales to Plato*, Macmillan & Co., Limited, London, p. 52.
591. Kirk, G. S., and Raven, J. E., 1957, *The Presocratic Philsophers*, Cambridge University Press, London, p. 274.
592. Burnet, J., 1920, *Greek Philosophy, Part I, Thales to Plato*, Macmillan & Co., Limited, London, p. 239.
593. Plato, 1937, *The Republic*, Book 7, in *The Dialogues of Plato*, translated into English by Benjamin Jowett (1817–1893), vol. 1, Random House, New York, p. 789 (529).
594. Ibid., Book 5, p. 739–740 (476).
595. Grote, G., 1888, *Plato and the Other Companions of Sokrates*, New Edition, vol. 4, John Murray, London, p. 51.
596. Plato, 1937, *The Republic*, Book 6, in *The Dialogues of Plato*, translated into English by Benjamin Jowett (1817–1893), vol. 1, Random House, New York, p. 766 (505).
597. Ibid., Book 6, p. 770 (509).
598. Aristotle, 1941, *Metaphysics*, Book 1, Chapter 6, translated by W. D. Ross, in *Basic Works of Aristotle*, Random House, New York, p. 700 (987a).
599. Ibid., p. 701 (987b).
600. Farrington, B., 1961, *Greek Science, Its Meaning For Us*, Penguin Books, Baltimore, Maryland, p. 110.
601. Gille, B., 1956, Machines, in *A History of Technology*, vol. 2, edited by Charles Singer, Oxford University Press, London, p. 632.
602. Laërtius, Diogenes, 1905, *The Lives and Opinions of Eminent Philosophers*, Book 8, translated by C. D. Yonge, London, George Bell and Sons, p. 370.
603. Gellius, A., 1795, *The Attic Nights of Aulus Gellius*, Book 10, Chapter 12, translated into English by the Rev. W. Beloe, vol. 2, J. Johnson, London, p. 223.
604. Aristotle, 1885, *The Politics of Aristotle*, vol. 1, Book 8, Chapter 6, translated by Benjamin Jowett (1817–1893), Oxford at the Clarendon Press, London, p. 254 (1340b).
605. Plutarch, 1952, Marcellus, in *The Lives of the Noble Grecians and Romans*, translated by John Dryden (1631–1700), *Great Books of the Western World*, vol. 14, William Benton, Chicago, p. 252.
606. Plato, 1937, *The Republic*, Book 7, in *The Dialogues of Plato*, translated into English by Benjamin Jowett (1817–1893), vol. 1, Random House, New York, p. 773 (514).
607. Lindberg, D. C., 1978, The Transmission of Greek and Arabic Learning to the West, in *Science in the Middle Ages*,

edited by David C. Lindberg, University of Chicago Press, Chicago, p. 53.
608. Haskins, C. H., 1927, *The Renaissance of the 12th Century*, Harvard University Press, Cambridge, Massachusetts, p. 344.
609. Grote, G., 1888, *Plato and the Other Companions of Sokrates*, New Edition, vol. 4, John Murray, London, p. 215.
610. Plato, 1937, Timaeus, in *The Dialogues of Plato*, translated into English by Benjamin Jowett (1817–1893), vol. 2, Random House, New York, p. 6–10. (21–25).
611. Ibid. p. 10 (25).
612. Cicero, 1829, *The Republic of Cicero*, Book 1, Chapter 10, translated by G. W. Featherstonhaugh, G. & C. Carvill, New York, p. 42.
613. Plato, 1937, Timaeus, in *The Dialogues of Plato*, translated into English by Benjamin Jowett (1817–1893), vol. 2, Random House, New York, p. 12 (28).
614. Ibid., p. 13 (29).
615. Burnet, J., 1920, *Greek Philosophy, Part 1: Thales to Plato*, Macmillan, London, p. 340.
616. Hesiod, 1920, Theogony (line 117), in *Hesiod, the Homeric Hymns and Homerica*, with an English translation by Hugh G. Evelyn-White, London, William Heinemann, p. 87.
617. Plato, 1937, Timaeus, in *The Dialogues of Plato*, translated into English by Benjamin Jowett (1817–1893), vol. 2, Random House, New York, p. 14 (30).
618. Plato, 1937, Timaeus, in *The Dialogues of Plato*, translated into English by Benjamin Jowett (1817–1893), vol. 2, Random House, New York, p. 28 (48).
619. Burnet, J., 1920, *Greek Philosophy, Part 1: Thales to Plato*, Macmillan, London, p. 341.
620. Grote, G., 1888, *Plato and the Other Companions of Sokrates*, New Edition, vol. 4, John Murray, London, p. 238.
621. Vlastos, G., 1975, *Plato's Universe*, University of Washington Press, Seattle, p. 97.
622. Plato, 1937, Timaeus, in *The Dialogues of Plato*, translated into English by Benjamin Jowett (1817–1893), vol. 2, Random House, New York, p. 14 (30).
623. Grote, G., 1888, *Plato and the Other Companions of Sokrates*, New Edition, vol. 4, John Murray, London, p. 222–223.
624. Ibid., p. 222.
625. Plato, 1937, Timaeus, in *The Dialogues of Plato*, translated into English by Benjamin Jowett (1817–1893), vol. 2, Random House, New York, p. 16 (33).
626. Ibid., p. 14 (30).
627. Ibid., p. 16 (34).
628. Ibid., p. 20 (39).
629. Ibid., p. 19 (38).
630. Ibid., p. 20 (38).
631. Heath, T., 1913, *Aristarchus of Samos, the Ancient Copernicus*, Oxford at the Clarendon Press, London, p. 190–212.
632. Ibid., p. 140.
633. Plato, 1937, Timaeus, in *The Dialogues of Plato*, translated into English by Benjamin Jowett (1817–1893), vol. 2, Random House, New York, p. 18 (36).
634. Heath, T., 1913, *Aristarchus of Samos, the Ancient Copernicus*, Oxford at the Clarendon Press, London, p. 159.
635. Plato, 1937, Timaeus, in *The Dialogues of Plato*, translated into English by Benjamin Jowett (1817–1893), vol. 2, Random House, New York, p. 21 (40).
636. Vlastos, G., 1975, *Plato's Universe*, University of Washington Press, Seattle, p. 54.
637. Plato, 1937, Timaeus, in *The Dialogues of Plato*, translated into English by Benjamin Jowett (1817–1893), vol. 2, Random House, New York, p. 22 (40).
638. Ibid., p. 22. (40–41).
639. Grote, G., 1888, *Plato and the Other Companions of Sokrates*, New Edition, vol. 4, John Murray, London, p. 232.
640. Plato, 1937, Timaeus, in *The Dialogues of Plato*, translated into English by Benjamin Jowett (1817–1893), vol. 2, Random House, New York, p. 23–24 (42–43).
641. Ibid., p. 23 (42).
642. Ibid.
643. Ibid., p. 67 (91).
644. Ibid.
645. Ibid.
646. Ibid. p. 68 (92).
647. Ibid.
648. Ibid., p. 29 (49).
649. Ibid., p. 35 (55).
650. Grote, G., 1888, *Plato and the Other Companions of Sokrates*, New Edition, vol. 4, John Murray, London, p. 241.
651. Plato, 1937, Timaeus, in *The Dialogues of Plato*, translated into English by Benjamin Jowett (1817–1893), vol. 2, Random House, New York, p. 36 (56).
652. Ibid., p. 34 (54).
653. Ibid., p. 54 (76).
654. Ibid., p. 50 (71).
655. Ibid. p. 26 (45).
656. Ibid., p. 48 (70).
657. Ibid., p. 55 (77).
658. Ibid.
659. Grote, G., 1888, *Plato and the Other Companions of Sokrates*, New Edition, vol. 4, John Murray, London, p. 259.
660. Ibid.
661. Plato, 1937, Timaeus, in *The Dialogues of Plato*, translated into English by Benjamin Jowett (1817–1893), vol. 2, Random House, New York, p. 59–62 (82–86).
662. Ibid., p. 63 (86).
663. Ibid., p. 25 (44).
664. Ibid., p. 68 (92).
665. Grote, G., 1888, *Plato and the Other Companions of Sokrates*, New Edition, vol. 4, John Murray, London, p. 262.
666. Ibid., p. 95.
667. Plato, 1937, *The Republic*, Book 1, in *The Dialogues of Plato*, translated into English by Benjamin Jowett (1817–1893), vol. 1, Random House, New York, p. 603 (339).
668. Ibid., Book 2, p. 629 (366).
669. Ibid., Book 2, p. 630 (367).
670. Ibid., Book 2, p. 631 (369).
671. Ibid., Book 2, p. 632–633 (369–371).
672. Grote, G., 1888, *Plato and the Other Companions of Sokrates*, New Edition, vol. 4, John Murray, London, p. 21.
673. Plato, 1937, *The Republic*, Book 2, in *The Dialogues of Plato*, translated into English by Benjamin Jowett (1817–1893), vol. 1, Random House, New York, p. 638 (375).
674. Ibid., Book 7, p. 784 (525).
675. Ibid., Book 7, p. 766 (505).
676. Ibid., Book 5, p. 737 (473).
677. Ibid., Book 6, p. 750 (489).
678. Ibid., Book 6, p. 756 (496).
679. Ibid., Book 6, p. 756 (495).
680. Ibid., Book 6, p. 758 (497).
681. Ibid., Book 3, p. 679 (414).
682. Ibid.
683. Ibid., Book 3, p. 679–680 (415).
684. Ibid., Book 3, p. 680 (415).
685. Ibid.
686. Ibid.
687. Ibid., Book 3 p. 678 (413).
688. Ibid., Book 3, p. 678 (414).
689. Ibid., Book 6, p. 752 (491).
690. Ibid., Book 6, p. 752 (492).
691. Grote, G., 1888, *Plato and the Other Companions of Sokrates*, New Edition, vol. 4, John Murray, London, p. 23–24.
692. Plato, 1937, *The Republic*, Book 2, in *The Dialogues of Plato*, translated into English by Benjamin Jowett (1817–1893), vol. 1, Random House, New York, p. 641 (377).
693. Grote, G., 1888, *Plato and the Other Companions of Sokrates*, New Edition, vol. 4, John Murray, London, p. 23–24.
694. Plato, 1937, *The Republic*, Book 2, in *The Dialogues of Plato*, translated into English by Benjamin Jowett (1817–1893), vol. 1, Random House, New York, p. 641 (377).
695. Ibid., Book 3, p. 651 (389).
696. Ibid.
697. Ibid., Book 3, p. 652 (389).
698. Ibid., Book 3, p. 665 (401).
699. Ibid.

700. Grote, G., 1888, *Plato and the Other Companions of Sokrates*, New Edition, vol. 4, John Murray, London, p. 23–27.
701. Plato, 1937, *The Republic*, Book 3, in *The Dialogues of Plato*, translated into English by Benjamin Jowett (1817–1893), vol. 1, Random House, New York, p. 663 (399).
702. Ibid.
703. Ibid., Book 5, p. 717 (455).
704. Ibid.
705. Ibid., Book 5, p. 717 (456).
706. Ibid., Book 7, p. 785–788 (525–529).
707. Ibid., Book 7, p. 789 (529).
708. Ibid., Book 7, p. 796–797 (537).
709. Ibid., Book 7, p. 793–794 (533–534).
710. Ibid., Book 7, p. 799 (539).
711. Ibid.
712. Ibid., Book 7, p. 799 (540).
713. Ibid., Book 5, p. 721 (459).
714. Ibid., Book 5, p. 722 (460).
715. Ibid., Book 5, p. 719 (457).
716. Ibid., Book 5, p. 727 (464).
717. Ibid., Book 5, p. 722 (460).
718. Ibid., Book 5, p. 723–724 (461).
719. Ibid., Book 5, p. 723 (460).
720. Ibid., Book 5, p. 727 (464).
721. Ibid., Book 8, p. 803 (546).
722. Ibid., Book 8, p. 803 (545).
723. Ibid.
724. Ibid., p. 802 (544).
725. Ibid., Book 8, p. 820–822 (562–564).
726. Ibid., Book 8, p. 822 (564).
727. Ibid., Book 2, p. 640–641 (376–378).
728. Ibid., Book 7, p. 785–800 (505–542).
729. Ibid., Book 3, p. 663 (399).
730. Ibid., Book 3, p. 658 (395).
731. Ibid., Book 5, p. 744 (480).
732. Ibid., Book 8, p. 651 (389).
733. Plato, 1937, *Laws*, Book 4, in *The Dialogues of Plato*, translated into English by Benjamin Jowett (1817–1893), vol. 2, Random House, New York, p. 485 (713).
734. Ibid., Book 4, p. 487 (715).
735. Ibid., Book 5, p. 506 (739).
736. Ibid.
737. Ibid., Book 5, p. 504 (737).
738. Ibid., Book 5, p. 506 (740).
739. Ibid., Book 5, p. 508 (742).
740. Ibid.
741. Ibid., Book 5, p. 510–511 (744–745).
742. Ibid., Book 6, p. 544 (787); compare with Book 6, p. 532–533 (772).
743. Ibid., Book 6, p. 533 (773).
744. Ibid., Book 10, p. 651 (909–910).
745. Ibid., Book 10, p. 650 (908).
746. Ibid., Book 10, p. 650 (908–909).
747. Ibid., Book 10, p. 650 (909).
748. Ibid.
749. Ibid., Book 12, p. 677–678 (942).
750. Ibid., Book 7, p. 555 (800).
751. Ibid., Book 7, p. 554 (798).
752. Ibid., Book 7, p. 553 (797).
753. Ibid., Book 2, p. 434 (656).
754. Ibid., Book 2, p. 435 (656).
755. Ibid., Book 1, p. 416 (634).
756. Ibid.
757. Ibid., Book 2, p. 441 (663).
758. Grote, G., 1888, *Plato and the Other Companions of Sokrates*, New Edition, vol. 4, John Murray, London, p. 381.
759. Plato, 1937, Laws, Book 12, in *The Dialogues of Plato*, translated into English by Benjamin Jowett (1817–1893), vol. 2, Random House, New York, p. 695 (961).
760. Ibid.
761. Ibid.
762. Grote, G., 1888, *Plato and the Other Companions of Sokrates*, New Edition, vol. 4, John Murray, London, p. 416.
763. Plato, 1937, *Laws*, Book 3, in *The Dialogues of Plato*, translated into English by Benjamin Jowett (1817–1893), vol. 2, Random House, New York, p. 475 (701).
764. Sarton, G., 1960, *A History of Science, Ancient Science Through the Golden Age of Greece*, Harvard University Press, Cambridge, p. 412–413.
765. Popper, K., 1966, *The Open Society and Its Enemies*, vol. 1, Fifth Edition, Routledge, London.
766. Ibid., p. 103.
767. Ibid., p. 194–195.
768. Ibid., p. 101.
769. Ibid., p. 102.
770. Plato, 1937, *The Republic*, Book 2, in *The Dialogues of Plato*, translated into English by Benjamin Jowett (1817–1893), vol. 1, Random House, New York, p. 631 (368–369).
771. Grote, G., 1899, *A History of Greece*, vol. 6, Harper & Brothers, New York, p. 192.
772. Ibid., vol. 7, p. 399.
773. Ibid., p. 402.
774. Xenophon, 1898, *The Government of Lacedaemon*, Chapter 14, Paragraph 1, in *Xenophon's Minor Works*, translated by J. S. Watson, George Bell & Sons, London, p. 228.
775. Grote, G., 1899, *A History of Greece*, vol. 8, Harper & Brothers, New York, p. 172.
776. Aristotle, 1885, *The Politics of Aristotle*, vol. 1, Book 2, Chapter 9, translated by Benjamin Jowett (1817–1893), Oxford at the Clarendon Press, London, p. 54 (1270a).
777. Grote, G., 1899, *A History of Greece*, vol. 9, Harper & Brothers, New York, p. 235.
778. Niebuhr, B. G., 1851, *The History of Rome*, vol. 3, translated by William Smith and Leonhard Schmitz, Taylor, Walton, and Maberly, London, p. 467–470.
779. Plutarch, 1952, Alexander, in *The Lives of the Noble Grecians and Romans*, translated by John Dryden (1631–1700), *Great Books of the Western World*, vol. 14, William Benton, Chicago, p. 542.
780. Blakesley, J. W., 1834, *A Life of Aristotle*, John W. Parker, London, p. 46.
781. Ibid., p. 543.
782. Grote, G., 1899, *A History of Greece*, vol. 9, Harper & Brothers, New York, p. 507.
783. Plutarch, 1952, Alexander, in *The Lives of the Noble Grecians and Romans*, translated by John Dryden (1631–1700), *Great Books of the Western World*, vol. 14, William Benton, Chicago, p. 543.
784. Ibid., p. 543–544.
785. Diodorus Siculus, 1814, *The Historical Library of Diodorus the Sicilian*, Book 16, Chapter 15, vol. 2, translated by G. Booth, W. M'Dowall for J. Davis, London, p. 156.
786. Grote, G., 1899, *A History of Greece*, vol. 9, Harper & Brothers, New York, p. 488.
787. Ibid., p. 490–491.
788. Ibid., p. 493.
789. Diodorus Siculus, 1814, *The Historical Library of Diodorus the Sicilian*, Book 16, Chapter 14, vol. 2, translated by G. Booth, W. M'Dowall for J. Davis, London, p. 153.
790. Grote, G., 1899, *A History of Greece*, vol. 9, Harper & Brothers, New York, p. 235.
791. Ibid., p. 496.
792. Diodorus Siculus, 1814, *The Historical Library of Diodorus the Sicilian*, Book 16, Chapter 15, vol. 2, translated by G. Booth, W. M'Dowall for J. Davis, London, p. 156.
793. Ibid., p. 157.
794. Ibid.
795. Plutarch, 1952, Alexander, in *The Lives of the Noble Grecians and Romans*, translated by John Dryden (1631–1700), *Great Books of the Western World*, vol. 14, William Benton, Chicago, p. 545.
796. Diodorus Siculus, 1814, *The Historical Library of Diodorus the Sicilian*, Book 17, Chapter 1, vol. 2, translated by G. Booth, W. M'Dowall for J. Davis, London, p. 161.
797. Ibid.
798. Arrian, 1893, *Arrian's Anabasis of Alexander and Indica*, Book 1, Chapter 1, translated by Edward James Chinnock, George Bell & Sons, London, p. 4.

799. Grote, G., 1899, *A History of Greece*, vol. 9, Harper & Brothers, New York, p. 516.
800. Diodorus Siculus, 1814, *The Historical Library of Diodorus the Sicilian*, Book 17, Chapter 1, vol. 2, translated by G. Booth, W. M'Dowall for J. Davis, London, p. 162.
801. Grote, G., 1899, *A History of Greece*, vol. 9, Harper & Brothers, New York, p. 517.
802. Diodorus Siculus, 1814, *The Historical Library of Diodorus the Sicilian*, Book 17, Chapter 1, vol. 2, translated by G. Booth, W. M'Dowall for J. Davis, London, p. 163.
803. Ibid., p. 164.
804. Ibid., p. 165–166.
805. Ibid., p. 166.
806. Ibid., p. 167
807. Ibid., p. 166.
808. Ibid., p. 168.
809. Ibid.
810. Ibid., p. 169.
811. Arrian, 1893, *Arrian's Anabasis of Alexander and Indica*, Book 1, Chapter 8, translated by Edward James Chinnock, George Bell & Sons, London, p. 24.
812. Diodorus Siculus, 1814, *The Historical Library of Diodorus the Sicilian*, Book 17, Chapter 1, vol. 2, translated by G. Booth, W. M'Dowall for J. Davis, London, p. 169.
813. Ibid., p. 170.
814. Grote, G., 1899, *A History of Greece*, vol. 9, Harper & Brothers, New York, p. 548.
815. Laërtius, Diogenes, 1905, *The Lives and Opinions of Eminent Philosophers*, Book 6, translated by C. D. Yonge, London, George Bell and Sons, p. 225.
816. Plutarch, 1952, Alexander, in *The Lives of the Noble Grecians and Romans*, translated by John Dryden (1631–1700), *Great Books of the Western World*, vol. 14, William Benton, Chicago, p. 546.
817. Arrian, 1893, *Arrian's Anabasis of Alexander and Indica*, Book 1, Chapter 11, translated by Edward James Chinnock, George Bell & Sons, London, p. 29.
818. Diodorus Siculus, 1814, *The Historical Library of Diodorus the Sicilian*, Book 17, Chapter 1, vol. 2, translated by G. Booth, W. M'Dowall for J. Davis, London, p. 172.
819. Arrian, 1893, *Arrian's Anabasis of Alexander and Indica*, Book 1, Chapter 11, translated by Edward James Chinnock, George Bell & Sons, London, p. 30.
820. Diodorus Siculus, 1814, *The Historical Library of Diodorus the Sicilian*, Book 17, Chapter 2, vol. 2, translated by G. Booth, W. M'Dowall for J. Davis, London, p. 173.
821. Grote, G., 1899, *A History of Greece*, vol. 10, Harper & Brothers, New York, p. 18.
822. Ibid., p. 26–27.
823. Ibid., p. 20.
824. Diodorus Siculus, 1814, *The Historical Library of Diodorus the Sicilian*, Book 17, Chapter 2, vol. 2, translated by G. Booth, W. M'Dowall for J. Davis, London, p. 173.
825. Grote, G., 1899, *A History of Greece*, vol. 10, Harper & Brothers, New York, p. 23.
826. Arrian, 1893, *Arrian's Anabasis of Alexander and Indica*, Book 1, Chapter 13, translated by Edward James Chinnock, George Bell & Sons, London, p. 35.
827. Ibid., Chapter 14, p. 37.
828. Ibid., Chapter 15, p. 38.
829. Ibid., p. 40.
830. Diodorus Siculus, 1814, *The Historical Library of Diodorus the Sicilian*, Book 17, Chapter 2, vol. 2, translated by G. Booth, W. M'Dowall for J. Davis, London, p. 174.
831. Ibid., p. 175.
832. Ibid.
833. Ibid.
834. Grote, G., 1899, *A History of Greece*, vol. 10, Harper & Brothers, New York, p. 34.
835. Ibid., p. 35.
836. Arrian, 1893, *Arrian's Anabasis of Alexander and Indica*, Book 1, Chapter 16, translated by Edward James Chinnock, George Bell & Sons, London, p. 41.
837. Grote, G., 1899, *A History of Greece*, vol. 10, Harper & Brothers, New York, p. 35.
838. Ibid., p. 38.
839. Ibid., p. 38–39.
840. Ibid., p. 39.
841. Arrian, 1893, *Arrian's Anabasis of Alexander and Indica*, Book 1, Chapter 18, translated by Edward James Chinnock, George Bell & Sons, London, p. 46.
842. Diodorus Siculus, 1814, *The Historical Library of Diodorus the Sicilian*, Book 17, Chapter 2, vol. 2, translated by G. Booth, W. M'Dowall for J. Davis, London, p. 176.
843. Arrian, 1893, *Arrian's Anabasis of Alexander and Indica*, Book 1, Chapter 18, translated by Edward James Chinnock, George Bell & Sons, London, p. 47.
844. Ibid.
845. Ibid., p. 48.
846. Ibid., p. 47–48.
847. Sarton, G., 1960, *A History of Science, Ancient Science Through the Golden Age of Greece*, Harvard University Press, Cambridge, p. 464.
848. Grote, G., 1899, *A History of Greece*, vol. 6, Harper & Brothers, New York, p. 147–151.
849. Arrian, 1893, *Arrian's Anabasis of Alexander and Indica*, Book 1, Chapter 18, translated by Edward James Chinnock, George Bell & Sons, London, p. 49.
850. Diodorus Siculus, 1814, *The Historical Library of Diodorus the Sicilian*, Book 17, Chapter 2, vol. 2, translated by G. Booth, W. M'Dowall for J. Davis, London, p. 176.
851. Ibid.
852. Arrian, 1893, *Arrian's Anabasis of Alexander and Indica*, Book 1, Chapter 20, translated by Edward James Chinnock, George Bell & Sons, London, p. 51.
853. Grote, G., 1899, *A History of Greece*, vol. 10, Harper & Brothers, New York, p. 43.
854. Ibid., p. 53.
855. Ibid.
856. Ibid., p. 56.
857. Ibid., p. 57.
858. Ibid.
859. Ibid., p. 64.
860. Arrian, 1893, *Arrian's Anabasis of Alexander and Indica*, Book 2, Chapter 6, translated by Edward James Chinnock, George Bell & Sons, London, p. 80.
861. Ibid., p. 80–81.
862. Grote, G., 1899, *A History of Greece*, vol. 10, Harper & Brothers, New York, p. 65.
863. Arrian, 1893, *Arrian's Anabasis of Alexander and Indica*, Book 2, Chapter 7, translated by Edward James Chinnock, George Bell & Sons, London, p. 82.
864. Ibid.
865. Ibid., Book 2, Chapter 10, p. 88.
866. Grote, G., 1899, *A History of Greece*, vol. 10, Harper & Brothers, New York, p. 68.
867. Polybius, 1889, *The Histories of Polybius*, Book 12, Chapter 17, vol. 2, translated by Evelyn S. Shuckburgh, Macmillan and Co., London, p. 95. The stade was an ancient unit of measurement approximately equal to 185 meters (Engels, D., 1985, The Length of Eratosthenes' Stade, *The American Journal of Philology*, no. 3, v. 106, p. 298–311.)
868. Plutarch, 1952, Alexander, in *The Lives of the Noble Grecians and Romans*, translated by John Dryden (1631–1700), *Great Books of the Western World*, vol. 14, William Benton, Chicago, p. 549.
869. Arrian, 1893, *Arrian's Anabasis of Alexander and Indica*, Book 2, Chapter 10, translated by Edward James Chinnock, George Bell & Sons, London, p. 90.
870. Ibid., Book 2, Chapter 11, p. 92.
871. Diodorus Siculus, 1814, *The Historical Library of Diodorus the Sicilian*, Book 17, Chapter 3, vol. 2, translated by G. Booth, W. M'Dowall for J. Davis, London, p. 186.
872. Arrian, 1893, *Arrian's Anabasis of Alexander and Indica*, Book 2, Chapter 11, translated by Edward James Chinnock, George Bell & Sons, London, p. 92–93.
873. Ibid., p. 93–94.
874. Grote, G., 1899, *A History of Greece*, vol. 10, Harper & Brothers, New York, p. 72.
875. Ibid., p. 74.

876. Arrian, 1893, *Arrian's Anabasis of Alexander and Indica*, Book 2, Chapter 14, translated by Edward James Chinnock, George Bell & Sons, London, p. 102.
877. Ibid., p. 103.
878. Ibid., Book 2, Chapter 17, p. 109.
879. Grote, G., 1899, *A History of Greece*, vol. 10, Harper & Brothers, New York, p. 81.
880. Arrian, 1893, *Arrian's Anabasis of Alexander and Indica*, Book 2, Chapter 17, translated by Edward James Chinnock, George Bell & Sons, London, p. 109–110.
881. Grote, G., 1899, *A History of Greece*, vol. 10, Harper & Brothers, New York, p. 83.
882. Ibid., p. 84.
883. Curtius Rufus, Quintus, 1747, *The History of the Wars of Alexander the Great*, translated by John Digby, Third Edition, vol. 1, Book 4, Chapter 2, A. Millar, London, p. 203.
884. Ibid. p. 201.
885. Ibid.
886. Ibid., p. 203.
887. Grote, G., 1899, *A History of Greece*, vol. 10, Harper & Brothers, New York, p. 85.
888. Curtius Rufus, Quintus, 1747, *The History of the Wars of Alexander the Great*, translated by John Digby, Third Edition, vol. 1, Book 4, Chapter 2, A. Millar, London, p. 204.
889. Arrian, 1893, *Arrian's Anabasis of Alexander and Indica*, Book 2, Chapter 21, translated by Edward James Chinnock, George Bell & Sons, London, p. 116.
890. Ibid., Book 2, Chapter 23, p. 119.
891. Diodorus Siculus, 1814, *The Historical Library of Diodorus the Sicilian*, Book 17, Chapter 4, vol. 2, translated by G. Booth, W. M'Dowall for J. Davis, London, p. 196–197.
892. Grote, G., 1899, *A History of Greece*, vol. 10, Harper & Brothers, New York, p. 87.
893. Arrian, 1893, *Arrian's Anabasis of Alexander and Indica*, Book 2, Chapter 25, translated by Edward James Chinnock, George Bell & Sons, London, p. 123.
894. Diodorus Siculus, 1814, *The Historical Library of Diodorus the Sicilian*, Book 17, Chapter 5, vol. 2, translated by G. Booth, W. M'Dowall for J. Davis, London, p. 203.
895. Arrian, 1893, *Arrian's Anabasis of Alexander and Indica*, Book 2, Chapter 26, translated by Edward James Chinnock, George Bell & Sons, London, p. 124.
896. Ibid., Chapter 27, p. 126.
897. Ibid.
898. Ibid.
899. Polybius, 1889, *The Histories of Polybius*, Book 16, Chapter 22, vol. 2, translated by Evelyn S. Shuckburgh, Macmillan and Co., London, p. 189–190.
900. Grote, G., 1899, *A History of Greece*, vol. 10, Harper & Brothers, New York, p. 90.
901. Arrian, 1893, *Arrian's Anabasis of Alexander and Indica*, Book 2, Chapter 27, translated by Edward James Chinnock, George Bell & Sons, London, p. 126–127.
902. Grote, G., 1899, *A History of Greece*, vol. 10, Harper & Brothers, New York, p. 92.
903. Arrian, 1893, *Arrian's Anabasis of Alexander and Indica*, Book 3, Chapter 1, translated by Edward James Chinnock, George Bell & Sons, London, p. 128.
904. Diodorus Siculus, 1814, *The Historical Library of Diodorus the Sicilian*, Book 17, Chapter 3, vol. 2, translated by G. Booth, W. M'Dowall for J. Davis, London, p. 190.
905. Grote, G., 1899, *A History of Greece*, vol. 10, Harper & Brothers, New York, p. 95.
906. Ibid., p. 94.
907. Arnold, T., 1868, *History of Rome, Three Volumes in One*, Chapter 30, D. Appleton & Company, New York, p. 275.
908. Diodorus Siculus, 1814, *The Historical Library of Diodorus the Sicilian*, Book 17, Chapter 5, vol. 2, translated by G. Booth, W. M'Dowall for J. Davis, London, p. 190.
909. Arrian, 1893, *Arrian's Anabasis of Alexander and Indica*, Book 3, Chapter 8, translated by Edward James Chinnock, George Bell & Sons, London, p. 143.
910. Diodorus Siculus, 1814, *The Historical Library of Diodorus the Sicilian*, Book 17, Chapter 5, vol. 2, translated by G. Booth, W. M'Dowall for J. Davis, London, p. 203.
911. Grote, G., 1899, *A History of Greece*, vol. 10, Harper & Brothers, New York, p. 102.
912. Arrian, 1893, *Arrian's Anabasis of Alexander and Indica*, Book 3, Chapter 12, translated by Edward James Chinnock, George Bell & Sons, London, p. 150.
913. Creasy, E. S., 1908, *The Fifteen Decisive Battles of the World from Marathon to Waterloo*, Everyman's Library Edition, E. P. Dutton, New York, p. 68.
914. Plutarch, 1952, Alexander, in *The Lives of the Noble Grecians and Romans*, translated by John Dryden (1631–1700), *Great Books of the Western World*, vol. 14, William Benton, Chicago, p. 556.
915. Creasy, E. S., 1908, *The Fifteen Decisive Battles of the World from Marathon to Waterloo*, Everyman's Library Edition, E. P. Dutton, New York, p. 74. Creasy notes that the exact date of the battle is known from the recorded fact that it "was fought eleven days after an eclipse of the moon."
916. Arrian, 1893, *Arrian's Anabasis of Alexander and Indica*, Book 3, Chapter 12, translated by Edward James Chinnock, George Bell & Sons, London, p. 149.
917. Creasy, E. S., 1908, *The Fifteen Decisive Battles of the World from Marathon to Waterloo*, Everyman's Library Edition, E. P. Dutton, New York, p. 77.
918. Arrian, 1893, *Arrian's Anabasis of Alexander and Indica*, Book 3, Chapter 13, translated by Edward James Chinnock, George Bell & Sons, London, p. 152–153.
919. Ibid., Book 3, Chapter 14, p. 153–154.
920. Grote, G., 1899, *A History of Greece*, vol. 10, Harper & Brothers, New York, p. 109.
921. Arrian, 1893, *Arrian's Anabasis of Alexander and Indica*, Book 3, Chapter 15, translated by Edward James Chinnock, George Bell & Sons, London, p. 155.
922. Creasy, E. S., 1908, *The Fifteen Decisive Battles of the World from Marathon to Waterloo*, Everyman's Library Edition, E. P. Dutton, New York, p. 80.
923. Grote, G., 1899, *A History of Greece*, vol. 10, Harper & Brothers, New York, p. 113.
924. Diodorus Siculus, 1814, *The Historical Library of Diodorus the Sicilian*, Book 17, Chapter 7, vol. 2, translated by G. Booth, W. M'Dowall for J. Davis, London, p. 211.
925. Grote, G., 1899, *A History of Greece*, vol. 10, Harper & Brothers, New York, p. 120.
926. Diodorus Siculus, 1814, *The Historical Library of Diodorus the Sicilian*, Book 17, Chapter 7, vol. 2, translated by G. Booth, W. M'Dowall for J. Davis, London, p. 214.
927. Ibid., p. 215.
928. Ibid.
929. Ibid., p. 216.
930. Ibid.
931. Plutarch, 1952, Alexander, in *The Lives of the Noble Grecians and Romans*, translated by John Dryden (1631–1700), *Great Books of the Western World*, vol. 14, William Benton, Chicago, p. 559.
932. Ibid., p. 564.
933. Curtius Rufus, Quintus, 1747, *The History of the Wars of Alexander the Great*, translated by John Digby, Third Edition, vol. 1, Book 6, Chapter 11, A. Millar, London, p. 368.
934. Grote, G., 1899, *A History of Greece*, vol. 10, Harper & Brothers, New York, p. 140.
935. Arrian, 1893, *Arrian's Anabasis of Alexander and Indica*, Book 3, Chapter 27, translated by Edward James Chinnock, George Bell & Sons, London, p. 179.
936. Grote, G., 1899, *A History of Greece*, vol. 10, Harper & Brothers, New York, p. 143.
937. Ibid., p. 124.
938. Arrian, 1893, *Arrian's Anabasis of Alexander and Indica*, Book 4, Chapter 7, translated by Edward James Chinnock, George Bell & Sons, London, p. 199.
939. Diodorus Siculus, 1814, *The Historical Library of Diodorus the Sicilian*, Book 17, Chapter 9, vol. 2, translated by G. Booth, W. M'Dowall for J. Davis, London, p. 226.
940. Grote, G., 1880, *Aristotle*, Second Edition, John Murray, London, p. 9.
941. Grote, G., 1899, *A History of Greece*, vol. 10, Harper & Brothers, New York, p. 96.

942. Plutarch, 1952, Alexander, in *The Lives of the Noble Grecians and Romans*, translated by John Dryden (1631–1700), *Great Books of the Western World*, vol. 14, William Benton, Chicago, p. 554.
943. Grote, G., 1899, *A History of Greece*, vol. 10, Harper & Brothers, New York, p. 157.
944. Plutarch, 1870, Laconic Apophthegms, or Remarkable Sayings of the Spartans, in *Plutarch's Morals, Translated from the Greek by Several Hands, Corrected and Revised by William W. Goodwin*, vol. 1, Little, Brown, and Company, Boston, p. 407.
945. Diodorus Siculus, 1814, *The Historical Library of Diodorus the Sicilian*, Book 17, Chapter 8, vol. 2, translated by G. Booth, W. M'Dowall for J. Davis, London, p. 221.
946. Arrian, 1893, *Arrian's Anabasis of Alexander and Indica*, Book 1, Chapter 15, translated by Edward James Chinnock, George Bell & Sons, London, p. 40.
947. Ibid., Book 4, Chapter 8, p. 201.
948. Plutarch, 1952, Alexander, in *The Lives of the Noble Grecians and Romans*, translated by John Dryden (1631–1700), *Great Books of the Western World*, vol. 14, William Benton, Chicago, p. 565.
949. Arrian, 1893, *Arrian's Anabasis of Alexander and Indica*, Book 4, Chapter 9, translated by Edward James Chinnock, George Bell & Sons, London, p. 203.
950. Grote, G., 1899, *A History of Greece*, vol. 10, Harper & Brothers, New York, p. 166.
951. Clarke, J., 1910, *Physical Science in the Time of Nero, Being a Translation of the Quaestiones Naturales of Seneca*, Book 6, Chapter 23, Macmillan, London, p. 254.
952. Arrian, 1893, *Arrian's Anabasis of Alexander and Indica*, Book 7, Chapter 14, translated by Edward James Chinnock, George Bell & Sons, London, p. 368.
953. Ibid.
954. Grote, G., 1899, *A History of Greece*, vol. 10, Harper & Brothers, New York, p. 187.
955. Diodorus Siculus, 1814, *The Historical Library of Diodorus the Sicilian*, Book 17, Chapter 8, vol. 2, translated by G. Booth, W. M'Dowall for J. Davis, London, p. 221.
956. Plutarch, 1952, Alexander, in *The Lives of the Noble Grecians and Romans*, translated by John Dryden (1631–1700), *Great Books of the Western World*, vol. 14, William Benton, Chicago, p. 550.
957. Ibid., p. 575.
958. Grote, G., 1899, *A History of Greece*, vol. 10, Harper & Brothers, New York, p. 187.
959. Curtius Rufus, Quintus, 1747, *The History of the Wars of Alexander the Great*, translated by John Digby, Third Edition, vol. 2, Book 10, Chapter 5, A. Millar, London, p. 186.
960. Grote, G., 1899, *A History of Greece*, vol. 10, Harper & Brothers, New York, p. 187.
961. *Encyclopædia Britannica*, 1972, "Aristotle," vol. 2, William Benton, Chicago, p. 390.
962. Aquinas, T., 1952, Summa Theologica, Treatise on God, Question 5, Article 4, in *Thomas Aquinas: I*, translated by the Fathers of the English Domincan Province, *Great Books of the Western World*, vol. 19, William Benton, Chicago, p. 26 (first published in 1911 by T. Baker in London and others).
963. Sarton, G., 1960, *A History of Science, Ancient Science Through the Golden Age of Greece*, Harvard University Press, Cambridge, p. 522.
964. Coleridge, S. T., 1888, *The Table Talk and Omniana of Samuel Taylor Coleridge*, arranged and edited by T. Ashe, George Bell and Sons, London, p. 100.
965. Alighieri, D., 1902, *The Divine Comedy of Dante Alighieri*, Canto IV, translated by Charles Eliot Norton (1827–1908), Houghton, Mifflin & Co., New York, p. 27.
966. Strabo, 1856, *The Geography of Strabo*, Fragment 35, vol. 1, translated by H. C. Hamilton and W. Falconer, Henry G. Bohn, London, p. 513.
967. Blakesley, J. W., 1834, *A Life of Aristotle*, John W. Parker, London, p. 12.
968. Grote, G., 1899, *A History of Greece*, vol. 3, Harper & Brothers, New York, p. 252.
969. Ross, W. D., 1930, *Aristotle*, Second Edition, Revised, Methuen & Co., London, p. 1.
970. Lewes, G. H., 1864, *Aristotle, a Chapter from the History of Science*, Smith, Elder and Co., London, p. 7.
971. Blakesley, J. W., 1834, *A Life of Aristotle*, John W. Parker, London, p. 14.
972. Lewes, G. H., 1864, *Aristotle, a Chapter from the History of Science*, Smith, Elder and Co., London, p. 7.
973. Laërtius, Diogenes, 1905, *The Lives and Opinions of Eminent Philosophers*, Book 5, translated by C. D. Yonge, London, George Bell and Sons, p. 184.
974. Blakesley, J. W., 1834, *A Life of Aristotle*, John W. Parker, London, p. 17.
975. Ross, W. D., 1930, *Aristotle*, Second Edition, Revised, Methuen & Co., London, p. 2.
976. Laërtius, Diogenes, 1905, *The Lives and Opinions of Eminent Philosophers*, Book 5, translated by C. D. Yonge, London, George Bell and Sons, p. 181.
977. Blakesley, J. W., 1834, *A Life of Aristotle*, John W. Parker, London, p. 27.
978. Aristotle, 1941, *Nicomachean Ethics*, Book 1, Chapter 6, translated by W. D. Ross, in *Basic Works of Aristotle*, Random House, New York, p. 939 (1096a).
979. Grote, G., 1899, *A History of Greece*, vol. 9, Harper & Brothers, New York, p. 341.
980. Grote, G., 1880, *Aristotle*, Second Edition, John Murray, London, p. 6.
981. Blakesley, J. W., 1834, *A Life of Aristotle*, John W. Parker, London, p. 35.
982. Strabo, 1856, *The Geography of Strabo*, Book 8, Chapter 1, Paragraph 57, vol. 2, translated by H. C. Hamilton and W. Falconer, Henry G. Bohn, London, p. 382.
983. Laërtius, Diogenes, 1905, *The Lives and Opinions of Eminent Philosophers*, Book 5, translated by C. D. Yonge, London, George Bell and Sons, p. 182.
984. Strabo, 1856, *The Geography of Strabo*, Book 8, Chapter 1, Paragraph 57, vol. 2, translated by H. C. Hamilton and W. Falconer, Henry G. Bohn, London, p. 382.
985. Ibid.
986. Diodorus Siculus, 1814, *The Historical Library of Diodorus the Sicilian*, Book 16, Chapter 9, vol. 2, translated by G. Booth, W. M'Dowall for J. Davis, London, p. 124.
987. Blakesley, J. W., 1834, *A Life of Aristotle*, John W. Parker, London, p. 44.
988. Grote, G., 1880, *Aristotle*, Second Edition, John Murray, London, p. 5.
989. Plutarch, 1952, Alexander, in *The Lives of the Noble Grecians and Romans*, translated by John Dryden (1631–1700), *Great Books of the Western World*, vol. 14, William Benton, Chicago, p. 543.
990. Ibid.
991. Blakesley, J. W., 1834, *A Life of Aristotle*, John W. Parker, London, p. 50.
992. Grote, G., 1880, *Aristotle*, Second Edition, John Murray, London, p. 5.
993. Ibid., p. 7.
994. Gellius, A., 1795, *The Attic Nights of Aulus Gellius*, Book 20, Chapter 5, translated into English by the Rev. W. Beloe, vol. 3, J. Johnson, London, p. 423–424.
995. Laërtius, Diogenes, 1905, *The Lives and Opinions of Eminent Philosophers*, Book 5, translated by C. D. Yonge, London, George Bell and Sons, p. 181.
996. Grote, G., 1880, *Aristotle*, Second Edition, John Murray, London, p. 16–17.
997. Ibid.
998. Laërtius, Diogenes, 1905, *The Lives and Opinions of Eminent Philosophers*, Book 5, translated by C. D. Yonge, London, George Bell and Sons, p. 193.
999. Thomson, G., and Missner, M., 2000, *On Aristotle*, Wadsworth, Belmont, California, p. 6–8.
1000. Strabo, 1856, *The Geography of Strabo*, Book 13, Chapter 1, Paragraph 54, vol. 2, translated by H. C. Hamilton and W. Falconer, Henry G. Bohn, London, p. 379–380.
1001. Plutarch, 1952, Sulla, in *The Lives of the Noble Grecians and Romans*, translated by John Dryden (1631–1700), *Great Books of the Western World*, vol. 14, William Benton, Chicago, p. 381.

1002. Grote, G., 1880, *Aristotle*, Second Edition, John Murray, London, p. 68.
1003. Ibid.
1004. Ross, W. D., 1930, *Aristotle*, Second Edition, Revised, Methuen & Co., London, p. 112–113.
1005. Darwin, F. (editor), 1887, *The Life and Letters of Charles Darwin*, vol. 2, D. Appleton, New York, p. 427.
1006. Aristotle, 1882, *On the Parts of Animals*, Book 1, Chapter 5, translated by W. Ogle, Kegan Paul, Trench & Co., London, p. 16–17 (645a).
1007. Aristotle, 1897, *Aristotle's History of Animals, in Ten Books*, Book 6, Chapter 13, Paragraph 4, translated by Richard Cresswell, George Bell and Sons, London, p. 155–156.
1008. Sarton, G., 1960, *A History of Science, Ancient Science Through the Golden Age of Greece*, Harvard University Press, Cambridge, p. 538.
1009. Aristotle, 1897, *Aristotle's History of Animals, in Ten Books*, Book 6, Chapter 3, Paragraph 6, translated by Richard Cresswell, George Bell, London, p. 143.
1010. Ibid., Book 9, Chapter 27, Paragraph 7, p. 262.
1011. Aristotle, 1912, *De Generatione Animalium* (*On the Generation of Animals*), Book 3, Chapter 10, translated by Arthur Platt, in *The Works of Aristotle Translated into English*, vol. 5, edited by J. A. Smith and W. D. Ross, Oxford University Press, London, p. 760b.
1012. Sarton, G., 1960, *A History of Science, Ancient Science Through the Golden Age of Greece*, Harvard University Press, Cambridge, p. 537.
1013. Aristotle, 1897, *Aristotle's History of Animals, in Ten Books*, Book 6, Chapter 14, Paragraph 2, translated by Richard Cresswell, George Bell, London, p. 157.
1014. Aristotle, 1882, *On the Parts of Animals*, Book 4, Chapter 5, translated by W. Ogle, Kegan Paul, Trench & Co., London, p. 104 (681b).
1015. Ibid.
1016. Aristotle, 1939, *On the Heavens* (*De Caelo*), Book 3, Chapter 7, translated by W. K. C. Guthrie, William Heinemann, London, p. 313 (306a).
1017. Ibid., Book 3, Chapter 7, p. 315 (306a).
1018. Sarton, G., 1960, *A History of Science, Ancient Science Through the Golden Age of Greece*, Harvard University Press, Cambridge, p. 535.
1019. Aristotle, 1912, *De Generatione Animalium* (*On the Generation of Animals*), Book 2, Chapter 1, translated by Arthur Platt, in *The Works of Aristotle Translated into English*, vol. 5, edited by J. A. Smith and W. D. Ross, Oxford University Press, London, p. 733b.
1020. Aristotle, 1910, *Historia Animalium* (*History of Animals*), Book 8, Chapter 1, translated by D'Arcy Wentworth Thompson, in *The Works of Aristotle Translated into English*, vol. 4, edited by J. A. Smith and W. D. Ross, Oxford University Press, London, p. 588b.
1021. Aristotle, 1882, *On the Parts of Animals*, Book 1, Chapter 5, translated by W. Ogle, Kegan Paul, Trench & Co., London, p. 17 (645b).
1022. Plato, 1937, Timaeus, in *The Dialogues of Plato*, translated into English by Benjamin Jowett (1817–1893), vol. 2, Random House, New York, p. 29–30 (49).
1023. Aristotle, 1941, *De Caelo* (*On the Heavens*), Book 3, Chapter 7, translated by J. L. Stocks, in *Basic Works of Aristotle*, Random House, New York, p. 451 (306a).
1024. Aristotle, 1923, *Meteorologica* (*Meteorology*), Book 1, Chapter 3, translated by E. W. Webster, Oxford at the Clarendon Press, London, p. 339a.
1025. Ibid.
1026. Aristotle, 1941, *De Generatione et Corruptione* (*On Generation and Corruption*), Book 2, Chapter 3, translated by Harold H. Joachim, in *Basic Works of Aristotle*, Random House, New York, p. 511 (330b). Also, see, Lindberg, D. C., 1992, *The Beginnings of Western Science*, University of Chicago Press, Chicago, p. 55.
1027. Aristotle, 1941, *Metaphysica* (*Metaphysics*), Book 3, Chapter 1, translated by W. D. Ross, in *Basic Works of Aristotle*, Random House, New York, p. 716 (995b).
1028. Aristotle, 1941, *De Generatione et Corruptione* (*On Generation and Corruption*), Book 1, Chapter 2, translated by Harold H. Joachim, in *Basic Works of Aristotle*, Random House, New York, p. 477–478 (317a). The discussion here is difficult to follow.
1029. Thomson, G., and Missner, M., 2000, *On Aristotle*, Wadsworth, Belmont, California, p. 38.
1030. Aristotle, 1923, *Meteorologica* (*Meteorology*), Book 1, Chapter 9, translated by E. W. Webster, Oxford at the Clarendon Press, London, p. 347a.
1031. Ibid., p. 350a.
1032. Ibid., p. 351b, 353a.
1033. Playfair, J., 1997, Biographical Account of the late Dr. James Hutton, F. R. S. Edin., in *James Hutton & Joseph Black, Biographies by John Playfair and Adam Ferguson*, from Volume V of *Transactions of the Royal Society of Edinburgh*, 1805, RSE Scotland Foundation, Edinburgh, p. 73.
1034. Aristotle, 1923, *Meteorologica (Meteorology)*, Book 2, Chapter 8, translated by E. W. Webster, Oxford at the Clarendon Press, London, p. 366b.
1035. Ibid.
1036. Ibid., Book 2, Chapter 3, p. 358b.
1037. Aristotle, 1913, *Mechanica* (*Mechanics*), Chapter 1, translated by E. S. Forster, in *The Works of Aristotle Translated into English*, vol. 6, edited by W. D. Ross, Oxford at the Clarendon Press, London, p. 847a.
1038. Winter, T. N., 2007, *The Mechanical Problems in the Corpus of Aristotle*: Faculty Publications, Classics and Religious Studies Department, University of Nebraska-Lincoln. <http://digitalcommons.unl.edu/classicsfacpub/68>
1039. Aristotle, 1941, *De Generatione et Corruptione* (*On Generation and Corruption*), Book 2, Chapter 10, translated by Harold H. Joachim, in *Basic Works of Aristotle*, Random House, New York, p. 525 (336a).
1040. Aristotle, 1941, *De Caelo* (*On the Heavens*), Book 2, Chapter 13, translated by J. L. Stocks, in *Basic Works of Aristotle*, Random House, New York, p. 430 (294b).
1041. Plato, 1937, Phaedo, in *The Dialogues of Plato*, translated into English by Benjamin Jowett (1817–1893), vol. 1, Random House, New York, p. 493 (109).
1042. Aristotle, 1939, *On the Heavens* (*De Caelo*), Book 2, Chapter 12, translated by W. K. C. Guthrie, William Heinemann, London, p. 205 (292a).
1043. Ibid.
1044. Ibid., Book 2, Chapter 14, p. 245 (296b).
1045. Heath, T., 1913, *Aristarchus of Samos, the Ancient Copernicus*, Oxford at the Clarendon Press, London, p. 193.
1046. Ibid., p. 195.
1047. Ibid., p. 196.
1048. Ibid., p. 217.
1049. Ibid., p. 196.
1050. Lloyd, G. E. R., 1970, *Early Greek Science: Thales to Aristotle*, W. W. Norton & Co., New York, p. 98.
1051. Heath, T., 1913, *Aristarchus of Samos, the Ancient Copernicus*, Oxford at the Clarendon Press, London, p. 217.
1052. Aristotle, 1941, *Metaphysica* (*Metaphysics*), Book 12, Chapter 8, translated by W. D. Ross, in *Basic Works of Aristotle*, Random House, New York, p. 883 (1074a).
1053. Burnet, J., 1920, *Greek Philosophy, Part 1: Thales to Plato*, Macmillan, London, p. 345.
1054. Aristotle, 1939, *On the Heavens* (*De Caelo*), Book 1, Chapter 3, translated by W. K. C. Guthrie, William Heinemann, London, p. 25 (270b).
1055. Heath, T., 1913, *Aristarchus of Samos, the Ancient Copernicus*, Oxford at the Clarendon Press, London, p. 227.
1056. Aristotle, 1939, *On the Heavens* (*De Caelo*), Book 1, Chapter 3, translated by W. K. C. Guthrie, William Heinemann, London, p. 25 (270b).
1057. Ibid., Book 2, Chapter 14, p. 247 (297a), 253 (297b–298a).
1058. Ibid., p. 253 (298a).
1059. Ibid., p. 255 (298b).
1060. Turcotte, D. L., and Schubert, G., 1982, *Geodynamics, Applications of Continuum Physics to Geological Problems*, John Wiley & Sons, New York, p. 429.
1061. Aristotle, 1923, *Meteorologica (Meteorology)*, Book 1,

Chapter 4, translated by E. W. Webster, Oxford at the Clarendon Press, London, p. 341b.

1062. Ibid., p. 344a.

1063. Aristotle, 1930, *Physica* (*Physics*), Book 2, Chapter 8, translated by Robert Purves Hardie and Russell Kerr Gaye, Clarendon Press, Oxford, p. 199b.

1064. Ibid., Book 2, Chapter 3, p. 194b.

1065. Grant, E., 1977, *Physical Science in the Middle Ages*, Cambridge University Press, Cambridge, p. 36–59.

1066. Aristotle, 1939, *On the Heavens* (*De Caelo*), Book 1, Chapter 2, translated by W. K. C. Guthrie, William Heinemann, London, p. 16–17 (269a–269b)

1067. Ibid., p. 17 (269b)

1068. Formigari, L., 1973, Chain of Being, in *Dictionary of the History of Ideas*, vol. 1, edited by Philip P. Wiener, Charles Scribner's Sons, New York, p. 325.

1069. Aristotle, 1930, *Physica* (*Physics*), Book 7, Chapter 5, translated by Robert Purves Hardie and Russell Kerr Gaye, Clarendon Press, Oxford, p. 250a.

1070. Ibid., Book 4, Chapter 8, p. 215b.

1071. Aristotle, 1939, *On the Heavens* (*De Caelo*), Book 3, Chapter 2, translated by W. K. C. Guthrie, William Heinemann, London, p. 281 (301b).

1072. Ibid., Book 1, Chapter 6, p. 49 (274a).

1073. Butterfield, H., 1957, *The Origins of Modern Science* (revised edition), Free Press, New York, p. 17–18.

1074. Ibid., p. 18.

1075. Ibid., p. 14.

1076. Ibid., p. 17–18.

1077. Cohen, M. R., and Drabkin, I. E., 1948, *A Source Book in Greek Science*, McGraw-Hill, New York, p. 211.

1078. Wolff, M., 1987, Philoponus and the Rise of Preclassical Dynamics, in *Philoponus and the Rejection of Aristotelean Science*, edited by Richard Sorabji, Cornell University Press, Ithaca, New York, p. 89.

1079. Ibid., p. 84.

1080. Sambursky, S., 1962, *The Physical World of Late Antiquity*, Princeton University Press, Princeton, New Jersey, p. 73.

1081. Moody, E. A., 2008, Buridan, Jean, in *Complete Dictionary of Scientific Biography*, edited by Charles Gillispie, vol. 2, Cengage Learning, New York, p. 606.

1082. Wolff, M., 1987, Philoponus and the Rise of Preclassical Dynamics, in *Philoponus and the Rejection of Aristotelean Science*, edited by Richard Sorabji, Cornell University Press, Ithaca, New York, p. 94.

1083. Kuhn, T., 1996, *The Structure of Scientific Revolutions* (Third Edition), University of Chicago Press, Chicago.

1084. Grant, E., 1977, *Physical Science in the Middle Ages*, Cambridge University Press, Cambridge, p. 83.

1085. Butterfield, H., 1957, *The Origins of Modern Science* (revised edition), Free Press, New York, p. 27.

1086. Aristotle, 1941, *Metaphysica* (*Metaphysics*), Book 4, Chapter 1, translated by W. D. Ross, in *Basic Works of Aristotle*, Random House, New York, p. 731 (1003a).

1087. Ibid., Book 4, Chapter 3, p. 736 (1005a).

1088. Wallace, W. A., 1982, Aristotle in the Middle Ages, in *Dictionary of the Middle Ages*, vol. 1, edited by Joseph R. Strayer, Charles Scribner's Sons, New York, p. 456.

1089. Ibid.

1090. Aristotle, 1941, *Metaphysica* (*Metaphysics*), Book 1, Chapter 1, translated by W. D. Ross, in *Basic Works of Aristotle*, Random House, New York, p. 689–690 (981a).

1091. Ibid., p. 690 (981a).

1092. Coleridge, S. T., 1888, *The Table Talk and Omniana of Samuel Taylor Coleridge*, arranged and edited by T. Ashe, George Bell and Sons, London, p. 99.

1093. Ibid., Book 7, Chapter 13, p. 805 (1039a).

1094. Thomson, G., and Missner, M., 2000, *On Aristotle*, Wadsworth, Belmont, California, p. 56.

1095. Gilson, E., 1955, *History of Christian Philosophy in the Middle Ages*, Random House, New York, p. 221.

1096. Whittaker, T., 1911, *Priests, Philosophers and Prophets*, Adam and Charles Black, London, p. 11.

1097. Burnet, J., 1920, *Early Greek Philosophy*, Third Edition, A. & C. Black, London, p. 112.

1098. Ibid., p. 119.

1099. Ibid.

1100. Aristotle, 1941, *Metaphysica* (*Metaphysics*), Book 12, Chapter 6, translated by W. D. Ross, in *Basic Works of Aristotle*, Random House, New York, p. 878 (1071b).

1101. Ibid., p. 879 (1072a).

1102. Ibid., p. 880 (1072b).

1103. Sarton, G., 1960, *A History of Science, Ancient Science Through the Golden Age of Greece*, Harvard University Press, Cambridge, p. 500.

1104. Aristotle, 1941, *Analytica Priora* (*Prior Analytics*), Book 1, Chapter 1, translated by A. J. Jenkinson, in *Basic Works of Aristotle*, Random House, New York, p. 66 (24b).

1105. Aristotle, 1941, *Analytica Posteriora* (*Posterior Analytics*), Book 2, Chapter 19, translated by A. J. Jenkinson, in *Basic Works of Aristotle*, Random House, New York, p. 185 (100b).

1106. Ibid., Book 1, Chapter 18, p. 136 (81b).

1107. Aristotle, 1906, *The Nicomachean Ethics of Aristotle*, Book 6, Chapter 3, Tenth Edition, translated by F. H. Peters, Kegan Paul, Trench, Trübner & Co., Ltd., London, p. 185 (1139b).

1108. Rand, A., 1982, *Philosophy: Who Needs It*, Bobbs-Merrill, Indianapolis, p. 4.

1109. Thomson, G., and Missner, M., 2000, *On Aristotle*, Wadsworth, Belmont, California, p. 66.

1110. Aristotle, 1941, *Ethica Nicomachea* (*Nicomachean Ethics*), Book 10, Chapter 8, translated by W. D. Ross, in *Basic Works of Aristotle*, Random House, New York, p. 1108 (1179a).

1111. Ibid., Book 2, Chapter 6, p. 959 (1106b).

1112. Ibid., p. 959 (1107a).

1113. Ibid., Book 1, Chapter 13, p. 950 (1102a).

1114. Ibid., Book 1, Chapter 8, p. 945 (1099b).

1115. Ibid., Book 9, Chapter 10, p. 1091 (1171a).

1116. Aristotle, 1898, *The Poetics of Aristotle*, Chapter 9, translated by Samuel Henry Butcher (1850–1910), Macmillan and Co., London, p. 35 (1451b).

1117. Aristotle, 1930, *Physica* (*Physics*), Book 2, Chapter 2, translated by Robert Purves Hardie and Russell Kerr Gaye, Clarendon Press, Oxford, p. 194a.

1118. Ibid., Book 2, Chapter 8, p. 199a.

1119. Sarton, G., 1960, *A History of Science, Ancient Science Through the Golden Age of Greece*, Harvard University Press, Cambridge, p. 574.

1120. Aristotle, 1885, *The Politics of Aristotle*, vol. 1, Book 1, Chapter 10, translated by Benjamin Jowett (1817–1893), Oxford at the Clarendon Press, London, p. 18–19 (1258a).

1121. Ibid., Book 1, Chapter 2, p. 4–5 (1253a).

1122. Ibid., p. 5 (1253a).

1123. Ibid., Book 2, Chapter 7, p. 46 (1267b).

1124. Ibid., Book 4, Chapter 11, p. 128 (1295b–1296a).

1125. Ibid.

1126. Ibid., Book 3, Chapter 7, p. 80 (1279b).

1127. Ibid., Book 4, Chapter 13, p. 132 (1297b).

1128. Ibid., Book 4, Chapter 11, p. 129 (1296a).

1129. Ibid., Book 2, Chapter 5, p. 35 (1263b).

1130. Ibid., Book 1, Chapter 14, p. 25–26 (1260b).

1131. Ibid., Book 3, Chapter 7, p. 80 (1279b).

1132. Ibid., Book 3, Chapter 18, p. 105 (1288a).

1133. Ibid., Book 3, Chapter 15, p. 99 (1286a).

1134. Ibid.

1135. Ibid., Book 4, Chapter 8, p. 122 (1293b).

1136. Ibid., Book 4, Chapter 2, p. 109 (1289b).

1137. Ibid., Book 4, Chapter 11, p. 128 (1296a).

1138. Ibid., Book 3, Chapter 7, p. 80 (1279b).

1139. Plato, 1937, *Laws*, in *The Dialogues of Plato*, translated into English by Benjamin Jowett (1817–1893), vol. 2, Random House, New York, p. 506 (739).

1140. Ibid.

1141. Aristotle, 1885, *The Politics of Aristotle*, vol. 1, Book 2, Chapter 5, translated by Benjamin Jowett (1817–1893), Oxford at the Clarendon Press, London, p. 35 (1263b).

1142. Ibid., Book 2, Chapter 3, p. 30 (1261b).

1143. Ibid., Book 2, Chapter 5, p. 35 1263b).

1144. Ibid. Book 6, Chapter 5, p. 197 (1320a).

1145. Ibid.

1146. Laërtius, Diogenes, 1905, *The Lives and Opinions of Eminent Philosophers*, Book 5, translated by C. D. Yonge, London, George Bell and Sons, p. 191.

1147. Aristotle, 1885, *The Politics of Aristotle*, vol. 1, Book 1, Chapter 5, translated by Benjamin Jowett (1817–1893), Oxford at the Clarendon Press, London, p. 7 (1254a).

1148. Ibid., Book 1, Chapter 5, p. 8–9 (1254b).

1149. Ibid., Book 1, Chapter 5, p. 9 (1255a).

1150. Ibid., Book 1, Chapter 6, p. 11 (1255b).

1151. Hesiod, 1920, Works and Days (line 78), in *Hesiod, the Homeric Hymns and Homerica*, with an English translation by Hugh G. Evelyn-White, London, William Heinemann, p. 7.

1152. Ibid., lines 90–96, p. 9.

1153. Freeman, K., 1966, *Ancilla to the Pre-Socratic Philosophers, a Complete Translation of the Fragments in Diels, Fragmente der Vorsokratiker*, Harvard University Press, Cambridge, p. 103.

1154. Grote, G., 1899, *A History of Greece*, vol. 2, Harper & Brothers, New York, p. 506; also, Plutarch, 1952, Solon, in *The Lives of the Noble Grecians and Romans*, translated by John Dryden (1631–1700), *Great Books of the Western World*, vol. 14, William Benton, Chicago, p. 73.

1155. Plutarch, 1952, Solon, in *The Lives of the Noble Grecians and Romans*, translated by John Dryden (1631–1700), *Great Books of the Western World*, vol. 14, William Benton, Chicago, p. 73.

1156. Thucydides, 1942, *The Peloponnesian War*, Book 3, Chapter 17, translated by Benjamin Jowett (1817–1893), in *The Greek Historians*, vol. 1, Random House, New York, p. 690.

1157. Aeschylus, 1922, Seven Against Thebes, lines 181–190, in *Aeschylus*, vol. 1, translated by Herbert Weir Smyth, William Heineman, London, p. 335.

1158. Plato, 1937, Timaeus, in *The Dialogues of Plato*, translated into English by Benjamin Jowett (1817–1893), vol. 2, Random House, New York, p. 67 (91).

1159. Plato, 1937, *The Republic*, Book 5, in *The Dialogues of Plato*, translated into English by Benjamin Jowett (1817–1893), vol. 1, Random House, New York, p. 716 (455).

1160. Ibid.

1161. Ibid., p. 717 (455).

1162. Ibid.

1163. Aristotle, 1885, *The Politics of Aristotle*, vol. 1, Book 1, Chapter 5, translated by Benjamin Jowett (1817–1893), Oxford at the Clarendon Press, London, p. 8 (1254b).

1164. Ibid., Book 1, Chapter 13, p. 24 (1260a).

1165. Grote, G., 1880, *Aristotle*, Second Edition, John Murray, London, p. 12.

1166. Aelianus, Claudius, 1665, *Claudius Aelianus his Various History*, Book 3, Chapter 36, translated by Thomas Stanley, Thomas Dring, London, p. 92.

1167. Blakesley, J. W., 1834, *A Life of Aristotle*, John W. Parker, London, p. 71.

1168. Laërtius, Diogenes, 1905, *The Lives and Opinions of Eminent Philosophers*, Book 5, translated by C. D. Yonge, London, George Bell and Sons, p. 184.

1169. Gellius, A., 1795, *The Attic Nights of Aulus Gellius*, Book 13, Chapter 5, translated into English by the Rev. W. Beloe, vol. 3, J. Johnson, London, p. 12.

1170. Laërtius, Diogenes, 1905, *The Lives and Opinions of Eminent Philosophers*, Book 5, translated by C. D. Yonge, London, George Bell and Sons, p. 185–186.

1171. Censorinus, 1900, *De Die Natale* (*The Natal Day*), Chapter 3, translated into English by William Maude, Cambridge Encyclopedia Co., New York, p. 13.

## Chapter 2

1. Diodorus Siculus, 1814, *The Historical Library of Diodorus the Sicilian*, Book 18, Chapter 2, vol. 2, translated by G. Booth, W. M'Dowall for J. Davis, London, p. 268.

2. Tacitus, 1876, *The History of Tacitus* (Third Edition), Book 4, Chapters 83, 84, translated into English by Alfred John Church and William Jackson Brodribb, Macmillan and Co., London, p. 189–190.

3. Mahaffy, J. P., 1899, *A History of Egypt Under the Ptolemaic Dynasty*, Charles Scribner's Sons, New York, p. 60.

4. Bevan, E., 1927, *A History of Egypt Under the Ptolemaic Dynasty*, Methuen & Co., London, p. 21.

5. Laërtius, Diogenes, 1905, *The Lives and Opinions of Eminent Philosophers*, Book 2, translated by C. D. Yonge, London, George Bell and Sons, p. 101.

6. Sarton, G., 1959, *A History of Science, Hellenistic Science and Culture in the Last Three Centuries* B.C., Harvard University Press, Cambridge, p. 29.

7. *Oxford English Dictionary Online*, Draft Revision June 2008.

8. Hesiod, 1920, Theogony (lines 60–80), in *Hesiod, the Homeric Hymns and Homerica*, with an English translation by Hugh G. Evelyn-White, London, William Heinemann, p. 83–84.

9. Clagett, M., 1994, *Greek Science in Antiquity*, Barnes and Noble, New York (reprint of 1955 edition), p. 33.

10. Strabo, 1856, *The Geography of Strabo*, Book 17, Chapter 1, Paragraph 8, vol. 3, translated by H. C. Hamilton and W. Falconer, Henry G. Bohn, London, p. 229.

11. Sarton, G., 1959, *A History of Science, Hellenistic Science and Culture in the Last Three Centuries* B.C., Harvard University Press, Cambridge, p. 34.

12. Pliny the Elder, 1855, *The Natural History of Pliny*, translated by John Bostock and H. T. Riley, book 8, chapter 21, vol. 3, Henry G. Bohn, London, p. 185.

13. Laërtius, Diogenes, 1905, *The Lives and Opinions of Eminent Philosophers*, Book 5, translated by C. D. Yonge, London, George Bell and Sons, p. 209.

14. Ibid., p. 210.

15. Strabo, 1856, *The Geography of Strabo*, Book 9, Chapter 1, Paragraph 20, vol. 2, translated by H. C. Hamilton and W. Falconer, Henry G. Bohn, London, p. 88–89.

16. Thackeray, H. St. J. (translator), 1904, *The Letter of Aristeas Translated into English*, Macmillan, London, p. 7.

17. Laërtius, Diogenes, 1905, *The Lives and Opinions of Eminent Philosophers*, Book 5, translated by C. D. Yonge, London, George Bell and Sons, p. 210.

18. Gellius, A., 1795, *The Attic Nights of Aulus Gellius*, Book 6, Chapter 17, translated into English by the Rev. W. Beloe, vol. 2, J. Johnson, London, p. 43.

19. Barnes, R., 2001, Cloistered Bookworms in the Chicken-Coop of the Muses, the Ancient Library of Alexandria, in *The Library of Alexandria*, edited by Roy MacLeod, I. B. Tauris, London, p. 64.

20. Strabo, 1856, *The Geography of Strabo*, Book 13, Chapter 1, Paragraph 54, vol. 2, translated by H. C. Hamilton and W. Falconer, Henry G. Bohn, London, p. 379.

21. Barnes, R., 2001, Cloistered Bookworms in the Chicken-Coop of the Muses, the Ancient Library of Alexandria, in *The Library of Alexandria*, edited by Roy MacLeod, I. B. Tauris, London, p. 61.

22. Athenaeus, 1854, *The Deipnosophists, or Banquet of the Learned*, Book 1, Chapter 4, translated by C. D. Yonge, vol. 1, Henry G. Bohn, London, p. 4.

23. Strabo, 1856, *The Geography of Strabo*, Book 13, Chapter 4, Paragraph 2, vol. 2, translated by H.C. Hamilton and W. Falconer: Henry G. Bohn, London, p. 400–401.

24. Vitruvius, 1960, *The Ten Books on Architecture*, Introduction to Book 7, translated by Morris Hicky Morgan (first published by Harvard University Press in 1914), Dover, New York, p. 196.

25. Pliny the Elder, 1855, *The Natural History of Pliny*, translated by John Bostock and H. T. Riley, book 8, chapter 21, vol. 3, Henry G. Bohn, London, p. 186.

26. Plutarch, 1952, Antony, in *The Lives of the Noble Grecians and Romans*, translated by John Dryden (1631–1700), *Great Books of the Western World*, vol. 14, William Benton, Chicago, p. 769.

27. Mahaffy, J. P., 1899, *A History of Egypt Under the Ptolemaic Dynasty*, Charles Scribner's Sons, New York, p. 64.

28. Athenaeus, 1854, *The Deipnosophists, or Banquet of the*

*Learned*, Book 5, Chapter 25, translated by C. D. Yonge, vol. 1, Henry G. Bohn, London, p. 313.

29. Ibid., Book 5, Chapter 36, p. 324.

30. Ibid., vol. 3, Book 12, Chapter 51, p. 858.

31. Ibid., vol. 3, Book 8, Chapter 37, p. 922.

32. Strabo, 1856, *The Geography of Strabo*, Book 17, Chapter 1, Paragraph 5, vol. 3, translated by H. C. Hamilton and W. Falconer, Henry G. Bohn, London, p. 224.

33. Barnes, R., 2001, Cloistered Bookworms in the Chicken-Coop of the Muses: the Ancient Library of Alexandria, in *The Library of Alexandria*, edited by Roy MacLeod, I. B. Tauris, London, p. 65.

34. Bevan, E., 1927, *A History of Egypt Under the Ptolemaic Dynasty*, Methuen & Co., London, p. 112.

35. Tcherikover, V., 1958, The Ideology of the Letter of Aristeas, *Harvard Theological Review*, no. 2, vol. 51, p. 59.

36. Thackeray, H. St. J. (translator), 1904, *The Letter of Aristeas Translated into English*, Macmillan, London, p. 15.

37. Ibid., p. 53.

38. Ibid., p. 52.

39. Philo, 1855, *On the Life of Moses*, Book 2, Chapter 7, in *The Works of Philo Judaeus*, translated by C. D. Yonge, vol. 3, Henry G. Bohn, London, p. 82.

40. Sarton, G., 1959, *A History of Science, Hellenistic Science and Culture in the Last Three Centuries* B.C., Harvard University Press, Cambridge, p. 25.

41. Strabo, 1856, *The Geography of Strabo*, Book 17, Chapter 1, Paragraph 6, vol. 3, translated by H. C. Hamilton and W. Falconer, Henry G. Bohn, London, p. 226–227.

42. Pliny the Elder, 1855, *The Natural History of Pliny*, translated by John Bostock and H. T. Riley, Book 36, Chapter 18, vol. 6, Henry G. Bohn, London, p. 339.

43. de Camp, L. S., 1965, The Darkhouse of Alexandria, *Technology and Culture*, vol. 6, no. 3, p. 423–427.

44. Pliny the Elder, 1855, *The Natural History of Pliny*, translated by John Bostock and H. T. Riley, Book 36, Chapter 18, vol. 6, Henry G. Bohn, London, p. 339.

45. Josephus, 1987, *The Wars of the Jews*, Book 4, Chapter 10, Paragraph 5, in *The Works of Josephus*, translated by William Whiston (1667–1752), Hendrickson Publishers, p. 693.

46. Diodorus Siculus, 1814, *The Historical Library of Diodorus the Sicilian*, Book 17, Chapter 5, vol. 2, translated by G. Booth, W. M'Dowall for J. Davis, London, p. 202.

47. Sharpley, H. (translator), 1906, *A Realist of the Aegean, being a verse-translation of the Mimes of Herodas*, David Nutt, London, p. 4.

48. Bevan, E., 1927, *A History of Egypt Under the Ptolemaic Dynasty*, Methuen & Co., London, p. 38.

49. Ibid., p. 87.

50. Polybius, 1889, *The Histories of Polybius*, Book 34, Chapter 14, vol. 2, translated by Evelyn S. Shuckburgh, Macmillan and Co., London, p. 492.

51. Bevan, E., 1927, *A History of Egypt Under the Ptolemaic Dynasty*, Methuen & Co., London, p. 111.

52. Ibid., p. 145.

53. Ibid., p. 149.

54. Ibid., p. 135.

55. Ibid., p. 144–157.

56. Mahaffy, J. P., 1899, *A History of Egypt Under the Ptolemaic Dynasty*, Charles Scribner's Sons, New York, p. 127–128.

57. Bevan, E., 1927, *A History of Egypt Under the Ptolemaic Dynasty*, Methuen & Co., London, p. 220.

58. Polybius, 1889, *The Histories of Polybius*, Book 5, Chapter 34, vol. 1, translated by Evelyn S. Shuckburgh, Macmillan and Co., London, p. 388–389.

59. Bevan, E., 1927, *A History of Egypt Under the Ptolemaic Dynasty*, Methuen & Co., London, p. 221.

60. Strabo, 1856, *The Geography of Strabo*, Book 17, Chapter 1, Paragraph 11, vol. 3, translated by H. C. Hamilton and W. Falconer, Henry G. Bohn, London, p. 231.

61. Diodorus Siculus, 1814, *The Historical Library of Diodorus the Sicilian*, Book 26, Chapter 109, vol. 2, translated by G. Booth, W. M'Dowall for J. Davis, London, p. 620.

62. Athenaeus, 1854, *The Deipnosophists, or Banquet of the Learned*, Book 12, Chapter 73, translated by C. D. Yonge, vol. 3, Henry G. Bohn, London, p. 879–880.

63. Nutton, V., 2004, *Ancient Medicine*, Routledge, London, p. 130.

64. Bevan, E., 1927, *A History of Egypt Under the Ptolemaic Dynasty*, Methuen & Co., London, p. 380–384.

65. Proclus, 1792, *The Philosophical and Mathematical Commentaries of Proclus, on the First Book of Euclid's Elements*, vol. 1, printed for the author, London, p. 101.

66. Heath, T., 1981, *A History of Greek Mathematics*, vol. 1 (first published in 1921 by the Clarendon Press, Oxford), Dover, New York, p. 355.

67. Ibid., p. 356.

68. Proclus, 1792, *The Philosophical and Mathematical Commentaries of Proclus, on the First Book of Euclid's Elements*, vol. 1, printed for the author, London, p. 101.

69. Heath, T., 1981, *A History of Greek Mathematics*, vol. 1 (first published in 1921 by the Clarendon Press, Oxford), Dover, New York, p. 357.

70. Euclid, 1956, *The Thirteen Books of Euclid's Elements*, Translated from the Text of Heiberg with Introduction and Commentary by Sir Thomas L. Heath, Second Edition, volumes 1, 2, and 3 (first published by Cambridge University Press in 1926), Dover, New York.

71. Ibid., vol. 1, p. vi.

72. Ibid., vol. 1, p. 202.

73. Ibid., vol. 1, p. 155.

74. Ibid., vol. 1, p. 202.

75. Ibid., vol. 1, p. 202.

76. Sarton, G., 1959, *A History of Science, Hellenistic Science and Culture in the Last Three Centuries* B.C., Harvard University Press, Cambridge, p. 39.

77. Heath, T., 1981, *A History of Greek Mathematics: Volume I, From Thales to Euclid* (first published in 1921 by the Clarendon Press, Oxford), Dover, New York, p. 357–358.

78. Laërtius, Diogenes, 1905, *The Lives and Opinions of Eminent Philosophers*, Book 5, translated by C. D. Yonge, London, George Bell and Sons, p. 202.

79. Ibid., p. 205.

80. Polybius, 1889, *The Histories of Polybius*, Book 12, Chapter 25(c), vol. 2, translated by Evelyn S. Shuckburgh, Macmillan and Co., London, p. 103.

81. Sarton, G., 1927, *Introduction to the History of Science*, vol. 1, Carnegie Institution of Washington, Washington, DC, p. 152.

82. Cicero, Marcus Tullius, 1744, Academics, in *The Morals of Cicero*, translated into English by William Guthrie, London, T. Waller, p. 315.

83. Ibid., p. 413–414.

84. Boas, M., 1949, Hero's Pneumatica: a Study of its Transmission and Influence, *Isis*, vol. 40, no. 1, p. 39.

85. e.g., see the discussion in Farrington, B., 1961, *Greek Science*: Penguin Books, Baltimore, Maryland, p. 169–185.

86. Gottschalk, H. B., 1965, Strato of Lampsacus, Some Texts, *Proceedings of the Leeds Philosophical and Literary Society*, vol. XI, part VI, p. 141.

87. Hero, 1851, *The Pneumatics of Hero of Alexandria*, translated by Bennet Woodcroft, Taylor Walton and Maberly, London, p. 1.

88. Ibid., p. 1–2.

89. Ibid., p. 6–7.

90. Ibid., p. 7–8.

91. Cohen, M. R., and Drabkin, I. E., 1948, *A Source Book in Greek Science*, McGraw-Hill, New York, p. 211.

92. Ibid.

93. Heath, T., 1913, *Aristarchus of Samos, the Ancient Copernicus*, Oxford at the Clarendon Press, London, p. 299.

94. Vitruvius, 1960, *The Ten Books on Architecture*, Book 1, Chapter 1, Paragraph 16, translated by Morris Hicky Morgan (first published by Harvard University Press in 1914), Dover, New York, p. 12.

95. Archimedes, 2002, *The Works of Archimedes*, edited by T. L. Heath, first published as *The Works of Archimedes* (1897, Cambridge University Press) and *The Method of Archimedes* (1912, Cambridge University Press), Dover, Mineola, New York, p. 221–222.

96. Burnet, J., 1920, *Greek Philosophy, Part I, Thales to Plato*, Macmillan & Co., Limited, London, p. 348.
97. Ibid., p. 347.
98. Heath, T., 1913, *Aristarchus of Samos, the Ancient Copernicus*, Oxford at the Clarendon Press, London, p. 175.
99. Ibid., p. 186.
100. Plutarch, 1874, Of the Face Appearing Within the Orb of the Moon, in *Plutarch's Morals, Translated from the Greek by Several Hands, Corrected and Revised by William W. Goodwin*, vol. 5, Little, Brown, and Company, Boston, p. 240.
101. Dreyer, J. L. E., 1953, *A History of Astronomy from Thales to Kepler* (first published in 1906 as *History of the Planetary Systems from Thales to Kepler*), Dover, New York, p. 147.
102. A complete English translation is given by: Heath, T., 1913, *Aristarchus of Samos, the Ancient Copernicus*, Oxford at the Clarendon Press, London, p. 351–414.
103. Aristotle, 1923, *Meteorologica* (*Meteorology*), Book 1, Chapter 8, translated by E. W. Webster, Oxford at the Clarendon Press, London, p. 345b.
104. Heath, T., 1913, *Aristarchus of Samos, the Ancient Copernicus*, Oxford at the Clarendon Press, London, p. 329.
105. Ibid., p. 311.
106. Archimedes, 2002, *The Works of Archimedes*, edited by T. L. Heath, first published as *The Works of Archimedes* (1897, Cambridge University Press) and *The Method of Archimedes* (1912, Cambridge University Press), Dover, Mineola, New York, p. 223.
107. Turcotte, D. L., and Schubert, G., 1982, *Geodynamics, Applications of Continuum Physics to Geological Problems*, John Wiley & Sons, New York, p. 430–431.
108. Heath, T., 1913, *Aristarchus of Samos, the Ancient Copernicus*, Oxford at the Clarendon Press, London, p. 338.
109. Dicks, D. R., 2007, Eratosthenes, in *Complete Dictionary of Scientific Biography*, edited by Charles Gillispie, vol. 4, Cengage Learning, New York, p. 388.
110. Strabo, 1856, *The Geography of Strabo*, Book 1, Chapter 2, Paragraph 2, vol. 1, translated by H. C. Hamilton and W. Falconer, Henry G. Bohn, London, p. 24.
111. Dicks, D. R., 2007, Eratosthenes, in *Complete Dictionary of Scientific Biography*, edited by Charles Gillispie, vol. 4, Cengage Learning, New York, p. 388.
112. Sarton, G., 1959, *A History of Science, Hellenistic Science and Culture in the Last Three Centuries B.C.*, Harvard University Press, Cambridge, p. 100–101.
113. Ibid., p. 101.
114. Ibid.
115. Archimedes, 2002, *The Works of Archimedes*, edited by T. L. Heath, first published as *The Works of Archimedes* (1897, Cambridge University Press) and *The Method of Archimedes* (1912, Cambridge University Press), Dover, Mineola, New York, p. 12 (of *The Method*).
116. Heath, T., 1981, *A History of Greek Mathematics: Volume I, From Thales to Euclid* (first published in 1921 by the Clarendon Press, Oxford), Dover, New York, p. 100.
117. Ibid., p. 244–270.
118. Ibid., p. 247–248.
119. Ibid., p. 244, 260.
120. Bowen, A. C., and Todd. R. B., 2004, *Cleomedes' Lectures on Astronomy*, University of California Press, Berkeley and Los Angeles, p. 78–85; also, Heath, T. L., 1932, *Greek Astronomy*, J. M. Dent & Sons, London, p. 109–112.
121. Aristotle, 1941, *De Caelo* (*On the Heavens*), Book 2, Chapter 14, translated by J. L. Stocks, in *Basic Works of Aristotle*, Random House, New York, p. 437 (298a).
122. Archimedes, 2002, *The Works of Archimedes*, edited by T. L. Heath, first published as *The Works of Archimedes* (1897, Cambridge University Press) and *The Method of Archimedes* (1912, Cambridge University Press), Dover, Mineola, New York, p. 222.
123. Dicks, D. R., 2007, Eratosthenes, in *Complete Dictionary of Scientific Biography*, edited by Charles Gillispie, vol. 4, Cengage Learning, New York, p. 390.
124. Strabo, 1856, *The Geography of Strabo*, Book 2, Chapter 7, Paragraph 7, vol. 1, translated by H. C. Hamilton and W. Falconer, Henry G. Bohn, London, p. 171.
125. Bunbury, E. H., 1883, *A History of Ancient Geography*, Second Edition, vol. 1, John Murray, London, p. 623.
126. Engels, D., 1985, The Length of Erastosthenes Stade, *American Journal of Philology*, vol. 106, no. 3, p. 298.
127. Sarton, G., 1959, *A History of Science, Hellenistic Science and Culture in the Last Three Centuries B.C.*: Harvard University Press, Cambridge, p. 105.
128. Bunbury, E. H., 1883, *A History of Ancient Geography*, Second Edition, vol. 1, John Murray, London, p. 623.
129. Dicks, D. R., 2007, Eratosthenes, in *Complete Dictionary of Scientific Biography*, edited by Charles Gillispie, vol. 4, Cengage Learning, New York, p. 390.
130. Ibid., p. 389.
131. Bunbury, E. H., 1883, *A History of Ancient Geography*, Second Edition, vol. 1, John Murray, London, p. 633.
132. Ibid., p. 632.
133. Strabo, 1856, *The Geography of Strabo*, Book 2, Chapter 1, Paragraph 1, vol. 1, translated by H. C. Hamilton and W. Falconer, Henry G. Bohn, London, p. 105.
134. Bunbury, E. H., 1883, *A History of Ancient Geography*, Second Edition, vol. 1, John Murray, London, p. 625.
135. Strabo, 1856, *The Geography of Strabo*, Book 1, Chapter 4, Paragraphs 2 and 5, vol. 1, translated by H. C. Hamilton and W. Falconer, Henry G. Bohn, London, p. 99, 101.
136. Ibid., Book 2, Chapter 5, Paragraph 34, vol. 1, p. 199.
137. Ibid., Book 1, Chapter 4, Paragraph 6, vol. 1, p. 101.
138. Ibid., Book 1, Chapter 4, Paragraph 6, vol. 1, p. 102.
139. According to Edward Grant, Columbus relied upon the book *Imago Mundi* (*Image of the World*), by Pierre d'Ailly (1351–1420 A.D.), which gave a value of 20,400 miles (32,831 kilometers) for the Earth's circumference, a figure eighteeen percent lower than the correct circumference of 40,030 kilometers (Grant, E., 1977, *Physical Science in the Middle Ages*, Cambridge University Press, Cambridge, p. 62–63.) And Pierre d'Ailly was evidently quoting from Roger Bacon's *Opus Majus*: "Aristotle says that the sea is small between the end of Spain on the west and the beginning of India on the east. Seneca in the fifth book on Natural History says that this sea is navigable in a very few days if the wind is favorable." (Bacon, R., 1928, *The Opus Majus of Roger Bacon*, vol. 1, translated by Robert Belle Burke, University of Pennsylvania Press, Philadelphia, p. 311).
140. Bowen, A. C., and Todd. R. B., 2004, *Cleomedes' Lectures on Astronomy*, University of California Press, Berkeley and Los Angeles, p. 80.
141. Bunbury, E. H., 1883, *A History of Ancient Geography*, Second Edition, vol. 2, John Murray, London, p. 95–96.
142. Strabo, 1856, *The Geography of Strabo*, Book 1, Chapter 4, Paragraph 9, vol. 1, translated by H. C. Hamilton and W. Falconer, Henry G. Bohn, London, p. 104.
143. Grote, G., 1899, *A History of Greece*, vol. 3, Harper & Brothers, New York, p. 176.
144. Thucydides, 1942, *The Peloponnesian War*, Book 6, Chapter 3, translated by Benjamin Jowett (1817–1893), in *The Greek Historians*, vol. 1, Random House, New York, p. 852.
145. Archimedes, 2002, *The Works of Archimedes*, edited by T. L. Heath, first published as *The Works of Archimedes* (1897, Cambridge University Press) and *The Method of Archimedes* (1912, Cambridge University Press), Dover, Mineola, New York, p. xvi.
146. Ibid., p. xv.
147. Ibid., p. xiv.
148. Ibid., p. 93.
149. Ibid., p. 1.
150. Ibid., p. cxlii–cliv.
151. Heath, T. L., 1981, *A History of Greek Mathematics*, vol. 2 (first published in 1921 by the Clarendon Press, Oxford), Dover, New York, p. 20.
152. Archimedes, 2002, *The Works of Archimedes*, edited by T. L. Heath, first published as *The Works of Archimedes* (1897, Cambridge University Press) and *The Method of Archimedes* (1912, Cambridge University Press), Dover, Mineola, New York, p. 151.
153. Ibid., p. 221.
154. Ibid., p. 232.

155. Anonymous, 1938, Introducing the Googol, Prof. Kasner in Kindergarten Talk Also Presents the Googolplex, *New York Times*, February 27, p. 41.
156. Archimedes, 2002, *The Works of Archimedes*, edited by T. L. Heath, first published as *The Works of Archimedes* (1897, Cambridge University Press) and *The Method of Archimedes* (1912, Cambridge University Press), Dover, Mineola, New York, p. 13 (of *The Method*).
157. Heath, T. L., 1981, *A History of Greek Mathematics*, vol. 2 (first published in 1921 by the Clarendon Press, Oxford), Dover, New York, p. 20.
158. Archimedes, 2002, *The Works of Archimedes*, edited by T. L. Heath, first published as *The Works of Archimedes* (1897, Cambridge University Press) and *The Method of Archimedes* (1912, Cambridge University Press), Dover, Mineola, New York, p. xl.
159. Ibid., p. 257.
160. Vitruvius, 1960, *The Ten Books on Architecture*, Book 9, Introduction, Paragraphs 9, 10, translated by Morris Hicky Morgan (first published by Harvard University Press in 1914), Dover, New York, p. 253–254.
161. Cicero, Marcus Tullius, 1853, On the Republic, Book 1, Chapter 14, in *The Treatises of M. T. Cicero*, translated by C. D. Yonge, H. G. Bohn, London, p. 295.
162. Ibid.
163. Ibid.
164. Ibid.
165. Ibid.
166. Price, D. D. S., 1974, Gears from the Greeks. The Antikythera Mechanism: a Calendar Computer from ca. 80 B.C.: *Transactions of the American Philosophical Society*, New Series, vol. 64, no. 7, p. 1–70.
167. Ibid., p. 8.
168. Freeth, T., et al., 2006, Decoding the Ancient Greek Astronomical Calculator Known as the Antikythera Mechanism, *Nature*, vol. 44, 30 November, p. 587.
169. Ibid.
170. Freeth, T., et al., 2008, Calendars with Olympiad Display and Eclipse Prediction on the Antikythera Mechanism, *Nature*, vol. 454, 31 July, p. 614.
171. Diodorus Siculus, 1814, *The Historical Library of Diodorus the Sicilian*, Book 5, Chapter 2, vol. 1, translated by G. Booth, W. M'Dowall for J. Davis, London, p. 321.
172. Athenaeus, 1854, *The Deipnosophists, or Banquet of the Learned*, Book 5, Chapter 43, translated by C. D. Yonge, vol. 1, Henry G. Bohn, London, p. 332.
173. Archimedes, 2002, *The Works of Archimedes*, edited by T. L. Heath, first published as *The Works of Archimedes* (1897, Cambridge University Press) and *The Method of Archimedes* (1912, Cambridge University Press), Dover, Mineola, New York, p. xxxvii.
174. Diodorus of Sicily, 1957, *Diodorus of Sicily*, Book 26, Chapter 18, translated into English by Francis R. Walton, vol. 11, Harvard University Press, Cambridge, p. 195.
175. Plutarch, 1952, Marcellus, in *The Lives of the Noble Grecians and Romans*, translated by John Dryden (1631–1700), *Great Books of the Western World*, vol. 14, William Benton, Chicago, p. 252.
176. Ibid.
177. Grote, G., 1899, *A History of Greece*, vol. 3, Harper & Brothers, New York, p. 159.
178. Mommsen, T., 1894, *The History of Rome*, Book 3, Chapter 1, translated by William Purdie Dickson, vol. 2, Richard Bentley, London, p. 136.
179. Ibid., p. 132.
180. Creasy, E. S., 1908, *The Fifteen Decisive Battles of the World from Marathon to Waterloo*, Everyman's Library Edition, E. P. Dutton, New York, p. 88–89.
181. Sarton, G., 1959, *A History of Science, Hellenistic Science and Culture in the Last Three Centuries B.C.*, Harvard University Press, Cambridge, p. 68–69.
182. Polybius, 1889, *The Histories of Polybius*, Book 1, Chapter 62, vol. 1, translated by Evelyn S. Shuckburgh, Macmillan and Co., London, p. 71.
183. Ibid., Book 1, Chapter 63, vol. 1, p. 71.
184. Ibid., Book 1, Chapter 63, vol. 1, p. 71–72.
185. Ibid., p. 72.
186. Niebuhr, B. G., 1849, *Lectures on the History of Rome*, Second Edition, vol. 2, Taylor, Walton, and Maberly, London, p. 17.
187. Ibid., p. 54.
188. Ibid., p. 54.
189. Ibid., p. 68.
190. Polybius, 1889, *The Histories of Polybius*, Book 2, Chapter 1, vol. 1, translated by Evelyn S. Shuckburgh, Macmillan and Co., London, p. 99.
191. Ibid., Book 3, Chapter 11, vol. 1, p. 175–176.
192. Livius, T., 1849, *The History of Rome*, Books Nine to Twenty-Six, Book 21, Chapter 4, translated by D. Spillan and Cyrus Edmonds, Henry G. Bohn, London, p. 703.
193. Ibid.
194. Ibid.
195. Ibid., Book 21, Chapter 1, p. 700.
196. Polybius, 1889, *The Histories of Polybius*, Book 3, Chapters 50–51, vol. 1, translated by Evelyn S. Shuckburgh, Macmillan and Co., London, p. 210–211.
197. Ibid., Book 3, Chapter 53, vol. 1, p. 213.
198. Ibid., Book 3, Chapters 56, 60, vol. 1, p. 217, 219.
199. Ibid., Book 3, Chapter 60, vol. 1, p. 219.
200. Ibid., Book 3, Chapter 56, vol. 1, p. 216.
201. Niebuhr, B. G., 1849, *Lectures on the History of Rome*, Second Edition, vol. 2, Taylor, Walton, and Maberly, London, p. 95.
202. Ibid., p. 96–97.
203. Polybius, 1889, *The Histories of Polybius*, Book 3, Chapter 80, vol. 1, translated by Evelyn S. Shuckburgh, Macmillan and Co., London, p. 239.
204. Ibid., Book 3, Chapter 82, vol. 1, p. 240–241.
205. Ibid., Book 3, Chapter 84, vol. 1, p. 242.
206. Ibid.
207. Livius, T., 1849, *The History of Rome*, Books Nine to Twenty-Six, Book 22, Chapter 7, translated by D. Spillan and Cyrus Edmonds, Henry G. Bohn, London, p. 773.
208. Niebuhr, B. G., 1849, *Lectures on the History of Rome*, Second Edition, vol. 2, Taylor, Walton, and Maberly, London, p. 101.
209. Livius, T., 1849, *The History of Rome*, Books Nine to Twenty-Six, Book 22, Chapter 1, translated by D. Spillan and Cyrus Edmonds, Henry G. Bohn, London, p. 766.
210. Plutarch, 1952, Marcellus, in *The Lives of the Noble Grecians and Romans*, translated by John Dryden (1631–1700), *Great Books of the Western World*, vol. 14, William Benton, Chicago, p. 247.
211. Ibid., p. 247–248.
212. Polybius, 1889, *The Histories of Polybius*, Book 3, Chapter 86, vol. 1, translated by Evelyn S. Shuckburgh, Macmillan and Co., London, p. 244–245.
213. Niebuhr, B. G., 1849, *Lectures on the History of Rome*, Second Edition, vol. 2, Taylor, Walton, and Maberly, London, p. 105.
214. Polybius, 1889, *The Histories of Polybius*, Book 3, Chapter 87, vol. 1, translated by Evelyn S. Shuckburgh, Macmillan and Co., London, p. 245.
215. Ibid., Book 3, Chapter 86, vol. 1, p. 245.
216. Ibid., Book 3, Chapter 89, vol. 1, p. 247.
217. Ibid., Book 3, Chapters 113–114, vol. 1, p. 269–270.
218. Ibid., Book 3, Chapter 115, vol. 1, p. 270.
219. Niebuhr, B. G., 1849, *Lectures on the History of Rome*, Second Edition, vol. 2, Taylor, Walton, and Maberly, London, p. 112.
220. Polybius, 1889, *The Histories of Polybius*, Book 3, Chapters 115–116, vol. 1, translated by Evelyn S. Shuckburgh, Macmillan and Co., London, p. 271–272.
221. Livius, T., 1849, *The History of Rome*, Books Nine to Twenty-Six, Book 22, Chapter 49, translated by D. Spillan and Cyrus Edmonds, Henry G. Bohn, London, p. 819.
222. Polybius, 1889, *The Histories of Polybius*, Book 3, Chapters 117, vol. 1, translated by Evelyn S. Shuckburgh, Macmillan and Co., London, p. 273.
223. Livius, T., 1849, *The History of Rome*, Books Nine to

Twenty-Six, Book 22, Chapter 49, translated by D. Spillan and Cyrus Edmonds, Henry G. Bohn, London, p. 818.

224. Polybius, 1889, *The Histories of Polybius*, Book 3, Chapter 117, vol. 1, translated by Evelyn S. Shuckburgh, Macmillan and Co., London, p. 273.

225. Livius, T., 1849, *The History of Rome*, Books Nine to Twenty-Six, Book 22, Chapter 51, translated by D. Spillan and Cyrus Edmonds, Henry G. Bohn, London, p. 821.

226. Polybius, 1889, *The Histories of Polybius*, Book 6, Chapter 58, vol. 1, translated by Evelyn S. Shuckburgh, Macmillan and Co., London, p. 508.

227. Niebuhr, B. G., 1849, *Lectures on the History of Rome*, Second Edition, vol. 2, Taylor, Walton, and Maberly, London, p. 120.

228. Polybius, 1889, *The Histories of Polybius*, Book 7, Chapter 8, vol. 1, translated by Evelyn S. Shuckburgh, Macmillan and Co., London, p. 514.

229. Livius, T., 1849, *The History of Rome*, Books Nine to Twenty-Six, Book 24, Chapter 4, translated by D. Spillan and Cyrus Edmonds, Henry G. Bohn, London, p. 900–901.

230. Ibid., Book 24, Chapter 5, p. 901.

231. Polybius, 1889, *The Histories of Polybius*, Book 7, Chapter 7, vol. 1, translated by Evelyn S. Shuckburgh, Macmillan and Co., London, p. 513.

232. Ibid.

233. Niebuhr, B. G., 1849, *Lectures on the History of Rome*, Second Edition, vol. 2, Taylor, Walton, and Maberly, London, p. 125.

234. Polybius, 1889, *The Histories of Polybius*, Book 7, Chapter 4, vol. 1, translated by Evelyn S. Shuckburgh, Macmillan and Co., London, p. 511.

235. Livius, T., 1849, *The History of Rome*, Books Nine to Twenty-Six, Book 24, Chapter 7, translated by D. Spillan and Cyrus Edmonds, Henry G. Bohn, London, p. 903–904.

236. Plutarch, 1952, Marcellus, in *The Lives of the Noble Grecians and Romans*, translated by John Dryden (1631–1700), *Great Books of the Western World*, vol. 14, William Benton, Chicago, p. 251.

237. Niebuhr, B. G., 1849, *Lectures on the History of Rome*, Second Edition, vol. 2, Taylor, Walton, and Maberly, London, p. 126.

238. Livius, T., 1849, *The History of Rome*, Books Nine to Twenty-Six, Book 24, Chapter 25, translated by D. Spillan and Cyrus Edmonds, Henry G. Bohn, London, p. 926.

239. Ibid., Book 24, Chapter 29, p. 930.

240. Ibid., Book 24, Chapter 30, p. 932.

241. Ibid.

242. Plutarch, 1952, Marcellus, in *The Lives of the Noble Grecians and Romans*, translated by John Dryden (1631–1700), *Great Books of the Western World*, vol. 14, William Benton, Chicago, p. 252.

243. Polybius, 1889, *The Histories of Polybius*, Book 8, Chapter 5, vol. 1, translated by Evelyn S. Shuckburgh, Macmillan and Co., London, p. 530.

244. Plutarch, 1952, Marcellus, in *The Lives of the Noble Grecians and Romans*, translated by John Dryden (1631–1700), *Great Books of the Western World*, vol. 14, William Benton, Chicago, p. 252–253.

245. Archimedes, 2002, T*he Works of Archimedes*, edited by T. L. Heath, first published as *The Works of Archimedes* (1897, Cambridge University Press) and *The Method of Archimedes* (1912, Cambridge University Press), Dover, Mineola, New York, p. xxi–xxii.

246. Lucian, 1913, Hippias, or The Bath, in *Lucian* with an English translation by A. M. Harmon, William Heinemann, London, p. 37.

247. Tzetzes, J., 1968, Book of Histories, Book 2, Chapters 103–144, in *Greek Mathematical Works*, Chapter 17, Archimedes, Part (a), translated into English by Ivor Thomas, vol. 2, Harvard University Press, Cambridge, p. 19–20.

248. Middleton, W. E. K., 1961, Archimedes, Kircher, Buffon, and the Burning-Mirrors, *Isis*, vol. 52, no. 4, p. 533.

249. Simms, D. L., 1977, Archimedes and the Burning Mirrors of Syracuse: *Technology and Culture*, vol. 18, no. 1, p. 22.

250. Middleton, W. E. K., 1961, Archimedes, Kircher, Buffon, and the Burning-Mirrors, *Isis*, vol. 52, no. 4, p. 540.

251. Modiano, M. S., 1973, How Archimedes Stole Sun to Burn Foe's Fleet: *New York Times*, 11 November, p. 16.

252. Plutarch, 1952, Marcellus, in *The Lives of the Noble Grecians and Romans*, translated by John Dryden (1631–1700), *Great Books of the Western World*, vol. 14, William Benton, Chicago, p. 253.

253. Homer, 1729, *The Illiad of Homer*, Book 1, verses 522–529, translated by Mr. Pope, vol. 1, printed by T. J. for B. L. & Company, London, p. 21.

254. Polybius, 1889, *The Histories of Polybius*, Book 8, Chapter 9, vol. 1, translated by Evelyn S. Shuckburgh, Macmillan and Co., London, p. 534.

255. Plutarch, 1952, Marcellus, in *The Lives of the Noble Grecians and Romans*, translated by John Dryden (1631–1700), *Great Books of the Western World*, vol. 14, William Benton, Chicago, p. 253.

256. Ibid., p. 254.

257. Polybius, 1889, *The Histories of Polybius*, Book 8, Chapter 9, vol. 1, translated by Evelyn S. Shuckburgh, Macmillan and Co., London, p. 534.

258. Plutarch, 1952, Marcellus, in *The Lives of the Noble Grecians and Romans*, translated by John Dryden (1631–1700), *Great Books of the Western World*, vol. 14, William Benton, Chicago, p. 254.

259. Ibid.

260. Polybius, 1889, *The Histories of Polybius*, Book 8, Chapter 37, vol. 1, translated by Evelyn S. Shuckburgh, Macmillan and Co., London, p. 561.

261. Livius, T., 1849, *The History of Rome*, Books Nine to Twenty-Six, Book 25, Chapter 24, translated by D. Spillan and Cyrus Edmonds, Henry G. Bohn, London, p. 993.

262. Ibid., Book 25, Chapter 25, p. 993–994.

263. Plutarch, 1952, Marcellus, in *The Lives of the Noble Grecians and Romans*, translated by John Dryden (1631–1700), *Great Books of the Western World*, vol. 14, William Benton, Chicago, p. 254–255.

264. Ibid., p. 254.

265. Cicero, M. T., 1758, *The Tusculan Disputations of Marcus Tullius Cicero*, Book 5, Chapter 23, printed for John Whiston and Benj. White, London, p. 296.

266. Niebuhr, B. G., 1849, *Lectures on the History of Rome*, Second Edition, vol. 2, Taylor, Walton, and Maberly, London, p. 128.

267. Polybius, 1889, *The Histories of Polybius*, Book 6, Chapter 52, vol. 1, translated by Evelyn S. Shuckburgh, Macmillan and Co., London, p. 502.

268. Creasy, E. S., 1908, *The Fifteen Decisive Battles of the World from Marathon to Waterloo*, Everyman's Library Edition, E. P. Dutton, New York, p. 95.

269. Niebuhr, B. G., 1849, *Lectures on the History of Rome*, Second Edition, vol. 2, Taylor, Walton, and Maberly, London, p. 132.

270. Livius, T., 1850, *The History of Rome*, Books Twenty-Seven to Thirty-Six, Book 27, Chapter 34, translated by Cyrus Edmonds, Henry G. Bohn, London, p. 1138.

271. Ibid., p. 1139.

272. Ibid.

273. Ibid., Book 27, Chapter 40, p. 1146.

274. Ibid., Book 27, Chater 43, p. 1149–1150.

275. Ibid., p. 1150.

276. Ibid., Book 27, Chapter 45, p. 1152.

277. Ibid., Book 27, Chapters, 45, 46, p. 1153.

278. Ibid., Book 27, Chapter 44, p. 1151.

279. Ibid.

280. Creasy, E. S., 1908, *The Fifteen Decisive Battles of the World from Marathon to Waterloo*, Everyman's Library Edition, E. P. Dutton, New York, p. 106–107.

281. Ibid., p. 107.

282. Livius, T., 1850, *The History of Rome*, Books Twenty-Seven to Thirty-Six, Book 27, Chapter 48, translated by Cyrus Edmonds, Henry G. Bohn, London, p. 1156.

283. Ibid., Book 27, Chapter 48, p. 1157.

284. Polybius, 1889, *The Histories of Polybius*, Book 11,

Chapter 1, vol. 2, translated by Evelyn S. Shuckburgh, Macmillan and Co., London, p. 50.

285. Ibid., Book 11, Chapter 3, vol. 2, p. 52.

286. Creasy, E. S., 1908, *The Fifteen Decisive Battles of the World from Marathon to Waterloo*, Everyman's Library Edition, E. P. Dutton, New York, p. 111.

287. Livius, T., 1850, *The History of Rome*, Books Twenty-Seven to Thirty-Six, Book 27, Chapters 50, 51, translated by Cyrus Edmonds, Henry G. Bohn, London, p. 1158, 1160.

288. Creasy, E. S., 1908, *The Fifteen Decisive Battles of the World from Marathon to Waterloo*, Everyman's Library Edition, E. P. Dutton, New York, p. 111–112.

289. Livius, T., 1850, *The History of Rome*, Books Twenty-Seven to Thirty-Six, Book 30, Chapter 4, translated by Cyrus Edmonds, Henry G. Bohn, London, p. 1286.

290. Polybius, 1889, *The Histories of Polybius*, Book 14, Chapter 4, vol. 2, translated by Evelyn S. Shuckburgh, Macmillan and Co., London, p. 127.

291. Ibid., Book 14, Chapter 4, vol. 2, p. 128.

292. Livius, T., 1850, *The History of Rome*, Books Twenty-Seven to Thirty-Six, Book 30, Chapter 6, translated by Cyrus Edmonds, Henry G. Bohn, London, p. 1289.

293. Polybius, 1889, *The Histories of Polybius*, Book 14, Chapter 5, vol. 2, translated by Evelyn S. Shuckburgh, Macmillan and Co., London, p. 129.

294. Livius, T., 1850, *The History of Rome*, Books Twenty-Seven to Thirty-Six, Book 30, Chapter 16, translated by Cyrus Edmonds, Henry G. Bohn, London, p. 1302–1303.

295. Ibid., Book 30, Chapter 16, p. 1303.

296. Ibid., Book 30, Chapter 20, p. 1308.

297. Polybius, 1889, *The Histories of Polybius*, Book 15, Chapter 9, vol. 2, translated by Evelyn S. Shuckburgh, Macmillan and Co., London, p. 144.

298. Ibid., Book 15, Chapter 12, vol. 2, p. 147.

299. Ibid., Book 15, Chapter 13, vol. 2, p. 147.

300. Ibid., p. 147–148.

301. Ibid., p. 148.

302. Ibid., Book 15, Chapter 14, vol. 2, p. 148

303. Ibid.

304. Ibid., p. 149.

305. Livius, T., 1850, *The History of Rome*, Books Twenty-Seven to Thirty-Six, Book 30, Chapter 36, translated by Cyrus Edmonds, Henry G. Bohn, London, p. 1329.

306. Ibid., Book 30, Chapter 37, p. 1330.

307. Ibid., Book 33, Chapter 46, p. 1486.

308. Ibid.

309. Ibid., Book 33, Chapter 47, p. 1487.

310. Ibid.

311. Ibid., Book 33, Chapter 48, p. 1488.

312. Livius, T., 1890, *The History of Rome*, Books Thirty-Seven to the End, Book 39, Chapter 51, translated by William A. M'Devitte, George Bell and Sons, London, p. 1848.

313. Appian, 1912, *Appian's Roman History*, Book 8, Chapter 10, Paragraph 67, translated into English by Horace White, vol. 1, William Heinemann, London, p. 509.

314. Ibid., p. 509–510.

315. Ibid., Book 8, Chapter 10, Paragraph 69, p. 514–515.

316. Livius, T., 1890, *The History of Rome*, Books Thirty-Seven to the End, Book 39, Chapter 40, translated by William A. M'Devitte, George Bell and Sons, London, p. 1837.

317. Plutarch, 1952, Marcus Cato, in *The Lives of the Noble Grecians and Romans*, translated by John Dryden (1631–1700), *Great Books of the Western World*, vol. 14, William Benton, Chicago, p. 278.

318. Ibid., p. 279.

319. Ibid., p. 284.

320. Livius, T., 1890, *The History of Rome*, Books Thirty-Seven to the End, Book 39, Chapter 44, translated by William A. M'Devitte, George Bell and Sons, London, p. 1841.

321. Plutarch, 1952, Marcus Cato, in *The Lives of the Noble Grecians and Romans*, translated by John Dryden (1631–1700), *Great Books of the Western World*, vol. 14, William Benton, Chicago, p. 284.

322. Frazer, J. G., 1920, *The Golden Bough*, vol. 3, *The Dying God* (first published in 1911), Macmillan, London, p. 75.

323. Plutarch, 1870, On Superstition, in *Plutarch's Morals, Translated from the Greek by Several Hands, Corrected and Revised by William W. Goodwin*, vol. 1, Little, Brown, and Company, Boston, p. 183.

324. Diodorus Siculus, 1814, *The Historical Library of Diodorus the Sicilian*, Book 20, Chapter 1, vol. 2, translated by G. Booth, W. M'Dowall for J. Davis, London, p. 422.

325. Bible, King James Version, I Kings 18:40.

326. Mommsen, T., 1894, *The History of Rome*, translated by William Purdie Dickson, vol. 3, Richard Bentley & Son, London, p. 238.

327. Plutarch, 1952, Marcus Cato, in *The Lives of the Noble Grecians and Romans*, translated by John Dryden (1631–1700), *Great Books of the Western World*, vol. 14, William Benton, Chicago, p. 290.

328. Mommsen, T., 1894, *The History of Rome*, translated by William Purdie Dickson, vol. 3, Richard Bentley & Son, London, p. 240.

329. Polybius, 1889, *The Histories of Polybius*, Book 38, Chapter 1, vol. 2, translated by Evelyn S. Shuckburgh, Macmillan and Co., London, p. 513.

330. Niebuhr, B. G., 1849, *Lectures on the History of Rome*, Second Edition, vol. 2, Taylor, Walton, and Maberly, London, p. 232–234.

331. Appian, 1912, *Appian's Roman History*, Book 8, Chapter 11, Paragraph 75, translated into English by Horace White, vol. 1, William Heinemann, London, p. 527.

332. Ibid., Book 8, Chapter 11, Paragraph 76, p. 527.

333. Mommsen, T., 1894, *The History of Rome*, translated by William Purdie Dickson, vol. 3, Richard Bentley & Son, London, p. 242.

334. Ibid., p. 243.

335. Appian, 1912, *Appian's Roman History*, Book 8, Chapter 11, Paragraph 76, translated into English by Horace White, vol. 1, William Heinemann, London, p. 529.

336. Ibid., Book 8, Chapter 11, Paragraph 77, p. 529.

337. Ibid., Book 8, Chapter 12, Paragraph 80, p. 537.

338. Ibid., Book 8, Chapter 12, Paragraph 81, p. 537–539.

339. Ibid., p. 539.

340. Ibid., Book 8, Chapter 12, Paragraph 86, p. 547.

341. Ibid., Book 8, Chapter 12, Paragraph 88, p. 551.

342. Ibid., Book 8, Chapter 13, Paragraph 92, p. 559–560.

343. Ibid., Book 8, Chapter 13, Paragraph 93, p. 561–563.

344. Appian, 1912, *Appian's Roman History*, Book 8, Chapter 14, Paragraph 95, translated into English by Horace White, vol. 1, William Heinemann, London, p. 565.

345. Mommsen, T., 1894, *The History of Rome*, translated by William Purdie Dickson, vol. 3, Richard Bentley & Son, London, p. 243.

346. Ibid., p. 254.

347. Appian, 1912, *Appian's Roman History*, Book 8, Chapter 18, Paragraph 117, translated into English by Horace White, vol. 1, William Heinemann, London, p. 609.

348. Ibid., Book 8, Chapter 18, Paragraph 118, p. 611.

349. Mommsen, T., 1894, *The History of Rome*, translated by William Purdie Dickson, vol. 3, Richard Bentley & Son, London, p. 256.

350. Appian, 1912, *Appian's Roman History*, Book 8, Chapter 19, Paragraph 127, translated into English by Horace White, vol. 1, William Heinemann, London, p. 629.

351. Mommsen, T., 1894, *The History of Rome*, translated by William Purdie Dickson, vol. 3, Richard Bentley & Son, London, p. 256.

352. Appian, 1912, *Appian's Roman History*, Book 8, Chapter 19, Paragraphs 128–129, translated into English by Horace White, vol. 1, William Heinemann, London, p. 631.

353. Ibid., Book 8, Chapter 19, Paragraph 131, p. 635–637.

354. Mommsen, T., 1894, *The History of Rome*, translated by William Purdie Dickson, vol. 3, Richard Bentley & Son, London, p. 256–257.

355. Ibid., p. 257.

356. Ibid., p. 258.

357. Dreyer, J. L. E., 1953, *A History of Astronomy from Thales to Kepler* (first published in 1906 as *History of the Planetary Systems from Thales to Kepler*), Dover, New York, p. 147.

358. Toomer, G. J., 2007, Hipparchus, in *Complete Dictionary of Scientific Biography*, edited by Charles Gillispie, vol. 15, Cengage Learning, New York, p. 208.
359. Ibid., p. 209.
360. Huxley, G. L., 2007, Sosigenes, in *Complete Dictionary of Scientific Biography*, edited by Charles Gillispie, vol. 12, Cengage Learning, New York, p. 547.
361. Heath, T., 1913, *Aristarchus of Samos, the Ancient Copernicus*, Oxford at the Clarendon Press, London, p. 221–222.
362. Ibid., p. 254–255.
363. Heath, T. L., 1932, *Greek Astronomy*, J. M. Dent & Sons, London, p. 93.
364. Dreyer, J. L. E., 1953, *A History of Astronomy from Thales to Kepler* (first published in 1906 as *History of the Planetary Systems from Thales to Kepler*), Dover, New York, p. 152.
365. Toomer, G. J., 2007, Hipparchus, in *Complete Dictionary of Scientific Biography*, edited by Charles Gillispie, vol. 15, Cengage Learning, New York, p. 211–215.
366. Ibid., p. 212.
367. Pliny the Elder, 1855, *The Natural History of Pliny*, translated by John Bostock and H. T. Riley, Book 2, Chapters 9 and 10, vol. 1, Henry G. Bohn, London, p. 37, 29.
368. Toomer, G. J., 2007, Hipparchus, in *Complete Dictionary of Scientific Biography*, edited by Charles Gillispie, vol. 15, Cengage Learning, New York, p. 215.
369. Ibid.
370. Ptolemy, C., 1952, *The Almagest*, Book 9, Chapter 2, translated by R. Catesby Taliaferro, *Great Books of the Western World*, vol. 16, William Benton, Chicago, p. 272.
371. Sarton, G., 1959, *A History of Science, Hellenistic Science and Culture in the Last Three Centuries B.C.*, Harvard University Press, Cambridge, p. 299.
372. Toomer, G. J., 2007, Hipparchus, in *Complete Dictionary of Scientific Biography*, edited by Charles Gillispie, vol. 15, Cengage Learning, New York, p. 218.
373. Ibid.
374. Heath, T. L., 1932, *Greek Astronomy*, J.M. Dent & Sons, London, p. 143–144.
375. Ibid., p. 119.
376. Pliny the Elder, 1855, *The Natural History of Pliny*, translated by John Bostock and H. T. Riley, Book 2, Chapter 24, vol. 1, Henry G. Bohn, London, p. 59.
377. Ptolemy, C., 1952, *The Almagest*, Book 8, Chapter 1, translated by R. Catesby Taliaferro, *Great Books of the Western World*, vol. 16, William Benton, Chicago, p. 258.
378. Toomer, G. J., 2007, Hipparchus, in *Complete Dictionary of Scientific Biography*, edited by Charles Gillispie, vol. 15, Cengage Learning, New York, p. 216.
379. Toomer, G. J., 2008, Ptolemy, in *Complete Dictionary of Scientific Biography*, edited by Charles Gillispie, vol. 11, Cengage Learning, New York, p. 186.
380. Ibid., p. 187, 202.
381. Ibid., p. 196.
382. Ptolemy, C., 1984, *Ptolemy's Almagest*, Book 1, Chapter 1, translated by G. J. Toomer, Springer-Verlag, New York, p. 37.
383. Ibid., Book 1, Chapter 2, p. 37.
384. Ibid., Book 1, Chapter 3, p. 39.
385. Ibid., Book 1, Chapter 4, p. 40.
386. Ibid., Book 1, Chapter 4, p. 41.
387. Heath, T. L., 1932, *Greek Astronomy*, J.M. Dent & Sons, London, p. 145.
388. Ibid., p. 147–148.
389. Ibid., p. 148.
390. Dreyer, J. L. E., 1953, *A History of Astronomy from Thales to Kepler* (first published in 1906 as *History of the Planetary Systems from Thales to Kepler*), Dover, New York, p. 191.
391. Toomer, G. J., 2008, Ptolemy, in *Complete Dictionary of Scientific Biography*, edited by Charles Gillispie, vol. 11, Cengage Learning, New York, p. 196.
392. Bayle, P., 1710, *An Historical and Critical Dictionary. By Monsieur Bayle. Translated into English, with Many Additions and Corrections, Made by the Author*, vol. 2, C. Harper and Others, London, p. 901.
393. Ptolemy, C., 1984, *Ptolemy's Almagest*, Book 3, Chapter 1, translated by G. J. Toomer, Springer-Verlag, New York, p. 140.
394. Ibid., Book 9, Chapter 2, p. 420.
395. Ibid., Book 3, Chapter 1, p. 136.
396. Koestler, A., 1963, *The Sleepwalkers*, Grosset & Dunlap, New York, p. 69.
397. Ptolemy, C., 1984, *Ptolemy's Almagest*, Book 13, Chapter 2, translated by G. J. Toomer, Springer-Verlag, New York, p. 600.
398. Ibid.
399. Ibid., p. 602.
400. Dreyer, J. L. E., 1953, *A History of Astronomy from Thales to Kepler* (first published in 1906 as *History of the Planetary Systems from Thales to Kepler*), Dover, New York, p. 196.
401. Toomer, G. J., 2008, Ptolemy, in *Complete Dictionary of Scientific Biography*, edited by Charles Gillispie, vol. 11, Cengage Learning, New York, p. 198.
402. Ptolemy, C., 1822, *Ptolemy's Tetrabiblos or Quadripartite: Being Four Books on the Influence of the Stars*, Book 1, Chapter 2, Davis and Dickson, London, p. 2.
403. Ibid., p. 3–4.
404. Ibid., p. 4.
405. Petronius, 1902, *The Satyricon of Petronius*, Charles Carrington, Paris, p. 96.
406. Ibid., p. 108–109.
407. Pliny the Elder, 1855, *The Natural History of Pliny*, translated by John Bostock and H. T. Riley, Book 2, Chapter 6, vol. 1, Henry G. Bohn, London, p. 26.
408. Clarke, J., 1910, *Physical Science in the Time of Nero, Being a Translation of the Quaestiones Naturales of Seneca*, Book 2, Chapter 32, Macmillan, London, p. 80–81.
409. Gibbon, E., 1909, *The History of the Decline and Fall of the Roman Empire*, Chapter 31, edited by J. B. Bury, vol. 3, Methuen & Co., London, p. 318.
410. Roberts, S. V., 1988, White House Confirms Reagans Follow Astrology, Up to a Point, *New York Times*, May 4, p. A-1.
411. Bunbury, E. H., 1883, *A History of Ancient Geography*, Second Edition, Chapter 28, vol. 2, John Murray, London, p. 549.
412. Ibid., p. 554.
413. Ibid.
414. Ibid., p. 551.
415. Ibid., p. 553.
416. *Encyclopædia Britannica*, Eleventh Edition, 1911, Ptolemy, vol. 22, Encyclopædia Britannica Company, New York, p. 623.
417. Bunbury, E. H., 1883, *A History of Ancient Geography*, Second Edition, Chapter 29, vol. 2, John Murray, London, p. 636.
418. Ptolemy, 2000, *Harmonics*, translated by Jon Solomon, Brill, Leiden, p. 8.
419. Ibid., p. 7–8.
420. Toomer, G. J., 2008, Ptolemy, in *Complete Dictionary of Scientific Biography*, edited by Charles Gillispie, vol. 11, Cengage Learning, New York, p. 200.
421. Neuburger, A., 1930, *The Technical Arts and Sciences of the Ancients*, Macmillan, New York, p. 152, 158.
422. Ibid., p. 163–164.
423. Charleston, R. J., and Angus-Butterworth, L. M., 1957, "Glass," in *A History of Technology*, vol. 3, edited by Charles Singer, Oxford University Press, London, p. 229.
424. Derry, T. K., and Williams, T. I., 1960, *A Short History of Technology*, Dover, New York, p. 112.
425. Charleston, R. J., and Angus-Butterworth, L. M., 1957, "Glass," in *A History of Technology*, vol. 3, edited by Charles Singer, Oxford University Press, London, p. 231.
426. Dreyer, J. L. E., 1953, *A History of Astronomy from Thales to Kepler* (first published in 1906 as *History of the Planetary Systems from Thales to Kepler*), Dover, New York, p. 202.
427. Newton, R. R., 1977, *The Crime of Claudius Ptolemy*, Johns Hopkins University Press, Baltimore.
428. Ibid., p. xiii.
429. Ibid., p. 378.

430. Ibid., p. 378–379.
431. Goldstein, B. R., 1978, Casting Doubt on Ptolemy, *Science*, 24 Feb., vol. 199, p. 872.
432. Gingerich, O., 1980, Was Ptolemy a Fraud? *Quarterly Journal of the Royal Astronomical Society*, vol. 21, p. 253–266.
433. Ibid., p. 253.
434. Ibid., p. 255.
435. Ibid., p. 262.
436. Ibid., p. 264.
437. Ibid.
438. Thurston, H., 2002, "Greek Mathematical Astronomy Reconsidered," *Isis*, vol. 93, no. 1, p. 63, 68.

## Chapter 3

1. Clagett, M., 1994, *Greek Science in Antiquity*, Barnes and Noble, New York (reprint of 1955 edition), p. 99.
2. Whitehead, A. N., 1935, *Science and the Modern World: Lowell Lectures*, 1925, Macmillan, New York, p. 9.
3. Singer, C., 1949, *A Short History of Science to the Nineteenth Century*, Clarendon Press, p. 94.
4. Sarton, G., 1960, *A History of Science, Ancient Science Through the Golden Age of Greece*, Harvard University Press, Cambridge, p. 561.
5. Russell, B., 1945, *A History of Western Philosophy* (1963 paperback edition), Simon and Schuster, New York, p. 278.
6. Crombie, A. C., 1995, *The History of Science from Augustine to Galileo*, Dover, New York, p. 30.
7. Stahl, W., 1962, *Roman Science*, University of Wisconsin Press, Madison, p. 3.
8. Stock, B., 1978, Science, Technology, and Economic Progress in the Early Middle Ages, in *Science in the Middle Ages*, edited by David C. Lindberg, University of Chicago Press, Chicago, p. 4.
9. Plutarch, 1952, Marcellus, in *The Lives of the Noble Grecians and Romans*, translated by John Dryden (1631–1700), *Great Books of the Western World*, vol. 14, William Benton, Chicago, p. 253.
10. Xenophon, 1898, *Oeconomicus*, Chapter 4, Paragraphs 2 and 3, in *Xenophon's Minor Works*, translated by J. S. Watson, George Bell & Sons, London, p. 86.
11. Gest, A. P., 1930, *Engineering, Our Debt to Greece and Rome* (1963 Edition), Cooper Square Publishers, New York, p. 19–20.
12. Steinman, D. B., and Watson, S. R., 1941, *Bridges and their Builders*, G. P. Putnam's Sons, New York, p. 37.
13. Pliny the Elder, 1855, *The Natural History of Pliny*, translated by John Bostock and H. T. Riley, Book 36, Chapter 16, vol. 6, Henry G. Bohn, London, p. 335–336.
14. Ibid., vol. 6, Book 36, Chapter 24, p. 347.
15. Ibid., vol. 6, Book 36, Chapter 24, p. 353–354.
16. Vitruvius, 1960, *The Ten Books on Architecture*, Book 2, Chapter 6 (Pozzolana), translated by Morris Hicky Morgan (first published by Harvard University Press in 1914), Dover, New York, p. 46–47.
17. Briggs, M.S., 1956, Building-Construction, in *A History of Technology*, vol. 2, edited by Charles Singer, Oxford University Press, London, p. 410.
18. Hodges, H., 1970, *Technology in the Ancient World*, Marboro Books, London, p. 233.
19. Briggs, M.S., 1956, Building-Construction, in *A History of Technology*, vol. 2, edited by Charles Singer, Oxford University Press, London, p. 405.
20. Gibbon, E., 1909, *The History of the Decline and Fall of the Roman Empire*, Chapter 2, edited by J. B. Bury, vol. 1, Methuen & Co., London, p. 48.
21. *Encyclopædia Britannica*, Eleventh Edition, 1911, Rome, vol. 23, Encyclopædia Britannica Company, New York, p. 604.
22. Loerke, W. C., 1990, A Rereading of the Interior Elevation of Hadrian's Rotunda, *The Journal of the Society of Architectural Historians*, vol. 49, no. 1, p. 26.
23. *Encyclopædia Britannica*, Eleventh Edition, 1911, Rome, vol. 23, Encyclopædia Britannica Company, New York, p. 604.
24. Von Hagen, V. W., 1967, *The Roads that Led to Rome*, World Publishing Company, Cleveland and New York, p. 8.
25. Derry, T. K., and Williams, T. I., 1993, *A Short History of Technology, From the Earliest Times to A.D. 1900* (reprint of 1960 edition published by Oxford University Press), Dover, New York, p. 168.
26. Goodchild, R. G., and Forbes, R. J., 1956, Road and Land Travel, in *A History of Technology*, vol. 2, edited by Charles Singer, Oxford University Press, London, p. 498.
27. Straub, H., 1952, *A History of Civil Engineering* (E. Rockwell, English translator), Leonard Hill, London, p. 3–6.
28. *Encyclopædia Britannica*, Eleventh Edition, 1911, Roads and Streets, vol. 23, Encyclopædia Britannica Company, New York, p. 388.
29. Ibid.
30. Ibid.
31. Codrington, T., 1903, *Roman Roads in Britain*, The Sheldon Press, London, p. 14.
32. De Camp, L. S., 1960, *The Ancient Engineers*, Dorset Press, New York, p. 186–187.
33. Statius, Publius Papinius, 1908, *The Silvae of Statius*, Book 4, Chapter 3, translated by D. A. Slater, Oxford at the Clarendon Press, Oxford, p. 148.
34. *Encyclopædia Britannica*, Eleventh Edition, 1911, Roads and Streets, vol. 23, Encyclopædia Britannica Company, New York, p. 388.
35. Vitruvius, 1960, *The Ten Books on Architecture*, Book 8, Chapter 1, translated by Morris H. Morgan (first published in 1914 by Harvard University Press), Dover, New York, p. 227.
36. Ibid., Book 8, Chapter 6, Paragraph 12, p. 247.
37. Forbes, R. J., 1956, Hydraulic Engineering and Sanitation, in *A History of Technology*, vol. 2, edited by Charles Singer, Oxford University Press, London, p. 670.
38. Finch, J. K., 1951, *Engineering and Western Civilization*, McGraw-Hill, New York, p. 17.
39. Gest, A. P., 1963, *Engineering, Our Debt to Greece and Rome* (first published in 1930), Cooper Square, New York, p. 62–107.
40. *Encyclopædia Britannica*, 1972, "Aqueduct," vol. 2, William Benton, Chicago, p. 156.
41. De Camp, L. S., 1960, *The Ancient Engineers*, Dorset Press, New York, p. 198.
42. Forbes, R. J., 1956, Hydraulic Engineering and Sanitation, in *A History of Technology*, vol. 2, edited by Charles Singer, Oxford University Press, London, p. 673.
43. Frontinus, Sextus Julius, 1899, *The Two Books on the Water Supply of the City of Rome*, translated by Clemens Herschel, Dana Estes Company, Boston, p. 50–51.
44. Neuburger, A., 1930, *The Technical Arts and Sciences of the Ancients*, Macmillan, New York, p. 257.
45. Ibid., p. 259–264.
46. Ibid., p. 366.
47. *Encyclopædia Britannica*, 1972, "Bath," vol. 3, William Benton, Chicago, p. 275.
48. De Camp, L. S., 1960, *The Ancient Engineers*, Dorset Press, New York, p. 182.
49. Neuburger, A., 1930, *The Technical Arts and Sciences of the Ancients*, Macmillan, New York, p. 366.
50. Ibid., p. 444.
51. Ibid.
52. Niebuhr, B. G., 1853, *Lectures on the History of Rome*, vol. 1, Third Edition, Taylor, Walton, and Maberly, London, p. 52–53.
53. *Encyclopædia Britannica*, Eleventh Edition, 1910, Aqueduct, vol. 2, Encyclopædia Britannica Company, New York, p. 243.
54. Steinman, D. B., and Watson, S. R., 1941, *Bridges and their Builders*, G. P. Putnam's Sons, New York, p. 52.
55. Rousseau, J. J., 1856, *The Confessions of J. J. Rousseau*, vol. 1, Calvin Blanchard, New York, p. 291–292.
56. Horace, 1892, *The Odes and Carmen Saeculare of Horace*, Book 3, Ode 30, translated by John Conington (1825–1869), George Bell and Sons, London, p. 102.
57. Steinman, D. B. and Watson, S. R., 1941, *Bridges and their Builders*, G. P. Putnam's Sons, New York, p. 50.

58. Ibid., p. 54.
59. Straub, H., 1952, *A History of Civil Engineering* (E. Rockwell, English translator), Leonard Hill, London, p. 11.
60. Steinman, D. B., and Watson, S. R., 1941, *Bridges and their Builders*. G. P. Putnam's Sons, New York, p. 40.
61. Caesar, 1919, *The Gallic War*, Book 4, Paragraph 17, translated by H. J. Edwards, William Heinemann, London, p. 201–203.

## Chapter 4

1. Ward-Perkins, J., 2008, Vitruvius Pollio, in *Complete Dictionary of Scientific Biography*, edited by Charles Gillispie, vol. 15, Cengage Learning, New York, p. 514.
2. Vitruvius, 1960, *The Ten Books on Architecture*, Book 1, Chapter 3, translated by Morris H. Morgan (first published in 1914 by Harvard University Press), Dover, New York, p. 16.
3. Ibid., Book 1, Chapter 1, p. 5–6.
4. Ibid., p. 8.
5. Ward-Perkins, J., 2008, Vitruvius Pollio, in *Complete Dictionary of Scientific Biography*, edited by Charles Gillispie, vol. 15, Cengage Learning, New York, p. 515.
6. Vitruvius, 1960, *The Ten Books on Architecture*, Book 1, Chapter 2, translated by Morris H. Morgan (first published in 1914 by Harvard University Press), Dover, New York, p. 13.
7. Ibid., p. 14.
8. Ibid., Book 2, Chapter 1, p. 38.
9. Ibid., Book 2, Chapter 6, p. 46–47.
10. Ibid., Book 2, Chapter 9, p. 59.
11. Ibid., Book 6, Introduction, p. 168.
12. Ibid., Book 6, Chapter 1, p. 173.
13. Ibid., Book 7, Chapter 5, p. 210.
14. Ibid., p. 211.
15. Forbes, R., J., 1956, Hydraulic Engineering and Sanitation, in *A History of Technology*, vol. 2, edited by Charles Singer, Oxford University Press, London, p. 672.
16. Vitruvius, 1960, *The Ten Books on Architecture*, Book 8, Chapter 6, translated by Morris H. Morgan (first published in 1914 by Harvard University Press), Dover, New York, p. 245.
17. Ibid., p. 246.
18. Ward-Perkins, J., 2008, Vitruvius Pollio, in *Complete Dictionary of Scientific Biography*, edited by Charles Gillispie, vol. 15, Cengage Learning, New York, p. 516–517.
19. Clarke, J., 1910, *Physical Science in the Time of Nero, Being a Translation of the Quaestiones Naturales of Seneca*, Macmillan, London, p. xxi.
20. *Encyclopædia Britannica*, Eleventh Edition, 1911, Seneca, vol. 20, Encyclopædia Britannica Company, New York, p. 215.
21. Clarke, J., 1910, *Physical Science in the Time of Nero, Being a Translation of the Quaestiones Naturales of Seneca*, Macmillan, London, p. xxii.
22. *Encyclopædia Britannica*, 1972, "Seneca, Lucius Annaeus," vol. 4, William Benton, Chicago, p. 695.
23. Clarke, J., 1910, *Physical Science in the Time of Nero, Being a Translation of the Quaestiones Naturales of Seneca*, Macmillan, London, p. xxiii.
24. Niebuhr, B. G., 1849, *Lectures on the History of Rome*, vol. 3, Second Edition, Taylor, Walton, and Maberly, London, p. 187.
25. Suetonius, C. T., 1906, *The Lives of the Twelve Caesars*, Claudius, Paragraph 26, George Bell & Sons, London, p. 319.
26. Ibid., Paragraph 46, p. 335.
27. Juvenal, 1918, *Satire 6, The Ways of Women*, Lines 115–130, in *Juvenal and Perseus*, translated by G. G. Ramsay, William Heinemann, London, p. 93.
28. Suetonius, C. T., 1906, *The Lives of the Twelve Caesars*, Claudius, Paragraph 46, George Bell & Sons, London, p. 335.
29. Tacitus, 1942, *Annals*, Book 11, Paragraph 27, in *The Complete Works of Tacitus*, translated by Alfred John Church (1824–1912) and William Jackson Brodribb (1829–1905), Modern Library, New York, p. 243.
30. Niebuhr, B. G., 1849, *Lectures on the History of Rome*, vol. 3, Second Edition, Taylor, Walton, and Maberly, London, p. 194.
31. Suetonius, C. T., 1906, *The Lives of the Twelve Caesars*, Claudius, Paragraph 46, George Bell & Sons, London, p. 335–336.
32. Ibid., Paragraph 26, p. 319.
33. Ibid., Paragraph 26, p. 319–320.
34. Niebuhr, B. G., 1849, *Lectures on the History of Rome*, vol. 3, Second Edition, Taylor, Walton, and Maberly, London, p. 190.
35. Tacitus, 1942, *Annals*, Book 12, Paragraph 66, in *The Complete Works of Tacitus*, translated by Alfred John Church (1824–1912) and William Jackson Brodribb (1829–1905), Modern Library, New York, p. 282.
36. Engel, D. J. F., 1940, The Correspondence between Lord Acton and Bishop Creighton, *Cambridge Historical Journal*, vol. 6, no. 3, p. 316.
37. Suetonius, C. T., 1906, *The Lives of the Twelve Caesars*, Nero, Paragraph 10, George Bell & Sons, London, p. 343.
38. Ibid.
39. Dio Cassius, 1906, *Dio's Rome*, Book 61, Paragraph 4, vol. 5, translated by Herbert Baldwin Foster, Pafraets Book Company, Troy, New York, p. 5.
40. Mommsen, T., 1996, *A History of Rome Under the Emperors*, Routledge, London, p. 175.
41. Dio Cassius, 1906, *Dio's Rome*, Book 61, Paragraphs 3, 4, vol. 5, translated by Herbert Baldwin Foster, Pafraets Book Company, Troy, New York, p. 4–5.
42. Ibid., Book 61, Paragraph 10, p. 13–14.
43. Tacitus, 1942, *Annals*, Book 13, Paragraph 12, in *The Complete Works of Tacitus*, translated by Alfred John Church (1824–1912) and William Jackson Brodribb (1829–1905), Modern Library, New York, p. 290.
44. Ibid., Book 13, Paragraph 14, p. 292.
45. Ibid., Book 13, Paragraphs 15, 16, p. 292–293.
46. Ibid., Book 13, Paragraph 17, p. 294.
47. Suetonius, C. T., 1906, *The Lives of the Twelve Caesars*, Nero, Paragraph 34, George Bell & Sons, London, p. 362–363.
48. Tacitus, 1942, *Annals*, Book 14, Paragraph 3, in *The Complete Works of Tacitus*, translated by Alfred John Church (1824–1912) and William Jackson Brodribb (1829–1905), Modern Library, New York, p. 321.
49. Ibid., Book 14, Paragraphs 3–5, p. 321–323.
50. Ibid., Book 14, Paragraph 8, p. 325.
51. Suetonius, C. T., 1906, *The Lives of the Twelve Caesars*, Nero, Paragraph 20, George Bell & Sons, London, p. 319.
52. Ibid., Paragraph 23, p. 353.
53. Tacitus, 1942, *Annals*, Book 13, Paragraph 45, in *The Complete Works of Tacitus*, translated by Alfred John Church (1824–1912) and William Jackson Brodribb (1829–1905), Modern Library, New York, p. 312.
54. Ibid., Book 13, Paragraph 46, p. 313.
55. Ibid., Book 13, Paragraph 45, p. 312.
56. Suetonius, C. T., 1906, *The Lives of the Twelve Caesars*, Nero, Paragraph 35, George Bell & Sons, London, p. 364.
57. Ibid., Paragraph 35, p. 365.
58. Tacitus, 1942, *Annals*, Book 14, Paragraph 64, in *The Complete Works of Tacitus*, translated by Alfred John Church (1824–1912) and William Jackson Brodribb (1829–1905), Modern Library, New York, p. 355–356.
59. Suetonius, C. T., 1906, *The Lives of the Twelve Caesars*, Nero, Paragraph 35, George Bell & Sons, London, p. 365.
60. Ibid., Paragraph 37, p. 366.
61. Ibid., Paragraph 35, p. 365.
62. Dio Cassius, 1906, *Dio's Rome*, Book 61, Paragraph 14, vol. 5, translated by Herbert Baldwin Foster, Pafraets Book Company, Troy, New York, p. 19.
63. Tacitus, 1942, *Annals*, Book 14, Paragraph 52, in *The Complete Works of Tacitus*, translated by Alfred John Church (1824–1912) and William Jackson Brodribb (1829–1905), Modern Library, New York, p. 348.
64. Ibid., Book 14, Paragraph 54, p. 349.
65. Ibid., Book 14, Paragraphs 55–56, p. 350.
66. Ibid., Book 15, Paragraph 45, p. 381.
67. Grant, E., 1977, *Physical Science in the Middle Ages*, Cambridge University Press, Cambridge, p. 7.
68. Clarke, J., 1910, *Physical Science in the Time of Nero*

*being a Translation of the Quaestiones Naturales of Seneca*, Book 1, Preface, Macmillan, London, p. 4.

69. Ibid., p. 7.

70. Ibid.

71. Ibid., Book 1, Section 1, Paragraph 4, p. 9.

72. Aristotle, 1923, *Meteorologica (Meteorology)*, Book 1, Chapter 4, translated by E. W. Webster, Oxford at the Clarendon Press, London, p. 344a.

73. Clarke, J., 1910, *Physical Science in the Time of Nero being a Translation of the Quaestiones Naturales of Seneca*, Book 1, Section 1, Macmillan, London, p. 11.

74. Ibid., Book 1, Sections 2, 3, p. 16–22.

75. Ibid., Book 1, Section 15, Paragraph 5, p. 40.

76. Ibid., Book 1, Section 16, Paragraph 5, p. 43.

77. Ibid., Book 1, Section 17, Paragraph 10, p. 47.

78. Singer, C., 1958, *From Magic to Science* (first published in 1928), Dover, New York, p. 15.

79. Clarke, J., 1910, *Physical Science in the Time of Nero being a Translation of the Quaestiones Naturales of Seneca*, Book 2, Section 59, Paragraph 1, Macmillan, London, p. 102.

80. Ibid., Book 3, Preface, Paragraph 1, p. 109.

81. Ibid., Book 3, Section 7, Paragraph 1, p. 117–118.

82. Ibid., Book 3, Section 7, Paragraph 3, p. 118.

83. Ibid., Book 3, Section 9, Paragraph 2, p. 119–120.

84. Ibid., Book 3, Section 10, Paragraph 1, p. 120.

85. Ibid., Book 3, Section 5, p. 116–117.

86. Plato, 1937, Phaedo, in *The Dialogues of Plato*, translated into English by Benjamin Jowett (1817–1893), vol. 1, Random House, New York, p. 495–496 (111).

87. Clarke, J., 1910, *Physical Science in the Time of Nero being a Translation of the Quaestiones Naturales of Seneca*, Book 3, Section 15, Paragraphs 2,3, Macmillan, London, p. 126.

88. Boas, G., 1973, Macrocosm and Microcosm, in *Dictionary of the History of Ideas*, vol. 3, edited by Philip P. Wiener, Charles Scribner's Sons, New York, p. 126.

89. Freeman, K., 1966, *Ancilla to the Pre-Socratic Philosophers, a Complete Translation of the Fragments in Diels, Fragmente der Vorsokratiker*, Harvard University Press, Cambridge, p. 99.

90. Plato, 1937, Philebus, in *The Dialogues of Plato*, translated into English by Benjamin Jowett (1817–1893), vol. 2, Random House, New York, p. 361–362 (29).

91. Plato, 1937, Timaeus, in *The Dialogues of Plato*, translated into English by Benjamin Jowett (1817–1893), vol. 2, Random House, New York, p. 14 (30).

92. Adams, F. D., 1954, *The Birth and Development of the Geological Sciences* (first published in 1938 by William & Wilkins in Baltimore), Dover, New York, p. 66.

93. Clarke, J., 1910, *Physical Science in the Time of Nero being a Translation of the Quaestiones Naturales of Seneca*, Book 3, Section 25, Paragraph 4, Macmillan, London, p. 138.

94. Ibid.

95. Ibid., Paragraph 6, p. 139.

96. Ibid., Book 3, Section 30, Paragraph 7, p. 156.

97. Ibid., Book 6, Section 3, Paragraph 1, p. 228.

98. Ibid., Book 6, Section 4, Paragraphs 2,3, p. 229–230.

99. Aristotle, 1923, *Meteorologica* (*Meteorology*), Book 2, Chapters 6–9, translated by E. W. Webster, Oxford at the Clarendon Press, London, p. 365a–369a.

100. Clarke, J., 1910, *Physical Science in the Time of Nero being a Translation of the Quaestiones Naturales of Seneca*, Book 6, Section 16, Paragraph 1, Macmillan, London, p. 244.

101. Ibid., Book 6, Section 16, Paragraph 4, p. 245.

102. Ibid., Book 7, Section 21, Paragraph 1, p. 293.

103. Ibid., Book 7, Section 22, Paragraph 1, p. 295, Section 25, Paragraph 1, p. 298.

104. Ibid., Book 7, Section 25, Paragraph 2, p. 298.

105. Ibid., Book 7, Section 26, Paragraph 1, p. 299.

106. Ibid., Book 7, Section 31, Paragraph 2, p. 305–306.

107. Ibid., Book 7, Section 31, Paragraph 2, p. 306.

108. Ibid., Book 7, Section 31, Paragraph 4, p. 306.

109. Ibid., Book 7, Section 32, Paragraph 4, p. 308.

110. Suetonius, C. T., 1906, *The Lives of the Twelve Caesars*, Nero, Paragraph 26, George Bell & Sons, London, p. 356.

111. Ibid., Paragraph 28, p. 357.

112. Ibid., Paragraph 29, p. 357–358.

113. Ibid., Paragraph 30, p. 358.

114. Ibid., Paragraph 38, p. 367–368.

115. Dio Cassius, 1906, *Dio's Rome*, Book 62, Paragraph 18, vol. 5, translated by Herbert Baldwin Foster, Pafraets Book Company, Troy, New York, p. 45.

116. Ibid.

117. Tacitus, 1942, *Annals*, Book 15, Paragraph 39, in *The Complete Works of Tacitus*, translated by Alfred John Church (1824–1912) and William Jackson Brodribb (1829–1905), Modern Library, New York, p. 378.

118. Dio Cassius, 1906, *Dio's Rome*, Book 62, Paragraph 18, vol. 5, translated by Herbert Baldwin Foster, Pafraets Book Company, Troy, New York, p. 45.

119. Tacitus, 1942, *Annals*, Book 15, Paragraph 44, in *The Complete Works of Tacitus*, translated by Alfred John Church (1824–1912) and William Jackson Brodribb (1829–1905), Modern Library, New York, p. 380–381.

120. Ibid., Book 15, Paragraph 60, p. 390.

121. Ibid., Book 15, Paragraph 63, p. 392.

122. Ibid., Book 15, Paragraph 64, p. 392.

123. Ibid., p. 393.

124. Dio Cassius, 1906, *Dio's Rome*, Book 63, Paragraph 22, vol. 5, translated by Herbert Baldwin Foster, Pafraets Book Company, Troy, New York, p. 75.

125. Ibid., Book 63, Paragraph 23, p. 76–77.

126. Suetonius, C. T., 1906, *The Lives of the Twelve Caesars*, Nero, Paragraph 44, George Bell & Sons, London, p. 373.

127. Ibid., Paragraph 47, p. 375.

128. Dio Cassius, 1906, *Dio's Rome*, Book 63, Paragraph 27, vol. 5, translated by Herbert Baldwin Foster, Pafraets Book Company, Troy, New York, p. 80.

129. Suetonius, C. T., 1906, *The Lives of the Twelve Caesars*, Nero, Paragraph 48, George Bell & Sons, London, p. 376.

130. Ibid., Paragraph 49, p. 377.

131. Ibid., Paragraph 49, p. 378.

132. *Encyclopædia Britannica*, Eleventh Edition, 1911, Pliny, The Elder, vol. 21, Encyclopædia Britannica Company, New York, p. 841.

133. Tacitus, 1942, *Annals*, Book 5, Paragraph 8, in *The Complete Works of Tacitus*, translated by Alfred John Church (1824–1912) and William Jackson Brodribb (1829–1905), Modern Library, New York, p. 192.

134. *Encyclopædia Britannica*, Eleventh Edition, 1911, Pliny, The Elder, vol. 21, Encyclopædia Britannica Company, New York, p. 842.

135. Ibid., p. 841–844.

136. Tacitus, 1942, *Annals*, Book 1, Paragraph 69, in *The Complete Works of Tacitus*, translated by Alfred John Church (1824–1912) and William Jackson Brodribb (1829–1905), Modern Library, New York, p. 46.

137. Pliny the Elder, 1855–1857, *The Natural History of Pliny*, translated by John Bostock and H. T. Riley, in 6 volumes, Henry G. Bohn, London.

138. Pliny the Consul, 1809, *The Letters of Pliny the Consul*, Book 3, Letter 5, translated by William Melmoth, E. Larkin, Boston, p. 127.

139. Pliny the Elder, 1855, *The Natural History of Pliny*, Book 1, Dedication, translated by John Bostock and H. T. Riley, vol. 1, Henry G. Bohn, London, p. 6.

140. Pliny the Consul, 1809, *The Letters of Pliny the Consul*, Book 3, Letter 5, translated by William Melmoth, E. Larkin, Boston, p. 127–130.

141. Pliny the Elder, 1855, *The Natural History of Pliny*, Book 2, Chapter 5, translated by John Bostock and H. T. Riley, vol. 1, Henry G. Bohn, London, p. 20.

142. Ibid., Book 2, Chapter 5, p. 25.

143. Ibid., Book 2, Chapter 5, p. 20–21.

144. Ibid., Book 2, Chapter 5, p. 21–22.

145. Gibbon, E., 1909, *The History of the Decline and Fall of the Roman Empire*, Chapter 13, edited by J. B. Bury, vol. 1, Methuen & Co., London, p. 394.

146. Sarton, G., 1924, Review of "A History of Magic and Experimental Science During the First Thirteen Centuries of Our Era," *Isis*, no. 1, vol. 6, p. 75.

147. Pliny the Elder, 1855, *The Natural History of Pliny*, Book 7, Chapter 2, translated by John Bostock and H. T. Riley, vol. 2, Henry G. Bohn, London, p. 130–131.
148. Ibid., Book 8, Chapter 30, p. 278–280.
149. Pliny the Elder, 1856, *The Natural History of Pliny*, Book 29, Chapter 32, translated by John Bostock and H. T. Riley, vol. 5, Henry G. Bohn, London, p. 405.
150. Pliny the Elder, 1857, *The Natural History of Pliny*, Book 37, Chapter 15, translated by John Bostock and H. T. Riley, vol. 6, Henry G. Bohn, London, p. 408.
151. Ibid., Book 37, Chapter 16, p. 409.
152. Ibid., Book 36, Chapter 24, p. 354.
153. Pliny the Elder, 1856, *The Natural History of Pliny*, Book 28, Chapter 34, translated by John Bostock and H. T. Riley, vol. 5, Henry G. Bohn, London, p. 283–284.
154. Ibid.
155. Ibid., p. 305.
156. Ibid., p. 304.
157. Pliny the Elder, 1855, *The Natural History of Pliny*, Book 8, Chapter 2, translated by John Bostock and H. T. Riley, vol. 2, Henry G. Bohn, London, p. 130–131.
158. Pliny the Elder, 1857, *The Natural History of Pliny*, Book 36, Chapter 34, translated by John Bostock and H. T. Riley, vol. 6, Henry G. Bohn, London, p. 361.
159. Pliny the Elder, 1855, *The Natural History of Pliny*, Book 2, Chapter 88, translated by John Bostock and H. T. Riley, vol. 1, Henry G. Bohn, London, p. 117.
160. Decker, R., and Decker, B., 1981, *Volcanoes*, W. H. Freeman, San Francisco, p. 15–27.
161. Fridriksson, S., 1987, Plant Colonization of a Volcanic Island, Surtsey, Iceland, *Arctic and Alpine Research*, vol. 19, no. 4, p. 425.
162. Pliny the Consul, 1809, *The Letters of Pliny the Consul*, Book 6, Letter 16, translated by William Melmoth, E. Larkin, Boston, p. 294–300.
163. Walsh, J. J., 1920, *Medieval Medicine*, A. & C. Black, Ltd., London, p. 18–19.
164. Coxe, J. R., 1846, Introduction, in *The Writings of Hippocrates and Galen*, Lindsay and Blakiston, Philadelphia, p. 29. This Introduction appears to be Coxe's translation of *Historie de la Medecine* by Daniel Le Clerc, published at A la Haye (The Hague) by Chez Isaac van der Kloot in 1729 (first edition in 1696).
165. Eusebius, 1851, *Ecclesiastical History*, Book 5, Chapter 28, translated by C. F. Cruse, Henry G. Bohn, London, p. 203.
166. Meyerhof, M., 1931, Science and Medicine, in *The Legacy of Islam* (First Edition), edited by T. Arnold and A. Guillaume, Oxford University Press, London, p. 316.
167. Sarton, G., 1927, *Introduction to the History of Science*, vol. 1, Carnegie Institution of Washington, Washington, DC, p. 302.
168. Sarton, G., 1954, *Galen of Pergamon*, University of Kansas Press, Lawrence, p. 11–12.
169. Ibid., p. 15.
170. Nutton, V., 2004, *Ancient Medicine*, Routledge, London, p. 216.
171. Coxe, J. R., 1846, Introduction, in *The Writings of Hippocrates and Galen*, Lindsay and Blakiston, Philadelphia, p. 28. This Introduction appears to be Coxe's translation of *Historie de la Medecine* by Daniel Le Clerc, published at A la Haye (The Hague) by Chez Isaac van der Kloot in 1729 (first edition in 1696).
172. Nutton, V., 2004, *Ancient Medicine*, Routledge, London, p. 217.
173. Galen, 1929, On the Passions, in *Greek Medicine, Being Extracts Illustrative of Medical Writers from Hippocrates to Galen*, translated by Arthur J. Brock, J.M. Dent & Sons, London, p. 171.
174. Ibid.
175. Sarton, G., 1954, *Galen of Pergamon*, University of Kansas Press, Lawrence, p. 16.
176. Nutton, V., 2004, *Ancient Medicine*, Routledge, London, p. 216.
177. Sarton, G., 1954, *Galen of Pergamon*, University of Kansas Press, Lawrence, p. 30.
178. Ibid., p. 16.
179. Galen, 1929, *On the Passions*, in *Greek Medicine, Being Extracts Illustrative of Medical Writers from Hippocrates to Galen*, translated by Arthur J. Brock, J.M. Dent & Sons, London, p. 172.
180. Ibid.
181. Ibid.
182. Galen, 1916, *On the Natural Faculties*, Book 3, Chapter 10, translated by Arthur John Brock, William Heinemann, London, p. 279.
183. Coxe, J. R., 1846, Introduction, in *The Writings of Hippocrates and Galen*, Lindsay and Blakiston, Philadelphia, p. 28. This Introduction appears to be Coxe's translation of *Historie de la Medecine* by Daniel Le Clerc, published at A la Haye (The Hague) by Chez Isaac van der Kloot in 1729 (first edition in 1696).
184. Sarton, G., 1954, *Galen of Pergamon*, University of Kansas Press, Lawrence, p. 16.
185. Ibid., p. 83–84.
186. Hippocrates, 1950, *The Medical Works of Hippocrates*, translated by John Chadwick and W. N. Mann, Charles C. Thomas, Springfield, Illinois, p. 194.
187. Antoninus, Marcus Aurelius, 1901, *The Thoughts of the Emperor M. Aurelius Antoninus*, Book 1, Chapter 17, translated by George Long, George Bell & Sons, London, p. 76.
188. Galen, 1929, *On the Utility of Parts*, Book 10, Chapter 12, in *Greek Medicine, Being Extracts Illustrative of Medical Writers from Hippocrates to Galen*, translated by Arthur J. Brock, J.M. Dent & Sons, London, p. 156.
189. Sarton, G., 1954, *Galen of Pergamon*, University of Kansas Press, Lawrence, p. 17–19.
190. Nutton, V., 2004, *Ancient Medicine*, Routledge, London, p. 217.
191. Sarton, G., 1954, *Galen of Pergamon*, University of Kansas Press, Lawrence, p. 18.
192. Nutton, V., 2004, *Ancient Medicine*, Routledge, London, p. 218.
193. Ibid.
194. Ibid., p. 223.
195. Ibid., p. 182–186.
196. Sarton, G., 1954, *Galen of Pergamon*, University of Kansas Press, Lawrence, p. 20.
197. Nutton, V., 2004, *Ancient Medicine*, Routledge, London, p. 224.
198. Ibid., p. 159.
199. Plutarch, 1952, Marcus Cato, in *The Lives of the Noble Grecians and Romans*, translated by John Dryden (1631–1700), *Great Books of the Western World*, vol. 14, William Benton, Chicago, p. 288.
200. Horace, 1889, *Horace: the Odes, Epodes, Satires, and Epistles*, Book 2, Epistle 1, Frederick Warne and Co., London, p. 348.
201. Nutton, V., 2004, *Ancient Medicine*, Routledge, London, p. 164.
202. Ibid., p. 165.
203. Ibid., p. 224.
204. Kudlien, F., 2008, Galen, in *Complete Dictionary of Scientific Biography*, edited by Charles Gillispie, vol. 5, Cengage Learning, New York, p. 229.
205. Kudlien, F., 2008, Galen, in *Complete Dictionary of Scientific Biography*, edited by Charles Gillispie, vol. 5, Cengage Learning, New York, p. 229.
206. Nutton, V., 2004, *Ancient Medicine*, Routledge, London, p. 224.
207. Sarton, G., 1954, *Galen of Pergamon*, University of Kansas Press, Lawrence, p. 21.
208. Nutton, V., 2004, *Ancient Medicine*, Routledge, London, p. 224.
209. Galen, 1929, *On Habits*, in *Greek Medicine, Being Extracts Illustrative of Medical Writers from Hippocrates to Galen*, translated by Arthur J. Brock, J.M. Dent & Sons, London, p. 186.
210. Galen, 1929, *Notes on His Own Books*, in *Greek Medicine, Being Extracts Illustrative of Medical Writers from Hip-*

pocrates to Galen, translated by Arthur J. Brock, J.M. Dent & Sons, London, p. 176.

211. Ibid.

212. Sarton, G., 1954, *Galen of Pergamon*, University of Kansas Press, Lawrence, p. 23.

213. Nutton, V., 2004, *Ancient Medicine*, Routledge, London, p. 225–229.

214. Ibid., p. 232.

215. Kudlien, F., 2008, Galen, in *Complete Dictionary of Scientific Biography*, edited by Charles Gillispie, vol. 5, Cengage Learning, New York, p. 227.

216. Nutton, V., 2004, *Ancient Medicine*, Routledge, London, p. 226–227.

217. Kudlien, F., 2008, Galen, in *Complete Dictionary of Scientific Biography*, edited by Charles Gillispie, vol. 5, Cengage Learning, New York, p. 228.

218. Coxe, J. R., 1846, *The Works of Galen*, in *The Writings of Hippocrates and Galen*, Lindsay and Blakiston, Philadelphia, p. 473.

219. Galen, 1929, *Notes on His Own Books*, in *Greek Medicine, Being Extracts Illustrative of Medical Writers from Hippocrates to Galen*, translated by Arthur J. Brock, J.M. Dent & Sons, London, p. 174.

220. Sarton, G., 1954, *Galen of Pergamon*, University of Kansas Press, Lawrence, p. 25.

221. Thorndike, L., 1923, *A History of Magic and Experimental Science*, vol. 1, Columbia University Press, New York, p. 120.

222. Kudlien, F., 2008, Galen, in *Complete Dictionary of Scientific Biography*, edited by Charles Gillispie, vol. 5, Cengage Learning, New York, p. 232.

223. Galen, 1929, *Notes on His Own Books*, in *Greek Medicine, Being Extracts Illustrative of Medical Writers from Hippocrates to Galen*, translated by Arthur J. Brock, J.M. Dent & Sons, London, p. 174.

224. Sarton, G., 1954, *Galen of Pergamon*, University of Kansas Press, Lawrence, p. 48, 59.

225. Thorndike, L., 1923, *A History of Magic and Experimental Science*, vol. 1, Columbia University Press, New York, p. 158.

226. Ibid., p. 157.

227. Ibid., p. 158.

228. Bacon, R., 1907, *On Experimental Science*, in *The Library of Original Sources*, vol. 4, edited by Oliver Joseph Thatcher, University Research Extension, New York, p. 369–370.

229. Thorndike, L., 1923, *A History of Magic and Experimental Science*, vol. 1, Columbia University Press, New York, p. 162.

230. Ibid., p. 166–167.

231. Ibid., p. 170.

232. Ibid., p. 173.

233. Ibid.

234. Ibid., p. 175.

235. Nutton, V., 2004, *Ancient Medicine*, Routledge, London, p. 241.

236. Longrigg, J., 2008, Herophilus, in *Complete Dictionary of Scientific Biography*, edited by Charles Gillispie, vol. 6, Cengage Learning, New York, p. 316.

237. Tertullian, 1870, *De Anima* (*On the Soul*), Chapter 10, in *Anti-Nicene Christian Library*, vol. 15, *The Writings of Tertullian*, vol. 2, edited by Alexander Roberts and James Donaldson, T. & T. Clark, Edinburgh, p. 431.

238. Celsus, A. C., 1831, *A Translation of the Eight Books of Aul. Corn. Celsus on Medicine*, Book 1, Preface, Second Edition, revised by G. F. Collier, Simpkin and Marshall, London, p. 5.

239. Nutton, V., 2004, *Ancient Medicine*, Routledge, London, p. 129–131.

240. Ibid., p. 132.

241. Longrigg, J., 2008, Herophilus, in *Complete Dictionary of Scientific Biography*, edited by Charles Gillispie, vol. 6, Cengage Learning, New York, p. 317.

242. Longrigg, J., 2008, Erasistratus, in *Complete Dictionary of Scientific Biography*, edited by Charles Gillispie, vol. 4, Cengage Learning, New York, p. 382.

243. Ibid., p. 383.

244. Nutton, V., 2004, *Ancient Medicine*, Routledge, London, p. 135.

245. Longrigg, J., 2008, Erasistratus, in *Complete Dictionary of Scientific Biography*, edited by Charles Gillispie, vol. 4, Cengage Learning, New York, p. 383.

246. Nutton, V., 2004, *Ancient Medicine*, Routledge, London, p. 203.

247. Wilson, L. G., 2008, Galen: Anatomy and Physiology, in *Complete Dictionary of Scientific Biography*, edited by Charles Gillispie, vol. 5, Cengage Learning, New York, p. 234.

248. Longrigg, J., 2008, Erasistratus, in *Complete Dictionary of Scientific Biography*, edited by Charles Gillispie, vol. 4, Cengage Learning, New York, p. 384.

249. Nutton, V., 2004, *Ancient Medicine*, Routledge, London, p. 136.

250. Longrigg, J., 2008, Erasistratus, in *Complete Dictionary of Scientific Biography*, edited by Charles Gillispie, vol. 4, Cengage Learning, New York, p. 385.

251. Ibid.

252. Nutton, V., 2004, *Ancient Medicine*, Routledge, London, p. 231.

253. Galen, 1929, *On Anatomical Procedure*, Book 1, Chapter 2, in *Greek Medicine, Being Extracts Illustrative of Medical Writers from Hippocrates to Galen*, translated by Arthur J. Brock, J.M. Dent & Sons, London, p. 161–162.

254. Nutton, V., 2004, *Ancient Medicine*, Routledge, London, 230–231.

255. Wilson, L. G., 2008, Galen: Anatomy and Physiology, in *Complete Dictionary of Scientific Biography*, edited by Charles Gillispie, vol. 5, Cengage Learning, New York, p. 234.

256. Galen, 1916, *On the Natural Faculties*, Book 3, Chapter 15, translated by Arthur John Brock, William Heinemann, London, p. 321.

257. Wilson, L. G., 2008, Galen: Anatomy and Physiology, in *Complete Dictionary of Scientific Biography*, edited by Charles Gillispie, vol. 5, Cengage Learning, New York, p. 234–235.

258. Galen, 1929, *On the Utility of Parts*, Book 6, Chapter 17, in *Greek Medicine, Being Extracts Illustrative of Medical Writers from Hippocrates to Galen*, translated by Arthur J. Brock, J.M. Dent & Sons, London, p. 159.

259. Nutton, V., 2004, *Ancient Medicine*, Routledge, London, p. 231.

260. Galen, 1916, *On the Natural Faculties*, Book 3, Chapter 4, translated by Arthur John Brock, William Heinemann, London, p. 243.

261. Galen, 1929, *On Anatomical Procedure*, Book 1, Chapter 2, in *Greek Medicine, Being Extracts Illustrative of Medical Writers from Hippocrates to Galen*, translated by Arthur J. Brock, J.M. Dent & Sons, London, p. 160.

262. Cuvier, G., 1997, Extract from a Memoir, read 6 October, 1798, in *Georges Cuvier, Fossil Bones, and Geological Catastrophes*, translated by Martin J. S. Rudwick, University of Chicago Press, Chicago, p. 36.

263. Galen, 1916, *On the Natural Faculties*, Book 1, Chapter 13, translated by Arthur John Brock, William Heinemann, London, p. 59.

264. Nutton, V., 2004, *Ancient Medicine*, Routledge, London, p. 201.

265. Sarton, G., 1954, *Galen of Pergamon*, University of Kansas Press, Lawrence, p. 31–34.

266. Galen, 1929, *Notes on His Own Books*, in *Greek Medicine, Being Extracts Illustrative of Medical Writers from Hippocrates to Galen*, translated by Arthur J. Brock, J.M. Dent & Sons, London, p. 178.

267. Celsus, A. C., 1831, *A Translation of the Eight Books of Aul. Corn. Celsus on Medicine*, Book 1, Preface, Second Edition, revised by G. F. Collier, Simpkin and Marshall, London, p. 3–8.

268. Ibid., p. 3.

269. Ibid., p. 5–6.

270. Galen, 1929, *On the Medical Sects*, Chapter 1, in *Greek Medicine, Being Extracts Illustrative of Medical Writers from*

*Hippocrates to Galen*, translated by Arthur J. Brock, J.M. Dent & Sons, London, p. 132.
271. Nutton, V., 2004, *Ancient Medicine*, Routledge, London, p. 206–207.
272. Galen, 1916, *On the Natural Faculties*, Book 1, Chapter 13, translated by Arthur John Brock, William Heinemann, London, p. 57.
273. Nutton, V., 2004, *Ancient Medicine*, Routledge, London, p,. 222.
274. Kudlien, F., 2008, Galen, in *Complete Dictionary of Scientific Biography*, edited by Charles Gillispie, vol. 5, Cengage Learning, New York, p. 230.
275. Galen, 1916, *On the Natural Faculties*, Book 1, Chapter 13, translated by Arthur John Brock, William Heinemann, London, p. 49.
276. Nutton, V., 2004, *Ancient Medicine*, Routledge, London, p. 222.
277. Galen, 1916, *On the Natural Faculties*, Book 2, Chapter 4, translated by Arthur John Brock, William Heinemann, London, p. 139.
278. Ibid., Book 2, Chapter 9, p. 207–209.
279. Aristotle, 1941, *De Generatione et Corruptione* (*On Generation and Corruption*), Book 2, Chapter 3, translated by Harold H. Joachim, in *Basic Works of Aristotle*, Random House, New York, p. 511 (330b).
280. Galen, 1916, *On the Natural Faculties*, Book 2, Chapter 9, translated by Arthur John Brock, William Heinemann, London, p. 209.
281. Ibid., Book 2, Chapter 8, p. 191.
282. Sarton, G., 1954, *Galen of Pergamon*, University of Kansas Press, Lawrence, p. 53.
283. Hippocrates, 1929, *Epidemics*, Book 3, Chapter 14, in *Greek Medicine, Being Extracts Illustrative of Medical Writers from Hippocrates to Galen*, translated by Arthur J. Brock, J.M. Dent & Sons, London, p. 79.
284. Hicks, R. D., 1910, *Stoic and Epicurean*, Charles Scribner's Sons, New York, p. 238.
285. Burnet, J., 1920, *Early Greek Philosophy*, Third Edition, A. & C. Black, London, p. 354.
286. Balme, D. M., 1973, Biological Conceptions in Antiquity, in *Dictionary of the History of Ideas*, vol. 1, edited by Philip P. Wiener, Charles Scribner's Sons, New York, p. 231.
287. Solmsen, F., 1957, The Vital Heat, the Inborn Pneuma and the Aether, *The Journal of Hellenic Studies*, vol. 77, p. 119.
288. *Encyclopædia Britannica*, Eleventh Edition, 1911, Stoics, vol. 25, Encyclopædia Britannica Company, New York, p. 944.
289. Hicks, R. D., 1910, *Stoic and Epicurean*, Charles Scribner's Sons, New York, p. 329.
290. Brock, A. J., 1929, Introduction, in *Greek Medicine, Being Extracts Illustrative of Medical Writers from Hippocrates to Galen*, J.M. Dent & Sons, London, p. 15.
291. Nutton, V., 2004, *Ancient Medicine*, Routledge, London, p. 202.
292. Sarton, G., 1954, *Galen of Pergamon*, University of Kansas Press, Lawrence, p. 49–51.
293. Ibid.
294. Ibid., p. 50.
295. Galen, 1929, *On the Utility of Parts*, Book 3, Chapter 10, in *Greek Medicine, Being Extracts Illustrative of Medical Writers from Hippocrates to Galen*, translated by Arthur J. Brock, J.M. Dent & Sons, London, p. 154.
296. Laërtius, Diogenes, 1905, *The Lives and Opinions of Eminent Philosophers*, Book 10, translated by C. D. Yonge, George Bell & Sons, London, p. 468–469.
297. Galen, 1929, *On the Utility of Parts*, Book 6, Chapter 17, in *Greek Medicine, Being Extracts Illustrative of Medical Writers from Hippocrates to Galen*, translated by Arthur J. Brock, J.M. Dent & Sons, London, p. 158.
298. Hicks, L. E., 1883, *A Critique of Design-Arguments*, Charles Scribner's Sons, New York, p. 70.
299. Ibid., p. 70–71.

## Chapter 5

1. *Encyclopædia Britannica*, Eleventh Edition, 1911, Stoics, vol. 25, Encyclopædia Britannica Company, New York, p. 942.
2. Hicks, R. D., 1910, *Stoic and Epicurean*, Charles Scribner's Sons, New York, p. 153.
3. Ibid., p. 154.
4. Laërtius, Diogenes, 1905, *The Lives and Opinions of Eminent Philosophers*, Book 10, translated by C. D. Yonge, George Bell & Sons, London, p. 424, 429.
5. Hicks, R. D., 1910, *Stoic and Epicurean*, Charles Scribner's Sons, New York, p. 155.
6. Ibid., p. 154.
7. Ibid., p. 154–155.
8. Ibid., p. 155.
9. Ibid., p. 156–157.
10. Ibid., p. 157.
11. Ibid., p. 159.
12. Ibid., p. 155.
13. Laërtius, Diogenes, 1905, *The Lives and Opinions of Eminent Philosophers*, Book 10, translated by C. D. Yonge, George Bell & Sons, London, p. 427.
14. Ibid., p. 433.
15. Bailey, C., 1928, *The Greek Atomists and Epicurus*, Oxford at the Clarendon Press, London, p. 228.
16. Laërtius, Diogenes, 1905, *The Lives and Opinions of Eminent Philosophers*, Book 10, translated by C. D. Yonge, George Bell & Sons, London, p. 436–479.
17. Hicks, R. D., 1910, *Stoic and Epicurean*, Charles Scribner's Sons, New York, p. 161.
18. Ibid., p. 160.
19. Ibid.
20. Laërtius, Diogenes, 1905, *The Lives and Opinions of Eminent Philosophers*, Book 10, translated by C. D. Yonge, George Bell & Sons, London, p. 428.
21. Ibid., p. 471.
22. Hicks, R. D., 1910, *Stoic and Epicurean*, Charles Scribner's Sons, New York, p. 157–158.
23. Ibid., p. 163.
24. Ibid., p. 311.
25. Laërtius, Diogenes, 1905, *The Lives and Opinions of Eminent Philosophers*, Book 10, translated by C. D. Yonge, George Bell & Sons, London, p. 468.
26. Hicks, R. D., 1910, *Stoic and Epicurean*, Charles Scribner's Sons, New York, p. 165.
27. Zeller, E., 1908, *Outlines of the History of Greek Philosophy*, translated by Sarah Frances Alleyne and Evelyn Abbott, Henry Holt, New York, p. 264.
28. Laërtius, Diogenes, 1905, *The Lives and Opinions of Eminent Philosophers*, Book 10, translated by C. D. Yonge, George Bell & Sons, London, p. 470.
29. Ibid., p. 471.
30. Bailey, C., 1928, *The Greek Atomists and Epicurus*, Oxford at the Clarendon Press, London, p. 224.
31. Hicks, R. D., 1910, *Stoic and Epicurean*, Charles Scribner's Sons, New York, p. 174.
32. Grote, G., 1880, *Aristotle*, Second Edition, John Murray, London, p. 656.
33. Laërtius, Diogenes, 1905, *The Lives and Opinions of Eminent Philosophers*, Book 10, translated by C. D. Yonge, George Bell & Sons, London, p. 479.
34. Hicks, R. D., 1910, *Stoic and Epicurean*, Charles Scribner's Sons, New York, p. 182.
35. Laërtius, Diogenes, 1905, *The Lives and Opinions of Eminent Philosophers*, Book 10, translated by C. D. Yonge, George Bell & Sons, London, p. 455.
36. Hicks, R. D., 1910, *Stoic and Epicurean*, Charles Scribner's Sons, New York, p. 208.
37. Ibid., p. 217.
38. Bailey, C., 1928, *The Greek Atomists and Epicurus*, Oxford at the Clarendon Press, London, p. 238.
39. Laërtius, Diogenes, 1905, *The Lives and Opinions of Eminent Philosophers*, Book 10, translated by C. D. Yonge, George Bell & Sons, London, p. 435.

40. Hicks, R. D., 1910, *Stoic and Epicurean*, Charles Scribner's Sons, New York, p. 215.
41. Laërtius, Diogenes, 1905, *The Lives and Opinions of Eminent Philosophers*, Book 10, translated by C. D. Yonge, George Bell & Sons, London, p. 435.
42. Ibid., p. 436.
43. Bailey, C., 1928, *The Greek Atomists and Epicurus*, Oxford at the Clarendon Press, London, p. 248–249.
44. Laërtius, Diogenes, 1905, *The Lives and Opinions of Eminent Philosophers*, Book 10, translated by C. D. Yonge, George Bell & Sons, London, p. 458.
45. Ibid., p. 458–466.
46. Lloyd, G. E. R., 1973, *Greek Science After Aristotle*, W. W. Norton, London, p. 26.
47. Chamberlin, T. C., 1890, The Method of Multiple Working Hypotheses, *Science*, vol. 15, no. 366, p. 93.
48. Hicks, R. D., 1910, *Stoic and Epicurean*, Charles Scribner's Sons, New York, p. 211.
49. Laërtius, Diogenes, 1905, *The Lives and Opinions of Eminent Philosophers*, Book 10, translated by C. D. Yonge, George Bell & Sons, London, p. 439.
50. Hicks, R. D., 1910, *Stoic and Epicurean*, Charles Scribner's Sons, New York, p. 222.
51. Laërtius, Diogenes, 1905, *The Lives and Opinions of Eminent Philosophers*, Book 10, translated by C. D. Yonge, George Bell & Sons, London, p. 443.
52. Ibid., p. 447–448.
53. Ibid., p. 439.
54. Ibid., p. 456.
55. Ibid., p. 438.
56. Ibid., p. 456–457.
57. Titus Lucretius Carus, 1743, *Of the Nature of Things*, Book 2, vol. 1, Daniel Brown, London, p. 185–187.
58. Laërtius, Diogenes, 1905, *The Lives and Opinions of Eminent Philosophers*, Book 10, translated by C. D. Yonge, George Bell & Sons, London, p. 468–469.
59. Plutarch, 1870, *Of Those Sentiments Concerning Nature with Which Philosophers were Delighted*, Book 1, Chapter 7, translated by John Dowel, in *Plutarch's Morals, Translated from the Greek by Several Hands, Corrected and Revised by William W. Goodwin*, vol. 3, Little, Brown, and Company, Boston, p. 122.
60. Laërtius, Diogenes, 1905, *The Lives and Opinions of Eminent Philosophers*, Book 10, translated by C. D. Yonge, George Bell & Sons, London, p. 459.
61. Titus Lucretius Carus, 1743, *Of the Nature of Things*, Book 2, vol. 1, Daniel Brown, London, p. 153.
62. Ibid., Book 1, vol. 1, p. 19.
63. Ibid., Book 2, vol. 1, p. 189.
64. Ibid., Book 1, vol. 1, p. 11.
65. Ibid., Book 3, vol. 1, p. 279.
66. Laërtius, Diogenes, 1905, *The Lives and Opinions of Eminent Philosophers*, Book 9, translated by C. D. Yonge, George Bell & Sons, London, p. 395.
67. Ibid., Book 10, p. 472.
68. Ibid.
69. Hicks, R. D., 1910, *Stoic and Epicurean*, Charles Scribner's Sons, New York, p. 180.
70. Laërtius, Diogenes, 1905, *The Lives and Opinions of Eminent Philosophers*, Book 10, translated by C. D. Yonge, George Bell & Sons, London, p. 467.
71. Ibid., p. 435.
72. Hicks, R. D., 1910, *Stoic and Epicurean*, Charles Scribner's Sons, New York, p. 163.
73. Laërtius, Diogenes, 1905, *The Lives and Opinions of Eminent Philosophers*, Book 10, translated by C. D. Yonge, George Bell & Sons, London, p. 469.
74. Ibid., Book 7, p. 259.
75. Hicks, R. D., 1910, *Stoic and Epicurean*, Charles Scribner's Sons, New York, p. 4.
76. Laërtius, Diogenes, 1905, *The Lives and Opinions of Eminent Philosophers*, Book 7, translated by C. D. Yonge, George Bell & Sons, London, p. 271.
77. Furley, D. J., 2008, Zeno of Citium, in *Complete Dictionary of Scientific Biography*, edited by Charles Gillispie, vol. 14, Cengage Learning, New York, p. 606.
78. *Encyclopædia Britannica*, Eleventh Edition, 1911, Stoics, vol. 25, Encyclopædia Britannica Company, New York, p. 942.
79. Laërtius, Diogenes, 1905, *The Lives and Opinions of Eminent Philosophers*, Book 7, translated by C. D. Yonge, George Bell & Sons, London, p. 260.
80. Furley, D. J., 2008, Zeno of Citium, in *Complete Dictionary of Scientific Biography*, edited by Charles Gillispie, vol. 14, Cengage Learning, New York, p. 606.
81. *Encyclopædia Britannica*, Eleventh Edition, 1911, Stoics, vol. 25, Encyclopædia Britannica Company, New York, p. 942.
82. Laërtius, Diogenes, 1905, *The Lives and Opinions of Eminent Philosophers*, Book 7, translated by C. D. Yonge, George Bell & Sons, London, p. 263–264.
83. Ibid., p. 264.
84. Ibid., p. 269.
85. Ibid., p. 267.
86. Ibid., p. 269.
87. Ibid., p. 266.
88. Ibid.
89. Ibid., p. 268.
90. Ibid.
91. Ibid.
92. Ibid., p. 267.
93. Ibid.
94. Ibid., p. 263.
95. Inwood, B., 2006, Cleanthes, in *Encyclopedia of Philosophy*, edited by Donald Borchert, vol. 2, Macmillan Reference, Detroit, p. 288.
96. Laërtius, Diogenes, 1905, *The Lives and Opinions of Eminent Philosophers*, Book 7, translated by C. D. Yonge, George Bell & Sons, London, p. 322.
97. Ibid., p. 323.
98. *Encyclopædia Britannica*, Eleventh Edition, 1911, Stoics, vol. 25, Encyclopædia Britannica Company, New York, p. 942.
99. Laërtius, Diogenes, 1905, *The Lives and Opinions of Eminent Philosophers*, Book 7, translated by C. D. Yonge, George Bell & Sons, London, p. 325.
100. *Encyclopædia Britannica*, Eleventh Edition, 1911, Stoics, vol. 25, Encyclopædia Britannica Company, New York, p. 943.
101. Laërtius, Diogenes, 1905, *The Lives and Opinions of Eminent Philosophers*, Book 7, translated by C. D. Yonge, George Bell & Sons, London, p. 327–328.
102. Ibid., p. 329–330.
103. Ibid., p. 331.
104. Ibid., p. 329.
105. Ibid., p. 327–328.
106. Ibid., p. 330.
107. *Encyclopædia Britannica*, Eleventh Edition, 1911, Stoics, vol. 25, Encyclopædia Britannica Company, New York, p. 948.
108. Hicks, R. D., 1910, *Stoic and Epicurean*, Charles Scribner's Sons, New York, p. 9.
109. Seneca, 1882, Of Benefits, Chapter 12, in *Seneca's Morals of a Happy Life, Benefits, Anger and Clemency*, translated by Roger L'Estrange, Belford, Clarke & Co., Chicago, p. 71.
110. Ibid.
111. Grote, G., 1888, *Plato and the Other Companions of Sokrates*, New Edition, vol. 4, John Murray, London, p. 102.
112. *Encyclopædia Britannica*, Eleventh Edition, 1911, Stoics, vol. 25, Encyclopædia Britannica Company, New York, p. 942.
113. Hicks, R. D., 1910, Stoic and Epicurean, Charles Scribner's Sons, New York, p. 54.
114. Ibid., p. 88.
115. Ibid., p. 115–116.
116. Ibid., p. 79.
117. Ibid.
118. Ibid., p. 140.
119. Plutarch, 1870, *Of Those Sentiments Concerning Nature with which Philosophers were Delighted*, Book 4, Chapter 9, in *Plutarch's Morals, Translated from the Greek by Several Hands, Corrected and Revised by William W. Goodwin*, vol. 3, Little, Brown, and Company, Boston, p. 165.
120. Laërtius, Diogenes, 1905, *The Lives and Opinions of Eminent Philosophers*, Book 7, translated by C. D. Yonge, George Bell & Sons, London, p. 326.

121. Hicks, R. D., 1910, *Stoic and Epicurean*, Charles Scribner's Sons, New York, p. 55.
122. *Encyclopædia Britannica*, Eleventh Edition, 1911, Stoics, vol. 25, Encyclopædia Britannica Company, New York, p. 943.
123. Hicks, R. D., 1910, *Stoic and Epicurean*, Charles Scribner's Sons, New York, p. 32–33.
124. *Encyclopædia Britannica*, Eleventh Edition, 1911, Stoics, vol. 25, Encyclopædia Britannica Company, New York, p. 944.
125. Clarke, J., 1910, *Physical Science in the Time of Nero being a Translation of the Quaestiones Naturales of Seneca*, Book 3, Section 9, Paragraph 1, Macmillan, London, p. 119–120.
126. Plutarch, 1870, *Of Those Sentiments Concerning Nature with which Philosophers were Delighted*, Book 1, Chapter 9, in *Plutarch's Morals, Translated from the Greek by Several Hands, Corrected and Revised by William W. Goodwin*, vol. 3, Little, Brown, and Company, Boston, p. 123.
127. Clarke, J., 1910, *Physical Science in the Time of Nero being a Translation of the Quaestiones Naturales of Seneca*, Book 3, Section 13, Paragraph 1, Macmillan, London, p. 124.
128. Hicks, R. D., 1910, *Stoic and Epicurean*, Charles Scribner's Sons, New York, p. 11.
129. Plutarch, 1889, The Contradictions of the Stoics, Paragraph 44, in *Plutarch's Miscellanies and Essays, Translated from the Greek by Several Hands, Corrected and Revised by William W. Goodwin*, vol. 4, Sixth Edition, Little, Brown, and Company, Boston, p. 470.
130. Plutarch, 1870, *Of Those Sentiments Concerning Nature with which Philosophers were Delighted*, Book 1, Chapter 18, in *Plutarch's Morals, Translated from the Greek by Several Hands, Corrected and Revised by William W. Goodwin*, vol. 3, Little, Brown, and Company, Boston, p. 127.
131. Kirk, G. S., and Raven, J. E., 1957, *The Presocratic Philsophers*, Cambridge University Press, London, p. 187.
132. Fairbanks, A. 1898, *The First Philosophers of Greece*, Kegan Paul, Trench, Trübner & Co., Ltd., London, p. 6.
133. Hicks, R. D., 1910, *Stoic and Epicurean*, Charles Scribner's Sons, New York, p. 10.
134. Bible, King James Version, John 1:1.
135. Clarke, J., 1910, *Physical Science in the Time of Nero being a Translation of the Quaestiones Naturales of Seneca*, Book 1, Preface, Macmillan, London, p. 7.
136. Ibid., Book 2, Section 45, p. 91–92.
137. Epictetus, 1904, *Discourses of Epictetus*, Book 2, Chapter 8, translated by George Long, D. Appleton, New York, p. 114.
138. Plutarch, 1870, *Of Those Sentiments Concerning Nature with which Philosophers were Delighted*, Book 1, Chapter 6, in *Plutarch's Morals, Translated from the Greek by Several Hands, Corrected and Revised by William W. Goodwin*, vol. 3, Little, Brown, and Company, Boston, p. 115–116.
139. Ibid., Book 1, Chapter 7, p. 122.
140. *Encyclopædia Britannica*, Eleventh Edition, 1911, Stoics, vol. 25, Encyclopædia Britannica Company, New York, p. 944.
141. Plutarch, 1889, Why the Oracles Cease to Give Answers, Paragraph 19, in *Plutarch's Miscellanies and Essays, Translated from the Greek by Several Hands, Corrected and Revised by William W. Goodwin*, vol. 4, Sixth Edition, Little, Brown, and Company, Boston, p. 24–25.
142. Hicks, R. D., 1910, *Stoic and Epicurean*, Charles Scribner's Sons, New York, p. 60.
143. Plutarch, 1870, *Of Those Sentiments Concerning Nature with which Philosophers were Delighted*, Book 4, Chapter 7, in *Plutarch's Morals, Translated from the Greek by Several Hands, Corrected and Revised by William W. Goodwin*, vol. 3, Little, Brown, and Company, Boston, p. 164.
144. Hicks, R. D., 1910, *Stoic and Epicurean*, Charles Scribner's Sons, New York, p. 16.
145. *Encyclopædia Britannica*, Eleventh Edition, 1911, Stoics, vol. 25, Encyclopædia Britannica Company, New York, p. 945.
146. Cicero, Marcus Tullius, 1775, *Of the Nature of the Gods*, Book 2, T. Davies, London, p. 134.
147. Hicks, R. D., 1910, *Stoic and Epicurean*, Charles Scribner's Sons, New York, p. 39, 42.
148. Antoninus, Marcus Aurelius, 1875, *The Thoughts of the Emperor M. Aurelius Antoninus*, Book 2, Paragraph 3, translated by George Long, Second Edition, George Bell & Sons, London, p. 78–79.
149. Ibid., Book 6, Paragraph 42, p. 126–127.
150. Fairbanks, A. 1898, *The First Philosophers of Greece*, Kegan Paul, Trench, Trübner & Co., Ltd., London, p. 62.
151. Hicks, R. D., 1910, *Stoic and Epicurean*, Charles Scribner's Sons, New York, p. 23.
152. Ibid., p. 26.
153. Clarke, J., 1910, *Physical Science in the Time of Nero being a Translation of the Quaestiones Naturales of Seneca*, Book 2, Chapters 35, 36, Macmillan, London, p. 84.
154. Plutarch, 1870, *Of Those Sentiments Concerning Nature with which Philosophers were Delighted*, Book 1, Chapter 28, in *Plutarch's Morals, Translated from the Greek by Several Hands, Corrected and Revised by William W. Goodwin*, vol. 3, Little, Brown, and Company, Boston, p. 130.
155. Hicks, R. D., 1910, *Stoic and Epicurean*, Charles Scribner's Sons, New York, p. 41.
156. Clarke, J., 1910, *Physical Science in the Time of Nero being a Translation of the Quaestiones Naturales of Seneca*, Book 2, Chapter 32, Paragraphs 4, 5, 6, Macmillan, London, p. 80.
157. Gellius, A., 1795, *The Attic Nights of Aulus Gellius*, Book 6, Chapter 1, translated into English by the Rev. W. Beloe, vol. 2, J. Johnson, London, p. 2–3.
158. Hicks, R. D., 1910, *Stoic and Epicurean*, Charles Scribner's Sons, New York, p. 77.
159. Seneca, 1882, *Of a Happy Life*, Chapter 8, in *Seneca's Morals of a Happy Life, Benefits, Anger and Clemency*, translated by Roger L'Estrange, Belford, Clarke & Co., Chicago, p. 178–179.
160. Ibid., p. 179.
161. Antoninus, Marcus Aurelius, 1875, *The Thoughts of the Emperor M. Aurelius Antoninus*, Book 10, Paragraph 6, translated by George Long, Second Edition, George Bell & Sons, London, p. 172.
162. Ibid., Book 5, Paragraph 8, p. 108.
163. Ibid., Book 9, Paragraph 42, p. 169.
164. Hicks, R. D., 1910, *Stoic and Epicurean*, Charles Scribner's Sons, New York, p. 107.
165. Epictetus, 1904, *Discourses of Epictetus*, Book 1, Chapter 11, translated by George Long, D. Appleton, New York, p. 34.
166. O'Leary, D. L., 1948, *How Greek Science Passed to the Arabs*, Routledge and Kegan Paul, London, p. 22–23.
167. *Encyclopædia Britannica*, Eleventh Edition, 1911, Plotinus, vol. 21, Encyclopædia Britannica Company, New York, p. 849.
168. Porphyry, 1949 or 1950, Life of Plotinus, in *Plotinus, the Ethical Treatises*, translated by Stephen McKenna, vol. 1 (first published in 1917), Charles T. Branford Company, Boston, p. 3.
169. Eusebius, 1851, *Ecclesiastical History*, Book 6, Chapter 19, translated by C. F. Cruse, Henry G. Bohn, London, p. 224.
170. Porphyry, 1949 or 1950, Life of Plotinus, in *Plotinus, the Ethical Treatises*, translated by Stephen McKenna, vol. 1 (first published in 1917), Charles T. Branford Company, Boston, p. 3.
171. Ibid., p. 17.
172. Ibid., p. 10.
173. Ibid., p. 12.
174. Ibid., p. 9.
175. Ibid., p. 25.
176. Ibid., p. 9.
177. Taylor, H. O., 1914, *The Mediaeval Mind*, Second Edition, vol. 1, Macmillan, London, p. 45.
178. O'Leary, D. L., 1948, *How Greek Science Passed to the Arabs*, Routledge and Kegan Paul, London, p. 27–28.
179. Ibid., p. 29.
180. *Encyclopædia Britannica*, Eleventh Edition, 1911, Neoplatonism, vol. 19, Encyclopædia Britannica Company, New York, p. 375.
181. Plotinus, 1911, Against the Gnostics, Paragraph 1 (*Ennead* 2.9), in *Select Works of Plotinus*, translated by Thomas Taylor, G. Bell and Sons, London, p. 44.
182. *Encyclopædia Britannica*, Eleventh Edition, 1911, Neo-

platonism, vol. 19, Encyclopædia Britannica Company, New York, p. 375.

183. Ibid.

184. Ibid.

185. Ibid.

186. Ibid.

187. Hyman, A., and Walsh, J. J., 1967, *Philosophy in the Middle Ages*, Harper & Row, New York, p. 285.

188. *Oxford English Dictionary Online*, Draft Revision March 2009.

189. Plotinus, 1911, On the Immortality of the Soul, Paragraph 3 (*Ennead* 4.7), in *Select Works of Plotinus*, translated by Thomas Taylor, G. Bell and Sons, London, p. 145.

190. *Encyclopædia Britannica*, Eleventh Edition, 1911, Neoplatonism, vol. 19, Encyclopædia Britannica Company, New York, p. 375.

191. Ibid.

192. Porphyry, 1949 or 1950, Life of Plotinus, in *Plotinus, the Ethical Treatises*, translated by Stephen McKenna, vol. 1 (first published in 1917), Charles T. Branford Company, Boston, p. 9.

193. Plotinus, 1949 or 1950, The Soul's Descent Into Body (*Ennead* 4.8), in *Plotinus, the Ethical Treatises*, translated by Stephen McKenna, vol. 2, Charles T. Branford Company, Boston, p. 143.

194. Porphyry, 1949 or 1950, Life of Plotinus, in *Plotinus, the Ethical Treatises*, translated by Stephen McKenna, vol. 1 (first published in 1917), Charles T. Branford Company, Boston, p. 24.

# Bibliography


Adams, F. D. 1954. *The Birth and Development of the Geological Sciences* (first published in 1938 by William & Wilkins in Baltimore). New York: Dover.

Adams, F. 1849. Preliminary Discourse. In *The Genuine Works of Hippocrates*, vol. 1. London: Sydenham Society.

Aelianus, Claudius. 1665. *Claudius Aelianus his Various History*. Trans. Thomas Stanley. London: Thomas Dring.

Aeschylus. 1922. Seven Against Thebes. In *Aeschylus*, vol. 1. Trans. Herbert Weir Smyth. London: William Heineman.

Akkermans, P. M. M. G., and G. M. Schwartz. 2003. *The Archeology of Syria*. Cambridge University Press.

Alighieri, D. 1902. *The Divine Comedy of Dante Alighieri*. Trans. Charles Eliot Norton. New York: Houghton, Mifflin.

Allman, G. J. 1889. *Greek Geometry from Thales to Euclid*. London: Longmans, Green.

Anonymous. 1938. Introducing the Googol, Prof. Kasner in Kindergarten Talk Also Presents the Googolplex. *New York Times*, February 27.

Antoninus, Marcus Aurelius. 1875. *The Thoughts of the Emperor M. Aurelius Antoninus*. Trans. George Long, Second Edition. London: George Bell.

Apostol, T. M. 2004. The Tunnel of Samos. *Engineering & Science*, vol. 67, no. 1.

Appian. 1912. *Appian's Roman History*. Trans. Horace White, vol. 1. London: William Heinemann.

Apuleius, L. 1909. *The Apologia and Florida of Apuleius of Madaura*. Trans. Harold Edgeworth Butler. London: Oxford at the Clarendon Press.

Aquinas, T. 1952. Summa Theologica, Treatise on God. In *Thomas Aquinas: I*. Trans. the Fathers of the English Dominican Province, *Great Books of the Western World*, vol. 19. Chicago: William Benton, (first published 1911 by T. Baker in London and others).

Archimedes. 2002. *The Works of Archimedes*. Ed. T. L. Heath, first published as *The Works of Archimedes* (1897, Cambridge University Press) and *The Method of Archimedes* (1912, Cambridge University Press). Mineola, New York: Dover.

Aristotle. 1882. *On the Parts of Animals*. Trans. W. Ogle. London: Kegan Paul, Trench.

_____. 1885. *The Politics of Aristotle*, vol. 1. Trans. Benjamin Jowett. London: Oxford at the Clarendon Press.

_____. 1897. *Aristotle's History of Animals, in Ten Books*. Trans. Richard Cresswell. London: George Bell.

_____. 1898. *The Poetics of Aristotle*. Trans. Samuel Henry Butcher. London: Macmillan.

_____. 1906. *The Nicomachean Ethics of Aristotle*, Tenth Edition. Trans. F. H. Peters. London: Kegan Paul, Trench, Trübner.

_____. 1910. *Historia Animalium* (*History of Animals*). Trans. D'Arcy Wentworth Thompson. In *The Works of Aristotle Translated into English*, vol. 4. Ed. J. A. Smith and W. D. Ross. London: Oxford University Press.

_____. 1912. *De Generatione Animalium* (*On the Generation of Animals*). Trans. Arthur Platt. In *The Works of Aristotle Translated into English*, vol. 5. Ed. J. A. Smith and W. D. Ross. London: London: Oxford University Press.

_____. 1913. *Mechanica* (*Mechanics*). Trans. E. S. Forster. In *The Works of Aristotle Translated into English*, vol. 6. Ed. W. D. Ross. London: Oxford at the Clarendon Press.

_____. 1923. *Meteorologica* (*Meteorology*). Trans. E. W. Webster. London: Oxford at the Clarendon Press.

_____. 1930. *Physica* (*Physics*). Trans. Robert Purves Hardie and Russell Kerr Gaye. London: Clarendon Press, Oxford.

_____. 1939. *On the Heavens* (*De Caelo*). Trans. W. K. C. Guthrie. London: William Heinemann.

_____. 1941. *Analytica Posteriora* (*Posterior Analytics*). Trans. A. J. Jenkinson. In *Basic Works of Aristotle*. New York: Random House.

_____. 1941. *Analytica Priora* (*Prior Analytics*). Trans. A. J. Jenkinson. In *Basic Works of Aristotle*. New York: Random House.

_____. 1941. *De Caelo* (*On the Heavens*). Trans. John Leofric Stocks. In *Basic Works of Aristotle*. New York: Random House.

_____. 1941. *De Generatione et Corruptione* (*On Generation and Corruption*). Trans. Harold H. Joachim. In *Basic Works of Aristotle*. New York: Random House.

_____. 1941. *Ethica Nicomachea* (*Nicomachean Ethics*). Trans. W. D. Ross. In *Basic Works of Aristotle*. New York: Random House.

_____. 1941. *Metaphysica* (*Metaphysics*). Trans. W. D. Ross. In *Basic Works of Aristotle*. New York: Random House.

_____. 1941. *Nicomachean Ethics*. Trans. W. D. Ross. In *Basic Works of Aristotle*. New York: Random House.

_____. 1941. *On Generation and Corruption*. Trans. Harold H. Joachim. In *Basic Works of Aristotle*. New York: Random House.

_____. 1941. *Physics*. Trans. R. P. Hardie and R. K. Gaye. In *Basic Works of Aristotle*. New York: Random House.

Arnold, T. 1868. *History of Rome, Three Volumes in One*. New York: D. Appleton.

Arrian. 1893. *Arrian's Anabasis of Alexander and Indica*. Trans. Edward James Chinnock. London: George Bell.

Athenaeus. 1854. *The Deipnosophists, or Banquet of the Learned*. Trans. C. D. Yonge, vol. 1. London: Henry G. Bohn.

Bacon, R. 1907. *On Experimental Science*. In *The Library of Original Sources*, vol. 4. Ed. Oliver Joseph Thatcher. New York: University Research Extension.

_____. 1928. *The Opus Majus of Roger Bacon*, vol. 1. Trans. Robert Belle Burke. Philadelphia: University of Pennsylvania Press.

Bailey, C. 1928. *The Greek Atomists and Epicurus*. London: Oxford at the Clarendon Press.

Bakewell, C. M. 1907. *Source Book in Ancient Philosophy*. New York: Scribner's.

Balme, D. M. 1973. Biological Conceptions in Antiquity. In *Dictionary of the History of Ideas*, vol. 1. Ed. Philip P. Wiener. New York: Scribner's.

Barnes, R. 2001. Cloistered Bookworms in the Chicken-Coop of the Muses: the Ancient Library of Alexandria. In *The Library of Alexandria*. Ed. Roy MacLeod. London: I. B. Tauris.

Bayle, P. 1710. *An Historical and Critical Dictionary. By Monsieur Bayle. Translated into English, with Many Additions and Corrections, Made by the Author*, vol. 2. London: C. Harper and Others.

Bevan, E. 1927. *A History of Egypt Under the Ptolemaic Dynasty*. London: Methuen.

Blakesley, J. W. 1834. *A Life of Aristotle*. London: John W. Parker.

Boas, G. 1948. Fact and Legend in the Biography of Plato. *The Philosophical Review*, vol. 57, no. 5.

_____. 1973. Macrocosm and Microcosm. In *Dictionary of the History of Ideas*, vol. 3. Ed. Philip P. Wiener. New York: Scribner's.

Boas, M. 1949. Hero's Pneumatica: a Study of its Transmission and Influence. *Isis*, vol. 40, no. 1.

Bowen, A. C., and R. B. Todd. 2004. *Cleomedes' Lectures on Astronomy*. Berkeley and Los Angeles: University of California Press.

Briggs, M.S. 1956. Building-Construction. In *A History of Technology*, vol. 2. Ed. Charles Singer. London: Oxford University Press.

Brock, A. J. 1929. Introduction. In *Greek Medicine, Being Extracts Illustrative of Medical Writers from Hippocrates to Galen*. London: J.M. Dent.

Bromehead, C. N. 1954. Mining and Quarrying. In *A History of Technology*, vol. 1. Ed. Charles Singer. London: Oxford University Press.

Bryce, T. R. 2002. The Trojan War: Is There Truth Behind the Legend? *Near Eastern Archaeology*, no. 3, vol. 65.

Bunbury, E. H. 1883. *A History of Ancient Geography*, Second Edition, vol. 1. London: John Murray.

Burnet, J. 1908. *Early Greek Philosophy*, Second Edition. London: Adam and Charles Black.

_____. 1920. *Early Greek Philosophy*, Third Edition. London: Adam & Charles Black.

_____. 1920. *Greek Philosophy, Part 1, Thales to Plato*. London: Macmillan.

Burns, A. 1971. The Tunnel of Eupalinus and the Tunnel Problem of Hero of Alexandria. *Isis*, vol. 62, no. 2.

Butterfield, H. 1957. *The Origins of Modern Science* (revised edition). New York: Free Press.

Caesar. 1919. *The Gallic War*. Trans. H. J. Edwards. London: William Heinemann.

Celsus, A. C. 1831. *A Translation of the Eight Books of Aul. Corn. Celsus on Medicine*, Second Edition, revised by G. F. Collier. London: Simpkin and Marshall.

Censorinus. 1900. *De Die Natale (The Natal Day)*. Trans. William Maude. New York: Cambridge Encyclopedia

Chamberlin, T. C. 1890. The Method of Multiple Working Hypotheses. *Science*, vol. 15, no. 366.

Charleston, R. J., and L. M. Angus-Butterworth. 1957. Glass. In *A History of Technology*, vol. 3. Ed. Charles Singer. London: Oxford University Press.

Cicero, M. T. 1829. *The Republic of Cicero*. Trans. G. W. Featherstonhaugh. New York: G. & C. Carvill.

_____. 1899. *Cicero's Tusculan Disputations*. Trans. C. D. Yonge. New York: Harper & Brothers.

_____. 1758. *The Tusculan Disputations of Marcus Tullius Cicero*. London: printed for John Whiston and Benj. White.

_____. 1744. Academics. In *The Morals of Cicero*, translated into English by William Guthrie. London: T. Waller.

_____. 1775. *Of the Nature of the Gods, in Three Books*. London: T. Davies.

_____. 1853. On the Republic. In *The Treatises of M. T. Cicero*. Trans. C. D. Yonge. London: H. G. Bohn.

Clagett, M. 1994. *Greek Science in Antiquity*. New York: Barnes and Noble (reprint of 1955 edition).

Clarke, J. 1910. *Physical Science in the Time of Nero being a Translation of the Quaestiones Naturales of Seneca*. London: Macmillan.

Codrington, T. 1903. *Roman Roads in Britain*. London: The Sheldon Press.

Cohen, M. R., and I. E. Drabkin. 1948. *A Source Book in Greek Science*. New York: McGraw-Hill.

Coleridge, S. T. 1888. *The Table Talk and Omniana of Samuel Taylor Coleridge*, arranged and edited by T. Ashe. London: George Bell and Sons.

Coxe, J. R. 1846. Introduction. In *The Writings of Hippocrates and Galen*. Philadelphia: Lindsay and Blakiston.

_____. 1846. *The Works of Galen*. In *The Writings of Hippocrates and Galen*. Philadelphia: Lindsay and Blakiston.

Creasy, E. S. 1908. *The Fifteen Decisive Battles of the World from Marathon to Waterloo*, Everyman's Library Edition. New York: E. P. Dutton.

Crombie, A. C. 1995. *The History of Science from Augustine to Galileo*. New York: Dover.

Crowfoot, G. M. 1954. Textiles, Basketry, and Mats. In

*A History of Technology*, vol. 1. Ed. Charles Singer. London: Oxford University Press.
Crump, T. 2001. *A Brief History of Science as Seen Through the Development of Scientific Instruments*. New York: Carroll & Graf.
Curtius Rufus, Quintus. 1747. *The History of the Wars of Alexander the Great*. Trans. John Digby, Third Edition, vol. 1. London: A. Millar.
Cuvier, G. 1997. Extract from a Memoir, read 6 October 1798. In *Georges Cuvier, Fossil Bones, and Geological Catastrophes*. Trans. Martin J. S. Rudwick. Chicago: University of Chicago Press.
Dalberg-Acton, J. E. E. 1907. Freedom in Antiquity. In *The History of Freedom and Other Essays*. Ed. John Neville Figgis and Reginald Vere Laurence. London: Macmillan.
Darwin, C. 1897. *The Descent of Man and Selection in Relation to Sex*. New York: D. Appleton.
Darwin, F. (editor). 1887. *The Life and Letters of Charles Darwin*, vol. 2. New York: D. Appleton.
Davisson, W. I., and J. E. Harper. 1972. *European Economic History, Volume 1, The Ancient World*. New York: Appleton-Century-Crofts.
de Camp, L. S. 1960. *The Ancient Engineers*. New York: Dorset Press.
_____. 1965. The Darkhouse of Alexandria. *Technology and Culture*, vol. 6, no. 3.
Decker, R., and Decker, B. 1981. *Volcanoes*. San Francisco: W. H. Freeman.
Deming, D. 2008. Design, Science and Naturalism. *Earth Science Reviews*, vol. 90.
Derry, T. K., and T. I. Williams. 1993. *A Short History of Technology, From the Earliest Times to A. D. 1900* (reprint of 1960 edition published by Oxford University Press). New York: Dover.
Dicks, D. R. 2007. Eratosthenes. In *Complete Dictionary of Scientific Biography*. Ed. Charles Gillispie, vol. 4. New York: Cengage Learning.
Dio Cassius. 1906. *Dio's Rome*, vol. 5. Trans. Herbert Baldwin Foster. Troy: Pafraets Book Company.
Diodorus of Sicily. 1957. *Diodorus of Sicily*, vol. 11. Trans. Francis R. Walton. Cambridge: Harvard University Press.
Diodorus Siculus. 1814. *The Historical Library of Diodorus the Sicilian*, vol. 1. Trans. G. Booth. London: W. M'Dowall for J. Davis.
Dover, K. J. 1989. *Greek Homosexuality, Updated and With a New Postscript*. Cambridge: Harvard University Press.
Dreyer, J. L. E. 1953. *A History of Astronomy from Thales to Kepler* (first published in 1906 as *History of the Planetary Systems from Thales to Kepler*). New York: Dover.
Durant, W. 1939. *The Life of Greece*. New York: Simon and Schuster.
Edwards, C. (translator). 1904. *The Hammurabi Code and the Sinaitic Legislation*. London: Watts.
*Encyclopædia Britannica*. 1972. Chicago: William Benton.
*Encyclopædia Britannica*, Eleventh Edition. 1910. New York: Encyclopædia Britannica.
Engel, D. J. F. 1940. The Correspondence between Lord Acton and Bishop Creighton. *Cambridge Historical Journal*, vol. 6, no. 3.
Engels, D. 1985. The Length of Erastosthenes Stade. *American Journal of Philology*, vol. 106, no. 3.
Epictetus. 1904. *Discourses of Epictetus*. Trans. George Long. New York: D. Appleton.
Euclid. 1956. *The Thirteen Books of Euclid's Elements*, Translated from the Text of Heiberg with Introduction and Commentary by Sir Thomas L. Heath, Second Edition, volumes 1, 2, and 3 (first published by Cambridge University Press in 1926). New York: Dover.
Eusebius. 1851. *Ecclesiastical History*. Trans. C. F. Cruse. London: Henry G. Bohn.
Fairbanks, A. 1898, *The First Philosophers of Greece*. London: Kegan Paul, Trench, Trübner.
Farrington, B. 1961. *Greek Science, Its Meaning For Us*. Baltimore, Maryland: Penguin Books.
Ferrier, J. F. 1854. *Institutes of Metaphysic*. Edinburgh and London: William Blackwood.
Finch, J. K. 1951. *Engineering and Western Civilization*. New York: McGraw-Hill.
Forbes, R., J. 1956. Hydraulic Engineering and Sanitation. In *A History of Technology*, vol. 2. Ed. Charles Singer. London: Oxford University Press.
Formigari, L. 1973. Chain of Being. In *Dictionary of the History of Ideas*, vol. 1. Ed. Philip P. Wiener. New York: Scribner's.
Frazer, J. G. 1920. *The Golden Bough*, vol. 3, *The Dying God* (first published in 1911). London: Macmillan.
_____. 1922. *The Golden Bough* (abridged edition). New York: Macmillan.
Freeman, K. 1966. *Ancilla to the Pre-Socratic Philosophers, a Complete Translation of the Fragments in Diels, Fragmente der Vorsokratiker*. Cambridge: Harvard University Press.
Freeth, T., et al. 2006. Decoding the Ancient Greek Astronomical Calculator Known as the Antikythera Mechanism. *Nature*, vol. 44, 30 November.
_____., et al. 2008. Calendars with Olympiad Display and Eclipse Prediction on the Antikythera Mechanism. *Nature*, vol. 454, 31 July.
Fridriksson, S. 1987. Plant Colonization of a Volcanic Island, Surtsey, Iceland. *Arctic and Alpine Research*, vol. 19, no. 4.
Frontinus, Sextus Julius. 1899. *The Two Books on the Water Supply of the City of Rome*. Trans. Clemens Herschel. Boston: Dana Estes.
Furley, D. J. 2008. Zeno of Citium. In *Complete Dictionary of Scientific Biography*. Ed. Charles Gillispie, vol. 14. New York: Cengage Learning.
Galen. 1916. *On the Natural Faculties*. Trans. Arthur John Brock. London: William Heinemann.
_____. 1929. *Notes on His Own Books*. In *Greek Medicine, Being Extracts Illustrative of Medical Writers from Hippocrates to Galen*. Trans. Arthur J. Brock. London: J.M. Dent.
_____. 1929. *On Anatomical Procedure*. In *Greek Medicine, Being Extracts Illustrative of Medical Writers from Hippocrates to Galen*. Trans. Arthur J. Brock. London: J.M. Dent.
_____. 1929. *On Habits*. In *Greek Medicine, Being Extracts Illustrative of Medical Writers from Hippocrates to Galen*. Trans. Arthur J. Brock. London: J.M. Dent.
_____. 1929. *On the Medical Sects*. In *Greek Medicine*,

*Being Extracts Illustrative of Medical Writers from Hippocrates to Galen*. Trans. Arthur J. Brock. London: J.M. Dent.

_____. 1929. *On the Passions*. In *Greek Medicine, Being Extracts Illustrative of Medical Writers from Hippocrates to Galen*. Trans. Arthur J. Brock. London: J.M. Dent.

_____. 1929. *On the Utility of Parts*. In *Greek Medicine, Being Extracts Illustrative of Medical Writers from Hippocrates to Galen*. Trans. Arthur J. Brock. London: J.M. Dent.

Gellius, A. 1795. *The Attic Nights of Aulus Gellius*, translated into English by the Rev. W. Beloe, vol. 2. London: J. Johnson.

Gest, A. P. 1930. *Engineering, Our Debt to Greece and Rome* (1963 Edition). New York: Cooper Square.

Gibbon, E. 1909. *The History of the Decline and Fall of the Roman Empire*. Ed. J. B. Bury, vol. 3. London: Methuen.

Gille, B. 1956. Machines. In *A History of Technology*, vol. 2. Ed. Charles Singer. London: Oxford University Press.

Gilson, E. 1955. *History of Christian Philosophy in the Middle Ages*. New York: Random House.

Gingerich, O. 1980. Was Ptolemy a Fraud? *Quarterly Journal of the Royal Astronomical Society*, vol. 21.

Goldstein, B. R. 1978. Casting Doubt on Ptolemy. *Science*, 24 Feb., vol. 199.

Goodchild, R. G., and R. J. Forbes. 1956. Road and Land Travel. In *A History of Technology*, vol. 2. Ed. Charles Singer. London: Oxford University Press.

Goodfield, J. 1964. The Tunnel of Eupalinus. *Scientific American*, vol. 210, no. 6.

Gottschalk, H. B. 1965. Strato of Lampsacus: Some Texts. *Proceedings of the Leeds Philosophical and Literary Society*, vol. XI, part VI.

Grant, E. 1977. *Physical Science in the Middle Ages*. Cambridge University Press, Cambridge.

_____. 2007. *A History of Natural Philosophy*. Cambridge University Press, Cambridge.

Grote, G. 1875. *Plato and the Other Companions of Sokrates*, vol. 1. London: John Murray.

_____. 1880. *Aristotle*, Second Edition. London: John Murray.

_____. 1885. *Plato, and the Other Companions of Sokrates*, vol. 1. London: John Murray.

_____. 1899. *A History of Greece*, vol. 1. New York: Harper & Brothers.

Hamilton, E. 1953. *Mythology*. New York: Mentor.

Harden, D. B. 1956. Glass and Glazes. In *A History of Technology*, vol. 2. Ed. Charles Singer. London: Oxford University Press.

Haskins, C. H. 1927. *The Renaissance of the 12th Century*. Cambridge, Massachusetts: Harvard University Press.

Heath, T. L. 1932. *Greek Astronomy*. London: J. M Dent.

_____. 1956. *The Thirteen Books of Euclid's Elements*, Second Edition, vol. 1 (first published in 1908 by Cambridge University Press). New York: Dover.

_____. 1981. *A History of Greek Mathematics*, vol. 2 (first published in 1921 by the Clarendon Press, Oxford). New York: Dover.

Heath, T. 1913. *Aristarchus of Samos, the Ancient Copernicus*. London: Oxford at the Clarendon Press.

_____. 1981. *A History of Greek Mathematics: Volume I, From Thales to Euclid* (first published in 1921 by the Clarendon Press, Oxford). New York: Dover.

Hero. 1851. *The Pneumatics of Hero of Alexandria*. Trans. Bennet Woodcroft. London: Taylor Walton and Maberly.

Herodotus. 1910. *The History of Herodotus*, vol. 1 (first published in 1858). Trans. George Rawlinson. New York: J. M. Dent.

Hesiod. 1920. *Hesiod, the Homeric Hymns and Homerica*, with an English translation by Hugh G. Evelyn-White. London: William Heinemann.

Hesiod. 1920. *Works and Days*. In *Hesiod, the Homeric Hymns and Homerica*, with an English translation by Hugh G. Evelyn-White. London: William Heinemann.

Hicks, L. E. 1883. *A Critique of Design-Arguments*. New York: Scribner's.

Hicks, R. D. 1910. *Stoic and Epicurean*. New York: Scribner's.

Hillman, G., et al. 2001. New Evidence of Lateglacial cereal cultivation at Abu Hureyra on the Euphrates. *The Holocene*, vol. 11, no. 4.

Hippocrates. 1846. *The Writings of Hippocrates and Galen*. Trans. John Redman Coxe. Philadelphia: Lindsay and Blakiston.

_____. 1849. *The Genuine Works of Hippocrates*, vol. 1. Trans. Francis Adams. London: Sydenham Society.

_____. 1886. *The Genuine Works of Hippocrates*, vol. 1. Trans. Francis Adams. New York: William Wood and Company.

_____. 1929. Epidemics. In *Greek Medicine, Being Extracts Illustrative of Medical Writers from Hippocrates to Galen*. Trans. Arthur J. Brock. London: J.M. Dent.

_____. 1950. *The Medical Works of Hippocrates*. Trans. John Chadwick and W. N. Mann. Springfield, Illinois: Charles C. Thomas.

Hippolytus. 1921. *Philosophumena, or the Refutation of All Heresies*, vol. 1. Trans. F. Legge. New York: Macmillan.

Hodges, H. 1970. *Technology in the Ancient World*. Marboro Books, 1992 edition.

Homer. 1729. *The Illiad of Homer*, vol. 1. Trans. Mr. Pope. London: printed by T. J. for B. L.

_____. 1920. To Asclepius. In *Hesiod, the Homeric Hymns and Homerica*. Trans. Hugh G. Evelyn-White. London: William Heinemann.

_____. 1922. *The Iliad of Homer*, revised edition. Trans. Andrew Lang, Walter Leaf, and Ernest Myers. London: Macmillan.

Horace. 1748. *The Satires, Epistles, and Art of Poetry of Horace Translated into English Prose*, Third Edition. London: printed for Joseph Davidson.

_____. 1889. *Horace: the Odes, Epodes, Satires, and Epistles*. London: Frederick Warne.

_____. 1892. *The Odes and Carmen Saeculare of Horace*. Trans. John Conington. London: George Bell and Sons.

Hubbert, M. K. 1963. Are We Retrogressing in Science? *Geological Society of America Bulletin*, vol. 74.

Huxley, G. L. 2007. Sosigenes. In *Complete Dictionary of Scientific Biography*. Ed. Charles Gillispie, vol. 12. New York: Cengage Learning.
Hyman, A., and J. J. Walsh. 1967. *Philosophy in the Middle Ages*. New York: Harper & Row.
Iamblichus. 1986. *Life of Pythagoras*. Trans. Thomas Taylor, (reprinted from the edition published in 1818 by J. M. Watkins in London). Rochester, Vermont: Inner Traditions.
Inwood, B. 2006. Cleanthes. In *Encyclopedia of Philosophy*. Ed. Donald Borchert, vol. 2. Detroit: Macmillan Reference.
Joly, R. 2008. Hippocrates of Cos. In *Complete Dictionary of Scientific Biography*. Ed. Charles Gillispie, vol. 6. New York: Cengage Learning.
Josephus. 1987. The Wars of the Jews. In *The Works of Josephus*. Trans. William Whiston. Hendrickson.
Julian. 1784. *Select Works of the Emperor Julian, translated from the Greek by John Duncombe*, vol. 2. London: J. Nichols.
Juvenal. 1918. Satire 6, The Ways of Women. In *Juvenal and Perseus*. Trans. G. G. Ramsay. London: William Heinemann.
Keegan, J. 1993. *A History of Warfare*. New York: Alfred A. Knopf.
Khashaba, D. R. 2002. Excursions into the Dialogues of Plato. *The Examined Life Online Philosophy Journal*, vol. 3, no. 12.
Kirk, G. S., and J. E. Raven. 1957. *The Presocratic Philosophers*. London: Cambridge University Press.
Koestler, A. 1963. *The Sleepwalkers*. New York: Grosset & Dunlap.
Kudlien, F. 1968. Early Greek Primitive Medicine. *Clio Medica*, vol. 3.
Kudlien, F. 2008. Galen. In *Complete Dictionary of Scientific Biography*. Ed. Charles Gillispie, vol. 5. New York: Cengage Learning.
Kuhn, T. 1996. *The Structure of Scientific Revolutions* (Third Edition). University of Chicago Press.
Laërtius, Diogenes. 1905. *The Lives and Opinions of Eminent Philosophers*. Trans. C. D. Yonge. London: George Bell and Sons.
Le Clerc, Daniel. *Historie de la Medecine*. A la Haye (The Hague): Chez Isaac van der Kloot, 1729 (first edition 1696).
Lewes, G. H. 1864. *Aristotle, a Chapter from the History of Science*. London: Smith, Elder.
Lindberg, D. C. 1978. The Transmission of Greek and Arabic Learning to the West. In *Science in the Middle Ages*. Ed. David C. Lindberg. Chicago: University of Chicago Press.
_____. 1992. *The Beginnings of Western Science*. Chicago: University of Chicago Press.
Livius, T. 1849. *The History of Rome*, Books Nine to Twenty-Six. Trans. D. Spillan and Cyrus Edmonds. London: Henry G. Bohn.
Lloyd, G. E. R. 1970. *Early Greek Science: Thales to Aristotle*. New York: W. W. Norton.
_____. 1973. *Greek Science After Aristotle*. London: W. W. Norton.
_____. 1979. *Magic, Reason and Experience*. Cambridge: Cambridge University Press.
Loerke, W. C. 1990. A Rereading of the Interior Elevation of Hadrian's Rotunda. *The Journal of the Society of Architectural Historians*, vol. 49, no. 1.
Longrigg, J. 2008. Erasistratus. In *Complete Dictionary of Scientific Biography*. Ed. Charles Gillispie, vol. 4. New York: Cengage Learning.
_____. 2008. Herophilus. In *Complete Dictionary of Scientific Biography*. Ed. Charles Gillispie, vol. 6. New York: Cengage Learning.
Lucian. 1913. Hippias, or The Bath. In *Lucian* with an English translation by A. M. Harmon. London: William Heinemann.
Madison, J. 1952. Federalist Paper 10. In *American State Papers, Great Books of the Western World*, vol. 43. Chicago: William Benton.
Mahaffy, J. P. 1899. *A History of Egypt Under the Ptolemaic Dynasty*. New York: Scribner's.
Maurice, F. 1930. The Size of the Army of Xerxes in the Invasion of Greece 480 B.C. *The Journal of Hellenic Studies*, vol. 50, part 2.
May, H. 2000. *On Socrates*. Belmont, California: Wadsworth.
McMurtrie, D. C. 1943. *The Book: the Story of Printing and Bookmaking*. New York: Oxford University Press.
Meyerhof, M. 1931. Science and Medicine. In *The Legacy of Islam* (First Edition). Ed. T. Arnold, and A. Guillaume, London: Oxford University Press.
Michell, H. 1952. *Sparta*. London: Cambridge University Press.
Middleton, W. E. K. 1961. Archimedes, Kircher, Buffon, and the Burning-Mirrors. *Isis*, vol. 52, no. 4.
Modiano, M. S. 1973. How Archimedes Stole Sun to Burn Foe's Fleet. *New York Times*, 11 November.
Mommsen, T. 1894. *The History of Rome*. Trans. William Purdie Dickson, vol. 2. London: Richard Bentley.
_____. 1996. *A History of Rome Under the Emperors*. London: Routledge.
Moody, E. A. 2008. Buridan, Jean. In *Complete Dictionary of Scientific Biography*. Ed. Charles Gillispie, vol. 2. New York: Cengage Learning.
Mourelatos, A. P. D. 2008. Empedocles. In *Complete Dictionary of Scientific Biography*. Ed. Charles Gillispie, vol. 4. New York: Cengage Learning.
Muller, H. J. 1961. *Freedom in the Ancient World*. New York: Harper & Brothers.
Needham, J. 1956. *Science and Civilisation in China*, vol. 2. London: Cambridge at the University Press.
Neuburger, A. 1930. *The Technical Arts and Sciences of the Ancients*. New York: Macmillan.
Newton, R. R. 1977. *The Crime of Claudius Ptolemy*. Baltimore: Johns Hopkins University Press.
Niebuhr, B. G. 1849. *Lectures on the History of Rome*, Second Edition, vol. 2. London: Taylor, Walton, and Maberly.
_____. 1851. *The History of Rome*, vol. 3, translated by William Smith and Leonhard Schmitz. London: Taylor, Walton, and Maberly.
_____. 1853. *Lectures on the History of Rome*, vol. 1, Third Edition. London: Taylor, Walton, and Maberly.
Nutton, V. 2004. *Ancient Medicine*. London: Routledge.
O'Leary, D. L. 1948. *How Greek Science Passed to the Arabs*. London: Routledge and Kegan Paul.

Petronius. 1902. *The Satyricon of Petronius*. Paris: Charles Carrington.
Philo. 1855. *On the Life of Moses*. In *The Works of Philo Judaeus*. Trans. C. D. Yonge, vol. 3. London: Henry G. Bohn.
Plato. 1891. Epistle 7. In *The Works of Plato*, Translated from the Text of Stallbaum. Trans. George Burges, vol. 4. London: G. Bell.
_____. 1937. *The Dialogues of Plato*, vol. 1. Trans. Benjamin Jowett. New York: Random House.
_____. 1955. *The Republic*. Trans. Henry Desmond Pritchard Lee. Baltimore, Maryland: Penguin.
Playfair, J. 1997. Biographical Account of the late Dr. James Hutton, F. R. S. Edin. In *James Hutton & Joseph Black, Biographies by John Playfair and Adam Ferguson*, from Volume V of *Transactions of the Royal Society of Edinburgh*, 1805. Edinburgh: RSE Scotland Foundation.
Pliny the Consul. 1809. *The Letters of Pliny the Consul*. Trans. William Melmoth. Boston: E. Larkin.
Pliny the Elder. 1855. *The Natural History of Pliny*. Trans. John Bostock and H. T. Riley, vol. 1. London: Henry G. Bohn.
Plotinus. 1911. *Select Works of Plotinus*. Trans. Thomas Taylor. London: G. Bell and Sons.
Plotinus, 1949 or 1950, The Soul's Descent Into Body (*Ennead* 4.8). In *Plotinus, the Ethical Treatises*. Trans. Stephen McKenna, vol. 2. Boston: Charles T. Branford Company.
Plutarch. 1718. *Banquet of the Seven Wisemen*. Trans. Roger Davis. In *Plutarch's Morals: in Five Volumes, translated from the Greek, by Several Hands*, Fifth Edition, vol. 2. London: W. Taylor.
_____. 1870. *Plutarch's Morals, Translated from the Greek by Several Hands, Corrected and Revised by William W. Goodwin*, vol. 1. Boston: Little, Brown.
_____. 1952. *The Lives of the Noble Grecians and Romans*. Trans. John Dryden. *Great Books of the Western World*, vol. 14. Chicago: William Benton.
_____. 1870. *On Superstition*. In *Plutarch's Morals, Translated from the Greek by Several Hands, Corrected and Revised by William W. Goodwin*, vol. 1. Boston: Little, Brown.
Polybius. 1889. *The Histories of Polybius*, vol. 1. Trans. Evelyn S. Shuckburgh. London: Macmillan.
Popper, K. 1966. *The Open Society and Its Enemies*, vol. 1, Fifth Edition. London: Routledge.
Popper, K. 1998. *The World of Parmenides, Essays on the Presocratic Enlightenment*. New York: Routlege.
Porphyry, 1949 or 1950, Life of Plotinus. In *Plotinus, the Ethical Treatises*. Trans. Stephen McKenna, vol. 1 (first published in 1917). Boston: Charles T. Branford.
_____. 1987. The Life of Pythagoras. In *The Pythagorean Sourcebook and Library*. Trans. Kenneth Sylvan Guthrie. Grand Rapids, Michigan: Phanes Press.
Porter, R. 2001. Medical Science. In *The Cambridge Illustrated History of Medicine*. Ed. Roy Porter. New York: Cambridge University Press.
Price, D. D. S. 1974. Gears from the Greeks. The Antikythera Mechanism: a Calendar Computer from ca. 80 B.C. *Transactions of the American Philosophical Society*, New Series, vol. 64, no. 7.
Proclus. 1792. *The Philosophical and Mathematical Commentaries of Proclus, on the First Book of Euclid's Elements*, vol. 1. London: printed for the author.
Ptolemy. 2000. *Harmonics*. Trans. Jon Solomon. Leiden: Brill.
Ptolemy, C. 1822. *Ptolemy's Tetrabiblos or Quadripartite: Being Four Books on the Influence of the Stars*. London: Davis and Dickson.
_____. 1952. *The Almagest*. Trans. R. Catesby Taliaferro, *Great Books of the Western World*, vol. 16. Chicago: William Benton.
_____. 1984. *Ptolemy's Almagest*. Trans. G. J. Toomer. New York: Springer-Verlag.
Rand, A. 1982. *Philosophy: Who Needs It*. Indianapolis: Bobbs-Merrill.
Reinach, S. 1904. *The Story of Art Throughout the Ages, an Illustrated Record*. New York: Scribner's.
Roberts, S. V. 1988. White House Confirms Reagans Follow Astrology, Up to a Point. *New York Times*, May 4.
Ross, W. D. 1930. *Aristotle*, Second Edition, Revised. London: Methuen.
Rousseau, J. J. 1856. *The Confessions of J. J. Rousseau*, vol. 1. New York: Calvin Blanchard.
Russell, B. 1945. *A History of Western Philosophy* (1963 paperback edition). New York: Simon and Schuster.
Saint Teresa of Avila. 1916. *The Life of St. Teresa of Jesus*, Fifth Edition. Trans. David Lewis. New York: Benziger Brothers.
Sambursky, S. 1962. *The Physical World of Late Antiquity*. Princeton, New Jersey: Princeton University Press.
Sarton, G. 1924. Review of "A History of Magic and Experimental Science During the First Thirteen Centuries of Our Era." *Isis*, no. 1, vol. 6.
_____. 1927. *Introduction to the History of Science*, vol. 1. Washington, DC: Carnegie Institution of Washington.
_____. 1936. The Unity and Diversity of the Mediterranean World. *Osiris*, vol. 2.
_____. 1954. *Galen of Pergamon*. Lawrence: University of Kansas Press.
_____. 1959. *A History of Science, Hellenistic Science and Culture in the Last Three Centuries B. C.* Cambridge: Harvard University Press.
_____. 1960. *A History of Science, Ancient Science Through the Golden Age of Greece*. Cambridge: Harvard University Press.
Semaw, S. 2000. The World's Oldest Stone Artefacts from Gona, Ethiopa. *Journal of Archaeological Science*, vol. 27.
Seneca. 1882. *Seneca's Morals of a Happy Life, Benefits, Anger and Clemency*. Trans. Roger L'Estrange. Chicago: Belford, Clarke.
Sharpley, H. (translator). 1906. *A Realist of the Aegean, being a verse-translation of the Mimes of Herodas*. London: David Nutt.
Shields, C. 2006. Learning about Plato from Aristotle. In *A Companion to Plato*. Ed. Hugh H. Benson. Malden, Massachusetts: Blackwell.
Shuckburgh, E. S. 1889. *The Histories of Polybius*, vol. 1. London: Macmillan.
Simms, D. L. 1977. Archimedes and the Burning Mir-

rors of Syracuse. *Technology and Culture*, vol. 18, no. 1.

Singer, C. 1949. *A Short History of Science to the Nineteenth Century*. Clarendon Press.

_____. 1958. *From Magic to Science* (first published in 1928). New York: Dover.

Solmsen, F. 1957. The Vital Heat, the Inborn Pneuma and the Aether. *The Journal of Hellenic Studies*, vol. 77.

Stahl, W. 1962. *Roman Science*. Madison: University of Wisconsin Press.

Statius, Publius Papinius. 1908. *The Silvae of Statius*. Trans. D. A. Slater. Oxford: Oxford at the Clarendon Press.

Steinman, D. B., and S. R. Watson. 1941. *Bridges and their Builders*. New York: G. P. Putnam's.

Stock, B. 1978. Science, Technology, and Economic Progress in the Early Middle Ages. In *Science in the Middle Ages*. Ed. David C. Lindberg. Chicago: University of Chicago Press.

Strabo. 1856. *The Geography of Strabo*. Trans. H. C. Hamilton and W. Falconer, vol. 1. London: Henry G. Bohn.

Straub, H. 1952. *A History of Civil Engineering*. Trans. E. Rockwell. London: Leonard Hill.

Suetonius, C. T. 1906. *The Lives of the Twelve Caesars*. London: George Bell & Sons.

Tacitus. 1876. *The History of Tacitus* (Third Edition). Trans. Alfred John Church and William Jackson Brodribb. London: Macmillan.

_____. 1942. *Annals*. In *The Complete Works of Tacitus*. Trans. Alfred John Church and William Jackson Brodribb. New York: Modern Library.

Taylor, H. O. 1914. *The Mediaeval Mind*, Second Edition, vol. 1. London: Macmillan.

Tcherikover, V. 1958. The Ideology of the Letter of Aristeas. *Harvard Theological Review*, no. 2, vol. 51.

Tertullian. 1870. *De Anima (On the Soul)*. In *Anti-Nicene Christian Library*, vol. 15, *The Writings of Tertullian*, vol. 2. Ed. Alexander Roberts and James Donaldson. Edinburgh: T. & T. Clark.

Thackeray, H. St. J. (translator). 1904. *The Letter of Aristeas Translated into English*. London: Macmillan.

Thomson, G., and M. Missner. 2000. *On Aristotle*. Belmont, California: Wadsworth.

Thorndike, L. 1923. *A History of Magic and Experimental Science*, vol. 1. New York: Columbia University Press.

Thucydides. 1942. *The Peloponnesian War*. Trans. Benjamin Jowett. In *The Greek Historians*, vol. 1. New York: Random House.

Thurston, H. 2002. Greek Mathematical Astronomy Reconsidered. *Isis*, vol. 93, no. 1.

Titus Lucretius Carus. 1743. *Of the Nature of Things*, vol. 1. London: Daniel Brown.

Toomer, G. J. 2007. Hipparchus. In *Complete Dictionary of Scientific Biography*. Ed. Charles Gillispie, vol. 15. New York: Cengage Learning.

_____. 2008. Ptolemy. In *Complete Dictionary of Scientific Biography*. Ed. Charles Gillispie, vol. 11. New York: Cengage Learning.

Turcotte, D. L., and Schubert, G. 1982. *Geodynamics, Applications of Continuum Physics to Geological Problems*. New York: John Wiley.

Turner, W. 1903. *History of Philosophy*. Boston: Ginn and Company.

Tzetzes, J. 1968. Book of Histories. In *Greek Mathematical Works*, vol. 2. Trans. Ivor Thomas. Cambridge: Harvard University Press.

Vitruvius. 1960. *The Ten Books on Architecture*. Trans. Morris Hicky Morgan (first published by Harvard University Press in 1914). New York: Dover.

Vlastos, G. 1975. *Plato's Universe*. Seattle: University of Washington Press.

Von Hagen, V. W. 1967. *The Roads that Led to Rome*. Cleveland and New York: World.

Wade, N. 2006. *Before the Dawn*. New York: Penguin Press.

Wallace, W. A. 1982. Aristotle in the Middle Ages. In *Dictionary of the Middle Ages*, vol. 1. Ed. Joseph R. Strayer. New York: Scribner's.

Walsh, J. J. 1920. *Medieval Medicine*. London: A. & C. Black.

Ward-Perkins, J. 2008. Vitruvius Pollio. In *Complete Dictionary of Scientific Biography*. Ed. Charles Gillispie, vol. 15. New York: Cengage Learning.

Whitehead, A. N. 1929. *Process and Reality*. Cambridge: Cambridge University Press.

_____. 1935. *Science and the Modern World*. Lowell Lectures, 1925. New York: Macmillan.

Whittaker, T. 1911. *Priests, Philosophers and Prophets*. London: Adam and Charles Black.

Wilson, L. G. 2008. Galen: Anatomy and Physiology. In *Complete Dictionary of Scientific Biography*. Ed. Charles Gillispie, vol. 5. New York: Cengage Learning.

Winter, T. N. 2007. *The Mechanical Problems in the Corpus of Aristotle*. Faculty Publications, Classics and Religious Studies Department, University of Nebraska-Lincoln. <http://digitalcommons.unl.edu/classicsfacpub/68>

Withington, E. T. 1921. The Asclepiadae and the Priests of Asclepius. In *Studies in the History and Method of Science*, vol. 2. Ed. Charles Singer. London: Oxford at the Clarendon Press.

Wolff, M. 1987. Philoponus and the Rise of Preclassical Dynamics. In *Philoponus and the Rejection of Aristotelean Science*. Ed. Richard Sorabji. Ithaca, New York: Cornell University Press.

Xenophon. 1894. Memorabilia of Socrates. In *The Anabasis, or Expedition of Cyrus, and the Memorabilia of Socrates*. Trans. J. S. Watson. New York: Harper & Brothers.

_____. 1898. *Xenophon's Minor Works*. Trans. J. S. Watson. London: George Bell & Sons.

_____. 1942. *Hellenica*. Trans. Henry G. Dakyns (first published in 1890). In *The Greek Historians*, vol. 2. New York: Random House.

Zeller, E. 1908. *Outlines of the History of Greek Philosophy*. Trans. Sarah Frances Alleyne and Evelyn Abbott. New York: Henry Holt.

# Index